T0135802

Mathematical Physics I. Dynamical Systems and Classical Mechanics. Lecture Notes

Matteo Petrera

Institut für Mathematik, MA 7-2
Technische Universität Berlin
Strasse des 17. Juni 136

Bibliographic information published by the Deutsche Nationalbibliothek

The Deutsche Nationalbibliothek lists this publication in the Deutsche Nationalbibliografie; detailed bibliographic data are available in the Internet at http://dnb.d-nb.de .

ISBN 978-3-8325-3569-8

Logos Verlag Berlin GmbH
Comeniushof, Gubener Str. 47,
10243 Berlin
Tel.: +49 (0)30 42 85 10 90
Fax: +49 (0)30 42 85 10 92
INTERNET: http://www.logos-verlag.de

Motivations

These Lecture Notes are based on a one-semester course taught in the Winter semester of 2011/2012 at the Technical University of Berlin, to bachelor undergraduate mathematics and physics students. The exercises at the end of each chapter have been either solved in class, during Tutorial's hours, or assigned as weekly homework.

These Lecture Notes are based on the references listed on the next pages. It is worthwhile to warn the reader that there are several excellent and exhaustive books and monographs on the topics that were covered in this course. A practical drawback of some of these books is that they are not really suited for a 4-hours per week, one-semester course. On the contrary, these notes contain only those topics that were actually explained in class. They are a kind of one-to-one copy of blackboard lectures. Some topics, some aspects of the theory and some proofs were left out because of time constraints.

A characteristic feature of these notes is that they present subjects in a synthetic and schematic way, thus following exactly the same pedagogical strategy used in class. Notions, concepts, statements and proofs are intentionally written and organized in a way that I found well suited for a systematic and effective understanding/learning process.

The aim is to provide students with practical tools that allow them to prepare themselves for their exams and not to substitute the role of an exhaustive book. This purpose has, of course, drawbacks and benefits at the same time. As a matter of fact, many students wish to have a "product" which is readable, compact and self-contained. In other words, something that is necessary and sufficient to get a good mark with a reasonable effort. This is - at least ideally - the positive side of good Lecture Notes. On the other hand, the risk is that their understanding might not be fluid and therefore too confined. Indeed, I always encourage my students to also consult more "standard" books like the ones quoted on the next pages.

Aknowledgments

The DFG (Deutsche Forschungsgemeinschaft) collaborative Research Center TRR 109 "Discretization in Geometry and Dynamics" is acknowledged. I am grateful to Yuri Suris and to the BMS (Berlin Mathematical School) for having given me the opportunity to teach this course.

Matteo Petrera
Institut für Mathematik, MA 7-2, Technische Universität Berlin
Strasse des 17. Juni 136
petrera@math.tu-berlin.de
November 14, 2013

Books and references used during the preparation of these Lecture Notes

Ch1 Initial Value Problems

✓ [Ch] C. Chicone, *Ordinary Differential Equations with Applications*, Springer, 2006.

✓ [Ge] G. Gentile, *Meccanica Lagrangiana e Hamiltoniana*, Lecture Notes (in Italian) available at `http://www.mat.uniroma3.it/users/gentile/2011/testo/testo.html`.

✓ [Te] G. Teschl, *Ordinary Differential Equations and Dynamical Systems*, preliminary version of the book available at `http://www.mat.univie.ac.at/~gerald/ftp/book-ode/ode.pdf`.

Ch2 Continuous Dynamical Systems

✓ [Be] N. Berglund, *Geometric Theory of Dynamical Systems*, Lecture Notes available at `http://arxiv.org/abs/math/0111177`.

✓ [BlKu] G.W. Bluman, S. Kumei, *Symmetries and Integration Methods for Differential Equations*, Springer, 1989.

✓ [BuNe] P. Buttà, P. Negrini, *Sistemi Dinamici*, Lecture Notes (in Italian) available at `http://www1.mat.uniroma1.it/people/butta/didattica/sisdin.pdf`.

✓ [Ch] C. Chicone, *Ordinary Differential Equations with Applications*, Springer, 2006.

✓ [FaMa] A. Fasano, S. Marmi, *Meccanica Analitica: una Introduzione*, Bollati Boringhieri, 2002.

✓ [Ge] G. Gentile, *Meccanica Lagrangiana e Hamiltoniana*, Lecture Notes (in Italian) available at `http://www.mat.uniroma3.it/users/gentile/2011/testo/testo.html`.

✓ M.W. Hirsch, S. Smale, *Differential Equations, Dynamical Systems and Linear Algebra*, Academic Press, 1974.

✓ [Ku] Yu. Kuznetsov, *Elements of Applied Bifurcation Theory*, Springer, 1995, and *Dynamical Systems Notes*, Lecture Notes available at `http://www.staff.science.uu.nl/~kouzn101/NLDV/index.html`.

✓ [Ol1] P.J. Olver, *Applied Mathematics Lecture Notes*, Lecture Notes available at `http://www.math.umn.edu/~olver/appl.html`.

✓ [McMe] P.D. McSwiggen, K.R. Meyer, *Conjugate Phase Portraits of Linear Systems*, American Mathematical Monthly, Vol. 115, No. 7, 2008.

✓ [MaRaAb] J.E. Marsden, T. Ratiu, R. Abraham, *Manifolds, Tensor Analysis and Applications*, Springer, 2001.

✓ [Te] G. Teschl, *Ordinary Differential Equations and Dynamical Systems*, preliminary version of the book available at `http://www.mat.univie.ac.at/~gerald/ftp/book-ode/ode.pdf`.

Ch3 Lagrangian and Hamiltonian Mechanics on Euclidean Spaces

✓ [Ar] V. Arnold, *Mathematical Methods of Classical Mechanics*, Springer, 1989.

✓ [FaMa] A. Fasano, S. Marmi, *Meccanica Analitica: una Introduzione*, Bollati Boringhieri, 2002.

✓ [Ge] G. Gentile, *Meccanica Lagrangiana e Hamiltoniana*, Lecture Notes (in Italian) available at `http://www.mat.uniroma3.it/users/gentile/2011/testo/testo.html`.

Ch4 Introduction to Hamiltonian Mechanics on Poisson Manifolds

✓ [AdMoVa] M. Adler, P. van Moerbeke, P. Vanhaecke, *Algebraic Integrability, Painlevé Geometry and Lie Algebras*, Springer, 2004.

✓ [Ar] V. Arnold, *Mathematical Methods of Classical Mechanics*, Springer, 1989.

✓ [FaMa] A. Fasano, S. Marmi, *Meccanica Analitica: una Introduzione*, Bollati Boringhieri, 2002.

✓ [Ho] D.D. Holm, *Geometric Mechanics*, Imperial College Press, 2008.

✓ [MaRa] J.E. Marsden, T. Ratiu, *Introduction to Mechanics and Symmetry. A Basic Exposition of Classical Mechanical Systems*, Springer, 1999.

✓ [Ol2] P.J. Olver, *Applications of Lie Groups to Differential Equations*, Springer, 1998.

Contents

1

Initial Value Problems

1.1 Introduction

▶ In classical mechanics, the *position* $x \in \mathbb{R}^3$ of a point particle of mass $m > 0$ is described by a continuous function of *time*, a real variable $t \in I \subseteq \mathbb{R}$, i.e., $x : I \to \mathbb{R}^3$.

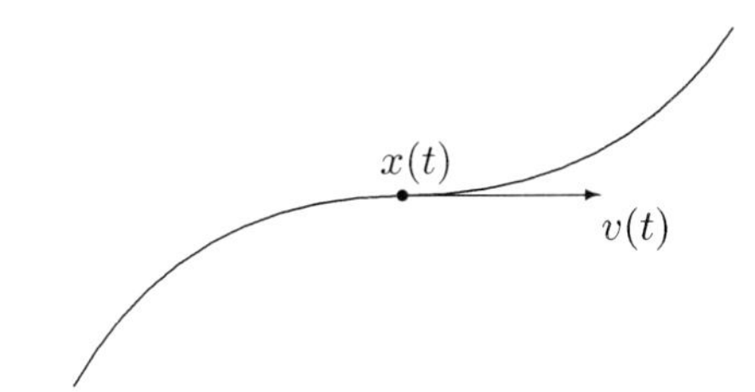

Fig. 1.1. Position and velocity vectors of a particle in $\mathbb{R}^2$ ([Te]).

- The first and the second derivatives of this function w.r.t. time define respectively the *velocity*, $v := \mathrm{d}x/\mathrm{d}t$, and the *acceleration*, $\mathrm{d}^2x/\mathrm{d}t^2$, of the particle.

- Assume that the particle is moving under the influence of an external vector field $f : \mathbb{R}^3 \to \mathbb{R}^3$, called *force*. Then the *Newton law of motion* states that at each point x the force acting on the particle is

$$f(x) = m\,\frac{\mathrm{d}^2x}{\mathrm{d}t^2} \qquad \forall\, t \in I. \tag{1.1}$$

- Formula (1.1) is a relation between x and its second derivative: it defines a system of three *second-order ordinary differential equations*. The variable x is called *dependent variable* and the variable t is called *independent variable*. Given a vector field f the problem is to find the general solution $x = x(t)$ satisfying (1.1).

- We can increase the number of dependent variables by considering the pair $(x, v) \in \mathbb{R}^6$. The advantage is that we get an equivalent *first-order* system of six ordinary differential equations:

$$\begin{cases} \dfrac{\mathrm{d}x}{\mathrm{d}t} = v, \\[2mm] \dfrac{\mathrm{d}v}{\mathrm{d}t} = \dfrac{f(x)}{m}. \end{cases}$$

▶ Informally, we can think of a *dynamical system* as the time evolution (continuous or discrete) of some physical system, such as the motion of two planets under the influence of the gravitational field, or the motion of a spinning top. One is often interested in knowing the fate of the system for long times. The theory of dynamical systems tries to predict the future of physical systems and understand the stability and limitations of these predictions. The first basic results on dynamical systems were found by I. Newton (1643-1727) but H. Poincaré (1854-1912) can be considered as the founder of the modern theory of dynamical systems.

► Dynamical systems naturally split into two classes, according to nature of the time variable t:

- If $t \in I \subseteq \mathbb{R}$ we have a *continuous dynamical system*, which is typically described by the *flow* of a system of first-order ordinary differential equations with prescribed initial conditions (*initial value problem*).
- If $t \in I \subseteq \mathbb{Z}$ we have a *discrete dynamical system*, which is typically described by the iteration of an invertible map.

► We here present two examples of dynamical systems, the first being with continuous time, the second with discrete time.

Example 1.1 (***Particle in*** $\mathbb{R}^3$ ***in a gravitational field***)

- In the vicinity of the surface of the earth, the gravitational force acting on a point particle of mass m, whose position is $x := (x_1, x_2, x_3) \in \mathbb{R}^3$, can be approximated by the vector field

$$f(x) := -m\, g\, e_3,$$

where $e_3 := (0,0,1) \in \mathbb{R}^3$ and $g > 0$ is the constant gravitational acceleration. Newton equations (1.1),

$$m\, \frac{\mathrm{d}^2 x}{\mathrm{d}t^2} = -m\, g\, e_3,$$

can be easily integrated to get the position vector of the particle

$$x(t) = x(0) + v(0)\, t - \frac{1}{2}\, g\, e_3\, t^2,$$

where $t = 0$ has be chosen as the initial time. The velocity vector is

$$\dot{x}(t) = v(t) = v(0) - g\, e_3\, t.$$

In this case we have a dynamical system defined in terms of a one-parameter (t is the parameter) family of maps (called *flow*) $\Phi_t : \mathbb{R} \times \mathbb{R}^6 \to \mathbb{R}^6$,

$$\Phi_t : (0, (x(0), v(0)) \mapsto (x(t, (x(0), v(0)), v(t, (x(0), v(0))) \,.$$

- Solutions of systems of ordinary differential equations of the type (1.1) cannot always be found by a straightforward integration. Indeed, a refinement of the model, which is valid not only in the vicinity of the surface of the earth, takes the real gravitational force

$$f(x) := -G\, m\, M\, \frac{x}{\|x\|^3},$$

which is a *central vector field*.

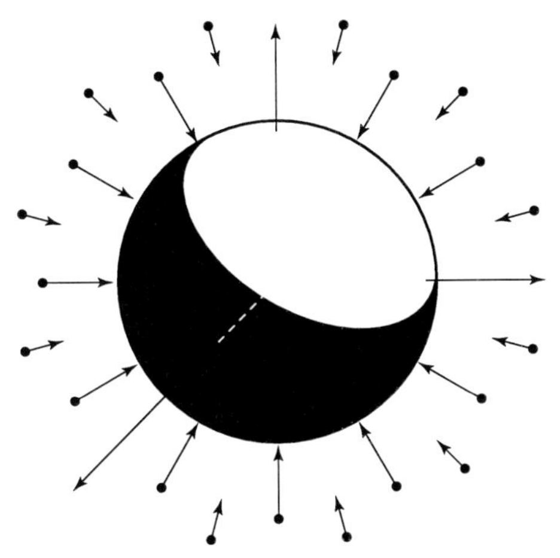

Fig. 1.2. The gravitational force field ([MaRaAb]).

Here M is the mass of the earth, $G > 0$ is the gravitational constant and $\|x\| := (x_1^2 + x_2^2 + x_3^2)^{1/2}$. Newton equations (1.1) now read

$$m\,\frac{\mathrm{d}^2 x}{\mathrm{d}t^2} = -G\,m\,M\,\frac{x}{\|x\|^3},$$

and it is no longer evident how to solve them. Moreover it is even unclear whether solutions exist at all.

Example 1.2 (*Logistic map*)

A discrete model for the evolution of a population $P_n : \mathbb{Z}_+ \to [0,\infty)$ of animals is

$$P_{n+1} = \alpha\,P_n - \beta\,P_n^2, \qquad n \in \mathbb{Z}_+,$$

where $\alpha > 0$ is the natality rate and $\beta > 0$ is a parameter which limits the growth. Note that if $\beta = 0$ then $P_n = \alpha^n P_0$ ("unrealistic" exponential growth). The rescaled variable $x_n := \beta\,P_n$ obeys the recurrence equation

$$x_{n+1} = \Phi(x_n, \alpha) := \alpha\,x_n(1 - x_n).$$

For $\alpha \in [0,4]$ then Φ maps $[0,1]$ into itself. The dynamics of the sequence $(x_n)_{n\in\mathbb{Z}_+}$, defined in terms of iterations of the parametric map Φ, depends drastically on the parameter α.

- $\alpha \in [0,1]$. The population will eventually die, irrespective of the initial population.
- $\alpha \in (1,3]$. The population approaches a stable equilibrium.
- $\alpha \in (3, 1+\sqrt{6})$. From almost all initial conditions the population will approach permanent oscillations between two values, which depend on α.
- $\alpha \in (1+\sqrt{6}, 3.54)$. From almost all initial conditions the population will approach permanent oscillations among four values, which depend on α.
- $\alpha > 3.54$. From almost all initial conditions the population will approach oscillations among 8 values, then 16, 32, ... (*period-doubling cascade*). At α approximately 3.57 is the onset of chaos, at the end of the period-doubling cascade. From almost all initial conditions we can no longer see any oscillations of finite period. Slight variations in the initial population yield dramatically different results over time.

1.2 Definition of an initial value problem (IVP)

► We start defining an ordinary differential equation.

Definition 1.1

1. *A continuous n-***th order ordinary differential equation (ODE)** *with independent variable $t \in I \subseteq \mathbb{R}$ and dependent (unknown) $\mathbb{R}$-valued variable $x : t \mapsto x(t)$ is a relation of the form*

$$F\left(t, x, x^{(1)}, \ldots, x^{(n)}\right) = 0, \qquad x^{(i)} \equiv \frac{\mathrm{d}^i x}{\mathrm{d}t^i}, \; i = 1, \ldots, n, \tag{1.2}$$

   where F is some continuous function of its arguments.

2. *A* **solution** *of (1.2) is a function $\phi : t \mapsto \phi(t)$, $\phi \in C^n(I_0, \mathbb{R})$, $I_0 \subseteq I$, such that*

$$F\left(t, \phi(t), \phi^{(1)}(t), \ldots, \phi^{(n)}(t)\right) = 0 \qquad \forall\, t \in I_0.$$

► We assume that (1.2) is solvable w.r.t. $x^{(n)}$ (*ODE in normal form*), that is

$$x^{(n)} = f\left(t, x, x^{(1)}, \ldots, x^{(n-1)}\right), \tag{1.3}$$

for some continuous function f. Thanks to the "Implicit function Theorem", this can be done at least locally in a neighborhood of some point $\left(t_0, x_0, x_0^{(1)}, \ldots, x_0^{(n)}\right)$ where there holds

$$\left.\frac{\partial F}{\partial x^{(n)}}\right|_{\left(t_0, x_0, x_0^{(1)}, \ldots, x_0^{(n)}\right)} \neq 0.$$

► Notational remarks:

- We introduce the *"dot" notation* for time-derivatives of the dependent variable x:

$$\dot{x} \equiv x^{(1)} \equiv \frac{\mathrm{d}x}{\mathrm{d}t} \qquad \left(\ddot{x} \equiv x^{(2)} \equiv \frac{\mathrm{d}^2 x}{\mathrm{d}t^2}, \ldots\right).$$

  From now on our dependent variable x is defined in M, which is an open set of the Euclidean space $\mathbb{R}^n$.

- Let M, N be open subsets, not necessarily of the same dimension, of the Euclidean space. We denote by $C^k(M, N)$, $k \geqslant 0$, the set of functions from M to N having continuous derivatives up to order k. We set $C(M, N) \equiv C^0(M, N)$, the set of continuous functions and $\mathscr{F}(M, N) \equiv C^\infty(M, N)$, the set of smooth functions.

Definition 1.2

1. *A continuous* **system of n first-order ODEs**, *with independent variable $t \in I \subseteq \mathbb{R}$ and dependent (unknown) $\mathbb{R}^n$-valued variable $x : t \mapsto x(t)$, is a set of n ODEs in normal form:*
$$\dot{x} = f(t, x), \tag{1.4}$$
where $f \in C(I \times M, \mathbb{R}^n)$. System (1.4) is **autonomous** *if f does not depend explicitly on t.*

2. *An* **initial value problem (IVP)** *(or* **Cauchy problem***) is the system (1.4) together with an initial value $x(t_0) = x_0$, $(t_0, x_0) \in I \times M$:*
$$\begin{cases} \dot{x} = f(t, x), \\ x(t_0) = x_0. \end{cases} \tag{1.5}$$

3. *A* **solution** *of (1.5) is a function $\phi : t \to \phi(t)$, $\phi \in C^1(I_0, \mathbb{R}^n)$, $I_0 \subseteq I$, such that*
$$\dot{\phi}(t) = f(t, \phi(t)) \qquad \forall t \in I_0,$$
satisfying $\phi(t_0) = x_0$.

▶ Notational remarks:

- From now on the solution of the IVP (1.5) will be denoted by $\phi(t, t_0, x_0)$ instead of $\phi(t)$. This is to emphasize explicitly the dependencies of the solution of (1.5) on its arguments. Note that $t \mapsto \phi(t, t_0, x_0)$ is a $\mathbb{R}^n$-valued *function of t*. The arguments (t_0, x_0) are fixed, and we may think of them as *parameters*. The initial condition reads $\phi(t_0, t_0, x_0) = x_0$. If $t_0 = 0$ we slightly simplify the notation omitting the second argument and thus denoting the solution by $\phi(t, x_0)$ with $\phi(0, x_0) = x_0$.

- If f depends on some parameters $\alpha \in A \subseteq \mathbb{R}^p$, $p \geqslant 1$, i.e., $f = f(t, x, \alpha)$, so does the solution of (1.5). In such a case the solution will be denoted by $\phi(t, t_0, x_0, \alpha)$.

- In what follows, we will be interested in regularities conditions of the solution of (1.5) w.r.t. the whole set of its arguments t, t_0, x_0, α. So we shall write $\phi(t, t_0, x_0, \alpha) \in C^k(I_0 \times I_0' \times M_0 \times A_0, \mathbb{R}^n)$ to say that the solution is of class C^k w.r.t. $t \in I_0$, $t_0 \in I_0' \subseteq I_0$, $x_0 \in M_0 \subseteq M$ and $\alpha \in A_0 \subseteq A$.

Example 1.3 (***Linear IVPs in $\mathbb{R}^n$***)

1. A *linear IVP* in $\mathbb{R}^n$ takes the form
$$\begin{cases} \dot{x} = A\, x, \\ x(0) = x_0, \end{cases}$$

where A is a constant $n \times n$ matrix. Its solution is given by

$$\phi(t, x_0) = e^{At} x_0, \qquad \phi(0, x_0) = x_0,$$

where e^{At} is the *matrix exponential*:

$$e^{At} := \sum_{k=0}^{\infty} \frac{t^k}{k!} A^k.$$

2. If $n = 1$, so that A is just a coefficient $a \in \mathbb{R}$, we simply have

$$\phi(t, x_0) = e^{at} x_0, \qquad \phi(0, x_0) = x_0.$$

► Any n-th order ODE (1.3) may be reduced to a system of n first-order ODEs (1.4) by introducing n new dependent variables $(y_1, y_2, \dots, y_n) := \left(x, x^{(1)}, \dots, x^{(n-1)}\right)$.

- Indeed, the variables $(y_1, y_2, \dots, y_n)$ obeys the system of ODEs

$$\begin{cases} \dot{y}_1 = y_2, \\ \vdots \\ \dot{y}_{n-1} = y_n, \\ \dot{y}_n = f(t, y_1, \dots, y_n). \end{cases}$$

- We can even define the set of variables $z := (t, y_1, y_2, \dots, y_n)$ in such a way that $(z_1, \dots, z_{n+1})$ satisfy the system of ODEs

$$\begin{cases} \dot{z}_1 = 1, \\ \dot{z}_2 = \dot{y}_1 = y_2 = z_3, \\ \vdots \\ \dot{z}_n = \dot{y}_{n-1} = y_n = z_{n+1}, \\ \dot{z}_{n+1} = f(z). \end{cases}$$

- This shows that, at least formally, *it suffices to consider the case of autonomous first-order systems of ODEs.*

1.3 Existence and uniqueness of solutions of IVPs

► Le us recall some facts from Analysis.

- Let $(X, \|\cdot\|_X)$ be a *Banach space*, i.e., a complete vector space X with *norm* $\|\cdot\|_X : X \to [0, \infty)$. Every Cauchy sequence in $(X, \|\cdot\|_X)$ is convergent.
- Let D be a (nonempty) closed subset of X and consider a map $K : D \to D$.

- A *fixed point* of K is an element $x \in D$ such that $K(x) = x$.
- The map K is a *contraction* if there exists $\eta \in [0,1)$ such that

$$\|K(x) - K(y)\|_X \leqslant \eta \, \|x - y\|_X \qquad \forall \, x, y \in D.$$

- The following claim holds true: *If K is a contraction then K has a unique fixed point $x \in D$* ("Contraction Principle" or "Banach fixed point Theorem").

Example 1.4 (*Banach spaces*)

1. $X = \mathbb{R}^n$ is a Banach space with norm given by the *Euclidean norm*

$$\|x\|_X = \|x\| := \sqrt{\langle x, x \rangle} := \sqrt{x_1^2 + \cdots + x_n^2}.$$

2. $X = C(I, \mathbb{R}^n)$, where I is a compact interval of $\mathbb{R}$, is a Banach space with norm given by

$$\|x\|_X = \sup_{t \in I} \|x(t)\|.$$

- Let $(t, x) \in I \times M$. A function $f \in C(I \times M, \mathbb{R}^n)$ is *locally Lipschitz continuous in x, uniformly w.r.t. t*, and we write $f \in \mathrm{Lip}\,(I \times M, \mathbb{R}^n)$, if, for every $(t_0, x_0) \in I \times M$, there exists a compact neighborhood V of (t_0, x_0) and $L \geqslant 0$ (*Lipschitz constant*) such that

$$\|f(t,x) - f(t,y)\| \leqslant L\|x - y\| \qquad \forall \, (t,x), \, (t,y) \in V. \tag{1.6}$$

If (1.6) holds true for all (t,x), $(t,y) \in I \times M$ then f is *globally Lipschitz continuous in x, uniformly w.r.t. t*.

- The following claim holds true: If $f \in C^1(M, \mathbb{R}^n)$ then $f \in \mathrm{Lip}\,(M, \mathbb{R}^n)$.

Example 1.5 (*One-variable real-valued Lipschitz continuous functions*)

Intuitively, a Lipschitz continuous function is limited in how fast it can change: for every pair of points on the graph of this function, the absolute value of the slope of the line connecting them is no greater than a definite real number, the Lipschitz constant.

1. $f(x) := \sin x$ is $C^\infty(\mathbb{R}, [-1,1])$. It is globally Lipschitz continuous because its derivative $\cos x$ is bounded above by 1 in absolute value.
2. $f(x) := |x|$ is not $C^1(\mathbb{R}, \mathbb{R})$. It is globally Lipschitz continuous by the reverse triangle inequality $||x| - |y|| \leqslant |x - y|$ for all $x, y \in \mathbb{R}$.
3. $f(x) := x^2$ is $C^\infty(\mathbb{R}, \mathbb{R})$. It is not globally Lipschitz continuous because it becomes arbitrarily steep as x approaches infinity. It is however locally Lipschitz continuous.

► The following fundamental Theorem provides sufficient conditions for the existence and uniqueness of solutions of the IVP (1.5).

Theorem 1.1 (*Picard-Lindelöf*)

Consider the IVP (1.5). If $f \in \mathsf{Lip}(I \times M, \mathbb{R}^n)$, then there exists, around each point $(t_0, x_0) \in I \times M$, an open set $I_0 \times M_0 \subset I \times M$, such that the IVP admits a unique local solution $\phi(t, t_0, x_0) \in C^1(I_0, \mathbb{R}^n)$.

Proof. We proceed by steps.

- Integrating both sides of the system of ODEs $\dot{x} = f(t,x)$ w.r.t. t we get an *integral equation* which formally defines the solution of (1.5):

$$x(t) = K(x)(t) := x_0 + \int_{t_0}^{t} f(s, x(s))\, \mathrm{d}s. \tag{1.7}$$

  Note that $x_0(t) := x_0$ is an approximating solution of (1.7) at least for small t.

- Plugging $x_0(t) := x_0$ into (1.7) we get another approximating solution

$$x_1(t) := K(x_0)(t) = x_0 + \int_{t_0}^{t} f(s, x_0(s))\, \mathrm{d}s,$$

  and so on,

$$x_2(t) := K^2(x_0)(t) = (K \circ K)(x_0)(t) = x_0 + \int_{t_0}^{t} f(s, x_1(s))\, \mathrm{d}s.$$

- The iteration of such a procedure, called *Picard iteration*, produces a a sequence of approximating solutions

$$x_m(t) := K^m(x_0)(t), \qquad m \in \mathbb{N}.$$

- This observation allows us to apply the "Contraction Principle" to the fixed point equation (1.7), which we compactly write $x = K(x)$. We set $t_0 = 0$ for notational simplicity and consider only the case $t \geqslant 0$ to avoid absolute values in the estimates which follow.

- We need a Banach space X and a closed subset $D \subseteq X$ such that $K : D \to D$ is a contraction. The natural choice is $X := C([0,T], \mathbb{R}^n)$ for some $T > 0$. Since $I \times M$ is open and $(0, x_0) \in I \times M$ we can choose $V := [0,T] \times \overline{M_0} \subset I \times M$, where $M_0 := \{x \in \mathbb{R}^n : \|x - x_0\| < \delta\}$, and define

$$P := \max_{(t,x) \in V} \|f(t,x)\|,$$

  where the maximum exists by continuity of f and compactness of V.

- Then we have:

$$\|K(x)(t) - x_0\| \leqslant \int_0^t \|f(s, x(s))\|\, \mathrm{d}s \leqslant t\,P,$$

  whenever $\{(t,x) : t \in [0,T]\} \subset V$.

- Hence for
$$t \leqslant T_0 := \min\left(T, \frac{\delta}{P}\right),$$
we have $T_0\,P \leqslant \delta$ and the graph of $K(x)$ restricted to $I_0 := [0, T_0]$ is again in V. Note that since $[0, T_0] \subseteq [0, T]$ the same constant P bounds $\|f\|$ on $V_0 := [0, T_0] \times \overline{M_0} \subset I \times M$.

- Therefore, if we choose $X := C([0, T_0], \mathbb{R}^n)$ as our Banach space with norm
$$\|x\|_X := \max_{t \in [0, T_0]} \|x(t)\|,$$
and $D := \{x \in X : \|x - x_0\|_X \leqslant \delta\}$ as our closed subset, then K is a map from D to D.

- The contraction property of K follows from the Lipschitz continuity property (1.6) of f. If $y \in D$ we have:
$$\begin{aligned} \|K(x)(t) - K(y)(t)\| &\leqslant \int_0^t \|f(s, x(s)) - f(s, y(s))\|\, \mathrm{d}s \\ &\leqslant L \int_0^t \|x(s) - y(s)\|\, \mathrm{d}s \\ &\leqslant L\,t \sup_{s \in [0,t]} \|x(s) - y(s)\|, \end{aligned}$$
provided that the graphs of both x and y lie in V_0. In other words:
$$\|K(x) - K(y)\|_X \leqslant \eta\, \|x - y\|_X, \qquad \forall\, x, y \in D,$$
with $\eta := L\,T_0$. Choosing $T_0 < L^{-1}$ we see that K is a contraction.

- The existence and the convergence of
$$\lim_{m \to \infty} x_m(t) = K^m(x_0)(t) =: \phi(t, x_0)$$
follows from the fact that X is a Banach space and its uniqueness follows from the "Contraction Principle".

- The fact that ϕ is differentiable w.r.t. t follows from the fact that $\dot{\phi}(t, x_0) = f(t, \phi(t, x_0))$, where $f \in C(I \times M, \mathbb{R}^n)$.

The Theorem is proved. ■

▶ Remarks:

- Reformulation of Theorem 1.1: *Let $V := [t_0, t_0 + T] \times M_0 \subset I \times M$ and P be the maximum of $\|f\|$ on V. Then the solution of the IVP (1.5) exists at least for $t \in [t_0, t_0 + T_0]$, where $T_0 := \min(T, \delta/P)$, and remains in M_0. The same holds for the interval $[t_0 - T, t_0]$. In these intervals the solution is of class C^1 w.r.t. t.*

- Theorem 1.1 can be interpreted as a *principle of determinism*: if we know the initial conditions of a system, then we can predict its future states. Although such a principle is mathematically validated by Theorem 1.1, its physical interpretation is not as clear as it might seem. The main reasons are:

    1. To find the explicit solution of an IVP (1.5) can be very complicated (if not impossible).
    2. To know the initial state exactly may be very difficult (if not impossible).

Corollary 1.1

Consider the IVP (1.5). If $f \in C^k(I \times M, \mathbb{R}^n), k \geqslant 1$, then there exists, around each point $(t_0, x_0) \in I \times M$, an open set $I_0 \times M_0 \subset I \times M$, such that the IVP admits a unique local solution $\phi(t, t_0, x_0) \in C^{k+1}(I_0, \mathbb{R}^n)$.

Proof. It is a consequence of the fact that $\dot{\phi}(t, t_0, x_0) = f(t, \phi(t, t_0, x_0))$, where $f \in C^k(I \times M, \mathbb{R}^n)$. ■

► The following result states that if f is continuous but not locally Lipschitz continuous in x, uniformly w.r.t. t, we still have existence of solutions, but me may lose their uniqueness. In other words, it provides sufficient conditions for existence of solutions of the IVP (1.5).

Theorem 1.2 (*Peano*)

Consider the IVP (1.5). If $f \in C(I \times M, \mathbb{R}^n)$, then there exists, around each point $(t_0, x_0) \in I \times M$, an open set $I_0 \times M_0 \subset I \times M$, such that the IVP admits a local solution $\phi(t, t_0, x_0) \in C^1(I_0, \mathbb{R}^n)$.

No Proof.

1.4 Dependence of solutions on initial values and parameters

► An IVP (1.5) is *well-posed* if:

1. There exists a unique local solution $\phi(t, t_0, x_0)$.
2. The solution admits a continuous (or even $C^k, k \geqslant 1$) dependence on the initial values (t_0, x_0).

If the IVP (1.5) is well-posed, one expects that small changes in the initial values will result in small changes of the solution.

► We start with the following technical Lemma.

Lemma 1.1 (*Gronwall inequality*)

Let $T > 0$. Let $\zeta : [0,T] \to \mathbb{R}$ be a non-negative function such that

$$\zeta(t) \leqslant \alpha + \int_0^t \beta(s)\,\zeta(s)\,\mathrm{ds}, \qquad \alpha \geqslant 0,\ t \in [0,T],$$

where $\beta(s) \geqslant 0$, $s \in [0,T]$. Then

$$\zeta(t) \leqslant \alpha \exp\left(\int_0^t \beta(s)\,\mathrm{ds}\right), \qquad t \in [0,T].$$

No Proof.

Theorem 1.3

Consider the following IVPs:

$$\begin{cases} \dot{x} = f(t,x), \\ x(t_0) = x_0, \end{cases} \qquad \begin{cases} \dot{y} = g(t,y), \\ y(t_0) = y_0, \end{cases} \tag{1.8}$$

where $f, g \in \mathsf{Lip}\,(I \times M, \mathbb{R}^n)$ (with Lipschitz constant $L \geqslant 0$) and $(t_0, x_0), (t_0, y_0) \in I \times M$. Let $\phi(t, t_0, x_0)$ and $\psi(t, t_0, y_0)$ be the unique local solutions of the IVPs (1.8). Then there holds

$$\|\phi(t,t_0,x_0) - \psi(t,t_0,y_0)\| \leqslant \|x_0 - y_0\| \mathrm{e}^{L|t-t_0|} + \frac{P}{L}\left(\mathrm{e}^{L|t-t_0|} - 1\right), \tag{1.9}$$

where

$$P := \sup_{(t,x)\in V} \|f(t,x) - g(t,x)\|,$$

with $V \subset I \times M$ being a set containing the graphs of $\phi(t,t_0,x_0)$ and $\psi(t,t_0,y_0)$.

Proof. Without any loss of generality we set $t_0 = 0$. We proceed by steps.

- We have $f \in \mathsf{Lip}\,(I \times M, \mathbb{R}^n)$, so that

$$\begin{aligned} \|f(s,\phi(s,x_0)) - g(s,\psi(s,y_0))\| &\leqslant \|f(s,\phi(s,x_0)) - f(s,\psi(s,y_0))\| \\ &+ \|f(s,\psi(s,y_0)) - g(s,\psi(s,y_0))\| \\ &\leqslant L\|\phi(s,x_0) - \psi(s,y_0)\| + P =: L\,\zeta(s) \geqslant 0. \end{aligned}$$

- We also have:

$$\|\phi(t,x_0) - \psi(t,y_0)\| \leqslant \|x_0 - y_0\| + \int_0^t \|f(s,\phi(s,x_0)) - g(s,\psi(s,y_0))\|\,\mathrm{ds},$$

that is

$$\|\phi(t,x_0) - \psi(t,y_0)\| = \zeta(t) - \frac{P}{L} \leqslant \|x_0 - y_0\| + \int_0^t L\,\zeta(s)\,\mathrm{ds},$$

or, equivalently,

$$\zeta(t) \leqslant \|x_0 - y_0\| + \frac{P}{L} + L\int_0^t \zeta(s)\,\mathrm{ds}.$$

- Use Lemma 1.1 with $\alpha := \|x_0 - y_0\| + P/L \geqslant 0$:

$$\zeta(t) \leqslant \alpha\,\mathrm{e}^{L|t|},$$

which is

$$\|\phi(t,x_0) - \psi(t,y_0)\| + \frac{P}{L} \leqslant \left(\|x_0 - y_0\| + \frac{P}{L}\right)\mathrm{e}^{L|t|},$$

which coincides with (1.9) for $t_0 = 0$.

The Theorem is proved. ■

Corollary 1.2

Consider the IVP (1.5). If $f \in \mathsf{Lip}\,(I \times M, \mathbb{R}^n)$ (with Lipschitz constant $L \geqslant 0$) and $(t_0,x_0), (t_0,y_0) \in I \times M$, then

$$\|\phi(t,t_0,x_0) - \phi(t,t_0,y_0)\| \leqslant \|x_0 - y_0\|\,\mathrm{e}^{L|t-t_0|}.$$

Proof. It is a consequence of Theorem 1.3 with $f = g$ (i.e., $P = 0$). ■

► We now denote by $\phi(t,s,\xi)$ the local solution of the IVP (1.5) to emphasize its dependence on the initial value $(s,\xi) \in I \times M$ which is now free to be varied within a neighborhood of (t_0,x_0). Note that with this notation we mean that ξ is the initial point and s is the initial time.

Theorem 1.4

Consider the IVP (1.5). If $f \in \mathsf{Lip}\,(I \times M, \mathbb{R}^n)$, then there exists, around each point $(t_0,x_0) \in I \times M$ a compact set $I_0 \times M_0 \subset I \times M$ such that the IVP admits a unique local solution $\phi(t,s,\xi) \in C(I_0 \times I_0 \times M_0, \mathbb{R}^n)$.

Proof. Using the same notation as in the Proof of Theorem 1.1 we can find a compact set $V := [t_0 - \varepsilon, t_0 + \varepsilon] \times \overline{B_0}$, where $B_0 := \{x \in \mathbb{R}^n : \|x - x_0\| < \delta\}$, such that $\phi(t,t_0,x_0)$ exists for $|t - t_0| \leqslant \varepsilon$ for $\varepsilon > 0$. Then we can find $V' := [t_0' - \varepsilon/2, t_0' + \varepsilon/2] \times \overline{C_0}$, where $C_0 := \{x \in \mathbb{R}^n : \|x - x_0'\| < \delta/2\}$, where $\phi(t,t_0',x_0')$ exists for

$|t - t_0'| \leqslant \varepsilon/2$ whenever $|t_0' - t_0| \leqslant \varepsilon/2$ and $\|x_0' - x_0\| \leqslant \delta/2$. Hence we can choose $I_0 := (t_0 - \varepsilon/4, t_0 + \varepsilon/4)$ and $M_0 := C_0$. The claim follows. ■

► In some cases the continuity w.r.t. (s, ξ) of the solution $\phi(t, s, \xi)$ assured by Theorem 1.4 is not good enough and we need differentiability w.r.t. the initial point ξ.

- Consider the IVP (1.5) with initial point $\xi \in M$. Suppose that $\phi(t, t_0, \xi)$ is differentiable w.r.t. ξ. Then the same is true for $\dot{\phi}(t, t_0, \xi)$. Let us write the IVP (1.5) as
$$\frac{\partial}{\partial t}\phi(t, t_0, \xi) = f(t, \phi(t, t_0, \xi)).$$
- Differentiating w.r.t. ξ and using the chain rule we get:
$$\begin{aligned}\frac{\partial}{\partial \xi}\left(\frac{\partial}{\partial t}\phi(t, t_0, \xi)\right) &= \left(\frac{\partial}{\partial \phi} f(t, \phi(t, t_0, \xi))\right)\frac{\partial}{\partial \xi}\phi(t, t_0, \xi)\\ &= \frac{\partial}{\partial t}\left(\frac{\partial}{\partial \xi}\phi(t, t_0, \xi)\right),\end{aligned}$$
where in the last step we interchanged the t and ξ partial derivatives.
- Defining the $n \times n$ matrix variable
$$\Theta(t, t_0, \xi) := \frac{\partial}{\partial \xi}\phi(t, t_0, \xi), \tag{1.10}$$
and the $n \times n$ matrix
$$A(t, t_0, \xi) := \frac{\partial}{\partial \phi} f(t, \phi(t, t_0, \xi)),$$
we see that Θ obeys the matrix differential equation
$$\dot{\Theta}(t, t_0, \xi) = A(t, t_0, \xi)\,\Theta(t, t_0, \xi), \tag{1.11}$$
which is called *first variational equation*. Note that the matrix A is, formally, nothing but the Jacobian matrix (w.r.t. x) of the function $f = f(t, x)$.
- Integration of (1.11) w.r.t. t yields the integral equation
$$\Theta(t, t_0, \xi) = \Theta(t_0, t_0, \xi) + \int_{t_0}^{t} A(s, t_0, \xi)\,\Theta(s, t_0, \xi)\,\mathrm{d}s,$$
where $\Theta(t_0, t_0, \xi) = \partial\phi(t_0, t_0, \xi)/\partial\xi = \mathbb{1}_n$ (here $\mathbb{1}_n$ is the identity $n \times n$ matrix) due to the fact that $\phi(t_0, t_0, \xi) = \xi$.

► The next Theorem shows, in particular, that if $f \in C^1(I \times M, \mathbb{R}^n)$ then the unique solution of the matrix differential equation (1.11), with $\Theta(t_0, t_0, \xi) = \mathbb{1}_n$, is exactly given by (1.10).

Theorem 1.5

Consider the IVP (1.5). If $f \in C^k(I \times M, \mathbb{R}^n)$, $k \geqslant 1$, then there exists, around each point $(t_0, x_0) \in I \times M$, an open set $I_0 \times M_0 \subset I \times M$ such that the IVP admits a unique local solution $\phi(t, s, \xi) \in C^k(I_0 \times I_0 \times M_0, \mathbb{R}^n)$.

No Proof.

► Finally, we assume that f depends on a set of p *parameters* $\alpha \in A \subseteq \mathbb{R}^p$ and consider the IVP

$$\begin{cases} \dot{x} = f(t, x, \alpha), \\ x(t_0) = x_0. \end{cases} \tag{1.12}$$

We are now interested in varying the initial condition (t_0, x_0) and α. Therefore, we denote by $\phi(t, s, \xi, \alpha)$ the local solution of the IVP (1.12) to emphasize these dependencies.

► The following result, which generalizes Theorem 1.5, establishes a C^k-dependence on (t_0, x_0) and α for the IVP (1.12). As for the case of initial conditions, it is natural to expect that small changes in the parameters will result in small changes of the solution.

Theorem 1.6

Consider the IVP (1.12). If $f \in C^k(I \times M \times A, \mathbb{R}^n)$, $k \geqslant 1$, then there exists, around each point $(t_0, x_0, \alpha) \in I \times M \times A$, an open set $I_0 \times M_0 \times A_0 \subset I \times M \times A$ such that the IVP admits a unique local solution $\phi(t, s, \xi, \alpha) \in C^k(I_0 \times I_0 \times M_0 \times A_0, \mathbb{R}^n)$.

No Proof.

1.5 Prolongation of solutions

► It is now clear that the local solution of an IVP (1.5) is locally defined on $I_0 \subset I$, where I is the definition domain of t for the function f. In particular, even though the IVP is defined for $t \in I = \mathbb{R}$, solutions might not exist for all $t \in \mathbb{R}$. This raises the question of existence of a *maximal interval* on which a *maximal solution* of (1.5) can be defined. In other words, we are now interested in understanding how much we can prolong the interval of existence I_0.

Definition 1.3

Let $\phi(t, t_0, x_0) \in C^1(I_0, \mathbb{R}^n)$ be the unique local solution of the IVP (1.5) defined in I_0.

1. *A solution $\widetilde{\phi}(t, t_0, x_0) \in C^1(\widetilde{I}_0, \mathbb{R}^n)$ of (1.5) is a* **prolongation** *of $\phi(t, t_0, x_0)$ if $I_0 \subset \widetilde{I}_0$ and $\widetilde{\phi}(t, t_0, x_0) = \phi(t, t_0, x_0)$ for all $t \in I_0$.*

2. *A solution $\phi(t, t_0, x_0)$ is called* **maximal solution** *if for any prolongation $\widetilde{\phi}(t, t_0, x_0)$ of $\phi(t, t_0, x_0)$ we have $I_0 = \widetilde{I}_0$. In this case I_0 is called* **maximal interval of existence.**

Theorem 1.7

Consider the IVP (1.5). If $f \in \mathrm{Lip}\,(I \times M, \mathbb{R}^n)$ then there exists a unique maximal solution defined on some open maximal interval (t_-, t_+), where $t_\pm$ depend on $(t_0, x_0) \in I \times M$.

No Proof.

▶ We know from Theorem 1.2 that if f is continuous but not locally Lipschitz continuous in x we still have existence of solutions, but we may lose uniqueness. Even without uniqueness, two given solutions of the IVP (1.5) can be glued together at some point.

Example 1.6 (*Non-uniqueness of the solution*)

Consider the following IVP in $\mathbb{R}$:

$$\begin{cases} \dot{x} = 3\,x^{2/3}, \\ x(0) = 0. \end{cases}$$

The function $f(x) := 3\,x^{2/3}$ is not Lipschitz continuous at $x = 0$. The IVP admits the a two-parameter family of solutions

$$\phi(t, 0) = \begin{cases} (t - t_1)^3 & t < t_1, \\ 0 & t_1 \leqslant t \leqslant t_2, \\ (t - t_2)^3 & t > t_2, \end{cases}$$

for all $t_1 < 0 < t_2$.

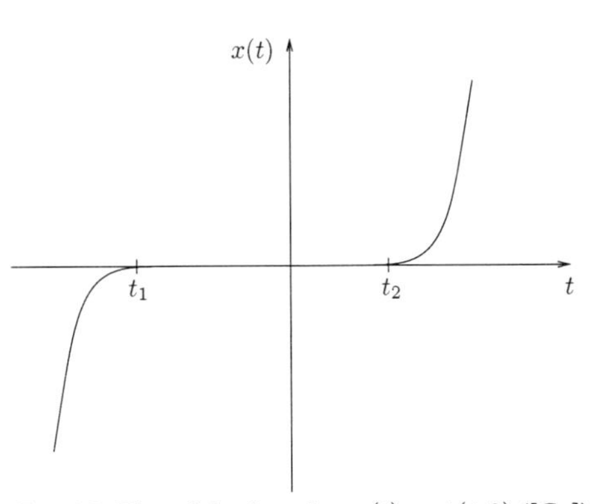

Fig. 1.3. Plot of the function $x(t) = \phi(t, 0)$ ([Ge]).

▶ The following claim holds.

Theorem 1.8

Consider the IVP (1.5). Assume that $f \in \mathrm{Lip}\,(I \times M, \mathbb{R}^n)$ and let (t_-, t_+), with $t_\pm$ depending on $(t_0, x_0) \in I \times M$, be the maximal interval of existence. If $t_+ < \infty$ then the solution must eventually leave every compact set D with $[t_0, t_+] \times D \subset I \times M$ as t approaches t_+. In particular, for $I \times M = \mathbb{R} \times \mathbb{R}^n$, the solution must tend to infinity as t approaches t_+.

No Proof.

► As a consequence of Theorem 1.8, it is possible to show that solutions exist for all $t \in \mathbb{R}$ if $f(t, x)$ grows at most linearly w.r.t. x.

Theorem 1.9

Let $I \times M = \mathbb{R} \times \mathbb{R}^n$ and assume that for every $T > 0$ there exist $S_1(T), S_2(T) \geqslant 0$ such that

$$\|f(t,x)\| \leqslant S_1(T) + S_2(T)\|x\|, \qquad (t,x) \in [-T, T] \times \mathbb{R}^n.$$

Then the solutions of the IVP (1.5) is defined on $I_0 = \mathbb{R}$.

No Proof.

Example 1.7 (*Global solutions and blow-up in finite time*)

1. Consider the following IVP in $\mathbb{R}$:

$$\begin{cases} \dot{x} = x, \\ x(0) = x_0. \end{cases}$$

Its solution is given by

$$\phi(t, x_0) = x_0\, \mathrm{e}^t,$$

which is globally defined for all $t \in \mathbb{R}$.

2. Consider the following IVP in $\mathbb{R}$:

$$\begin{cases} \dot{x} = x^2, \\ x(0) = x_0 > 0. \end{cases} \tag{1.13}$$

Its solution is given by

$$\phi(t, x_0) = \frac{x_0}{1 - x_0 t}.$$

The solution only exists on the interval $(-\infty, 1/x_0)$ and we have a blow up in finite time as $t \to 1/x_0$. There is no way to extend this solution for $t > 1/x_0$. Therefore solutions might only exist locally in t, even for perfectly nice f.

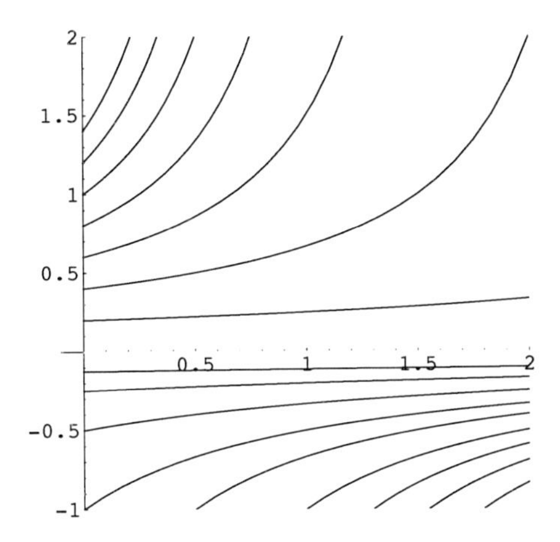

Fig. 1.4. Solutions of the IVP (1.13) ([Ol]).

► We conclude this Chapter with the following claim (see Theorem 1.1, Corollary 1.1, Theorems 1.5 and 1.8): *Every IVP*

$$\begin{cases} \dot{x} = f(t,x), \\ x(t_0) = x_0, \end{cases}$$

where $f \in C^k(I \times M, \mathbb{R}^n)$, $k \geqslant 1$, *has a unique local solution* $\phi(t, t_0, x_0)$ *that is of class* C^k *w.r.t.* (t, t_0, x_0). *Moreover, the solution can be extended in time until it either reaches the boundary of M or blows up to infinity.*

1.6 Exercises

Ch1.E1 (a) Let $f, g : \mathbb{R} \to \mathbb{R}$ be two globally Lipschitz continuous functions. Prove that their composition is globally Lipschitz continuous on $\mathbb{R}$.

(b) Prove that the real function

$$f(x) := \frac{1}{1+x^2}$$

is globally Lipschitz continuous.

(c) Use (a) and (b) to prove that the real function

$$f(x) := \frac{1}{1+\sin^2 x}$$

is globally Lipschitz continuous.

(d) Prove that the scalar IVP ($t \geqslant 0$)

$$\begin{cases} \dot{x} = \dfrac{e^{-t}}{1+\sin^2 x}, \\ x(0) = 1, \end{cases}$$

has a unique solution.

Ch1.E2 Consider the following IVP in $\mathbb{R}$:

$$\begin{cases} \dot{x} = |x|^{p/q}, \\ x(0) = 0, \end{cases}$$

with $p, q \in \mathbb{N} \setminus \{0\}$.

(a) Prove that it has a unique solution if $p > q$.

(b) Prove that it has an infinite number of solutions if $p < q$.

(c) What can you say if $p = q$?

Ch1.E3 Consider the following IVP in $\mathbb{R}$:

$$\begin{cases} \dot{x} = \dfrac{x^2}{x^2+\varepsilon}\sqrt{|x|}, \\ x(0) = 0, \end{cases}$$

with $\varepsilon > 0$. What can you say about existence and uniqueness of its solutions? Is the solution unique if $\varepsilon = 0$?

Ch1.E4 Consider the following IVP in $\mathbb{R}$:

$$\begin{cases} \dot{x} = x^3 - x, \\ x(0) = \dfrac{1}{2}. \end{cases}$$

(a) What can you say about existence and uniqueness of its solutions?

(b) Without solving the ODE, calculate $\lim_{t\to+\infty}\phi(t,1/2)$, where $\phi(t,1/2)$ is the solution.

Ch1.E5 Prove that the second-order scalar IVP

$$\begin{cases} \ddot{x} = 2\,\dot{x} - x + t - 2, \\ (x(0), \dot{x}(0)) = (x_0, v_0), \end{cases}$$

has a unique solution for any choice of the initial values.

Ch1.E6 Consider the following IVP in $\mathbb{R}$:

$$\begin{cases} \dot{x} = x^{\alpha}, \\ x(0) = 0, \end{cases}$$

with $\alpha > 0$.

(a) Does it admit a unique solution? Find the solution(s).

(b) Is there any essential modification of your conclusions if the initial condition is $x(0) = 1$?

Ch1.E7 Consider the following IVPs in $\mathbb{R}$:

$$\begin{cases} \dot{x} = 2\,t\,x, \\ x(0) = 1, \end{cases} \qquad \begin{cases} \dot{x} = t + x, \\ x(0) = 1. \end{cases}$$

For both of them construct the sequence of Picard iterations and obtain the explicit solution.

Ch1.E8 Consider the following ODE in $\mathbb{R}$:

$$\ddot{x} - 2\,t\,\dot{x} - 2\,x = 0.$$

(a) Find the solution by using a power series expansion around $t = 0$.

(b) Solve the IVP

$$\begin{cases} \ddot{x} = 2\,t\,\dot{x} + 2\,x, \\ (x(0), \dot{x}(0)) = (1, 0). \end{cases}$$

Ch1.E9 Prove that the solution of the scalar ODE

$$\dot{x} = 1 + x^{14}$$

diverges to infinity in a finite time, irrespective of the initial conditions $x(0) \in \mathbb{R}$.

Ch1.E10 Consider the following IVP in $\mathbb{R}$:

$$\begin{cases} \dot{x} = \frac{1}{2}\left(x^2 - 1\right), \\ x(0) \in \mathbb{R}. \end{cases}$$

(a) Find the solution.

(b) Find the intervals $(t_-, t_+) \subseteq \mathbb{R}$ on which the solution is defined.

(c) What happens if the solution approaches t_- from the right and t_+ from the left?

Ch1.E11 Consider the following ODE in $\mathbb{R}$:

$$\dot{x} = -2\,t\,x^2.$$

(a) Find the solution (this solution will depend on one arbitrary real constant, say $\alpha \in \mathbb{R}$).

(b) Depending on α, find the maximal solutions.

Ch1.E12 Consider the following IVP in $\mathbb{R}$:

$$\begin{cases} \dot{x} = 1 + \sqrt{|x|}, \\ x(0) = 0. \end{cases}$$

Find the solution and the intervals of existence.

Ch1.E13 Consider the following IVP in $\mathbb{R}$:

$$\begin{cases} \dot{x} = x - x^2, \\ x(t_0) = x_0. \end{cases}$$

(a) Find the solution.

(b) Find the intervals $(t_-, t_+) \subseteq \mathbb{R}$ on which the solution is defined.

(c) What happens if the solution approaches t_- from the right and t_+ from the left?

Ch1.E14 Consider the following IVP in $\mathbb{R}$:

$$\begin{cases} \dot{x} = x - \frac{4}{x}, \\ x(0) = -1. \end{cases}$$

(a) Find the solution.

(b) Find the maximal interval centered at $x = 0$ where the unique solution is defined and continuously differentiable.

Ch1.E15 Let A be a constant real $n \times n$ matrix. Define

$$\cosh(A) := \frac{e^A + e^{-A}}{2}, \qquad \sinh(A) := \frac{e^A - e^{-A}}{2}.$$

(a) Show that for all $t \in \mathbb{R}$ there holds:

$$\frac{d}{dt}\cosh(tA) = A\sinh(tA), \qquad \frac{d}{dt}\sinh(tA) = A\cosh(tA).$$

(b) Verify that the function

$$x(t) := \cosh(tA)\,a + \sinh(tA)\,b, \qquad a, b \in \mathbb{R}^n,\ t \in \mathbb{R},$$

satisfies the following second-order ODE in $\mathbb{R}^n$:

$$\ddot{x} - A^2 x = 0.$$

Ch1.E16 Consider the following IVP in $\mathbb{R}$:

$$\begin{cases} \dot{x} = x + f(t), \\ x(0) \in \mathbb{R}, \end{cases}$$

where $f \in C^1(\mathbb{R}, \mathbb{R})$.

(a) Find the solution.

(b) Use the result in (a) to prove that the IVP of n ODEs

$$\begin{cases} \dot{x}_k = x_{k-1} + x_k, \\ x_k(0) = 1, \end{cases}$$

where $k = 1, \ldots, n$ and $x_0 \equiv 0$, admits the solution

$$\phi_k(t, 1) = e^t \sum_{j=0}^{k-1} \frac{t^j}{j!}, \qquad k = 1, \ldots, n.$$

Compute the limit $\lim_{n \to \infty} \phi_n(t, 1)$.

2

Continuous Dynamical Systems

2.1 Introduction

► A dynamical system is the mathematical formalization of the notion of *deterministic process.*

- The future and past states of many physical, chemical, biological, ecological, economical, and even social systems can be predicted to a certain extent by knowing their present state and the laws governing their time evolution.
- Provided these laws do not change in time, the behavior of such systems can be considered as completely defined by their initial state.

► The main ingredients to define a dynamical system are:

1. A complete metric set M, called *phase space*, containing all possible states of the system. According to the dimension of M, the dynamical system is either finite- or infinite-dimensional. In the finite-dimensional case, M is usually an open subset of $\mathbb{R}^n$ or a smooth manifold (which, roughly speaking, is an abstract surface that locally looks like a linear space, see Definition 4.1).
2. A *time variable* t, which can be continuous ($t \in \mathbb{R}, \mathbb{R}_+$) or discrete ($t \in \mathbb{Z}, \mathbb{Z}_+$).
3. A law of time evolution of states in M defined in terms of *time evolution operator*. The time evolution provides a change in the state of the system. The time evolution operator must satisfy certain conditions to be defined later.

Example 2.1 (*Mechanical systems and the planar pendulum*)

- In the canonical Hamiltonian formulation of classical mechanics, the state of an isolated system with n degrees of freedom is characterized by a $2n$-dimensional real vector:

$$(q,p) := (q_1, \dots, q_n, p_1, \dots, p_n) \in M \subseteq \mathbb{R}^{2n},$$

where q_i is the i-th coordinate and p_i is the corresponding momentum. The time evolution of (q,p) is described by a system of $2n$ first-order ODEs (*Hamilton equations*):

$$\begin{cases} \dot{q}_i = \dfrac{\partial \mathscr{H}(q,p)}{\partial p_i}, \\ \dot{p}_i = -\dfrac{\partial \mathscr{H}(q,p)}{\partial q_i}, \end{cases}$$

with $i = 1, \dots, n$. Here the differentiable function $\mathscr{H} : M \to \mathbb{R}$ is the *Hamiltonian function* of the system.

- Assume $n = 1$ and

$$\mathscr{H}(q,p) := \frac{p^2}{2} + \cos q, \qquad (q,p) \in M := [0, 2\pi) \times \mathbb{R}.$$

Then Hamilton equations are

$$\begin{cases} \dot{q} = p, \\ \dot{p} = -\sin q. \end{cases} \tag{2.1}$$

They are equivalent to the Newton equation

$$\ddot{q} = -\sin q.$$

The *(Hamiltonian) flow* of Hamilton equations is

$$\Phi_t : (0, (q(0), p(0))) \longmapsto (q(t), p(t)),$$

where $(0, (q(0), p(0)))$ is the initial state at $t = 0$ and $(q(t), p(t))$ is the solution of (2.1), defining the state at time $t > 0$. Physically, the flow defines the time evolution of the angular displacement q from the vertical position of a unit-mass planar *pendulum* and the time evolution of the angular velocity $\dot{q} = p$.

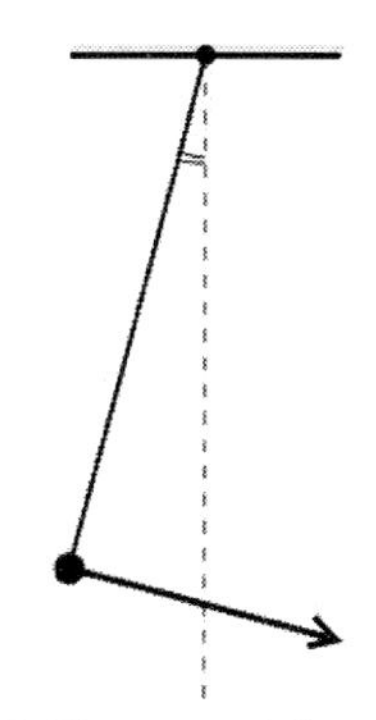

Fig. 2.1. Planar pendulum ([Ku]).

2.2 Definition of a dynamical system

► Let us recall the following facts from Group Theory:

- A *group* is a set $\mathbf{G}$ together with an operation $* : \mathbf{G} \times \mathbf{G} \to \mathbf{G}$, called *group law* of $\mathbf{G}$, that combines any two elements g and h to form another element $g * h \in \mathbf{G}$. The pair $(\mathbf{G}, *)$ must satisfy four axioms:

  1. *Closure*: $g * h \in \mathbf{G}$ for all $g, h \in \mathbf{G}$.
  2. *Associativity*: $(g * h) \cdot k = g * (h * k)$ for all $g, h, k \in \mathbf{G}$.
  3. *Existence of identity element*: there exists a unique element $e \in \mathbf{G}$, such that for every element $g \in \mathbf{G}$, the equation $g * e = e * g = g$ holds.
  4. *Existence of inverse element*: for each $g \in \mathbf{G}$, there exists an element $h \in \mathbf{G}$ such that $g * h = h * g = e$.

- A *semigroup* is a set $\mathbf{G}$ together with an operation $* : \mathbf{G} \times \mathbf{G} \to \mathbf{G}$, such that only the group axioms 1. and 2. hold true. Thus $\mathbf{G}$ need not have an identity element and its elements need not have inverses within $\mathbf{G}$.

- Let M be a set and $\mathbf{G}$ be a group. The (left) *group action* of $\mathbf{G}$ on a set M is a map
$$\Phi_g : \mathbf{G} \times M \to M : (g, x) \mapsto \Phi_g(x),$$
such that the following axioms hold:

  1. There holds:
$$\Phi_{g*h}(x) = \Phi_g(\Phi_h(x)) \equiv (\Phi_g \circ \Phi_h)(x) \qquad \forall g, h \in \mathbf{G},\ x \in M. \tag{2.2}$$
  2. *Existence of identity element*: there exists a unique element $e \in \mathbf{G}$, such that
$$\Phi_e(x) = x \qquad \forall x \in M. \tag{2.3}$$

  From these two axioms, it follows that for every $g \in \mathbf{G}$, the map Φ_g is a bijective map from M to M (its inverse being the map which maps x to $\Phi_{g^{-1}}(x)$).

▶ We now give a general definition of a (finite-dimensional) dynamical system.

Definition 2.1

Let M be either (a subset of) the Euclidean space or a real finite-dimensional smooth manifold. We say that M is the **phase space** *. Let $\mathbf{G}$ be a number set; an element $g \in \mathbf{G}$ is a* **time variable**. *A* **dynamical system** *is a triple $\{\Phi_g, \mathbf{G}, M\}$ defined in terms of a (semi)group action*
$$\Phi_g : \mathbf{G} \times M \to M : (g, x) \mapsto \Phi_g(x), \tag{2.4}$$
which is called **time evolution operator**.

1. *If $\mathbf{G}$ is a group then $\{\Phi_g, \mathbf{G}, M\}$ is an* **invertible dynamical system**.
2. *If $\mathbf{G}$ is a subset of $\mathbb{R}$ (containing 0) then $\{\Phi_g, \mathbf{G}, M\}$ is a* **continuous dynamical system**.
3. *If $\mathbf{G}$ is a subset of $\mathbb{Z}$ (containing 0) then $\{\Phi_g, \mathbf{G}, M\}$ is a* **discrete dynamical system**.

▶ Note that in Definition 2.1 $\mathbf{G}$ is a subset of $\mathbb{R}$ or $\mathbb{Z}$ containing 0. The group law of $\mathbf{G}$ is the addition, i.e., $g * h := g + h$, and $\mathbf{G}$ is abelian, i.e., $g + h = h + g$ for all $g, h \in \mathbf{G}$. Therefore (2.2) reads
$$\Phi_{g+h}(x) = \Phi_g(\Phi_h(x)) = \Phi_h(\Phi_g(x)) = \Phi_{h+g}(x) \qquad \forall g, h \in \mathbf{G},\ x \in M.$$

▶ So far time evolution operators are defined only in a formal way. The typical examples of invertible continuous and discrete dynamical systems are:

- $\{\Phi_t, \mathbb{R}, M\}$, where $M \subset \mathbb{R}^n$ (is compact). Here Φ_t is the global (i.e., $\mathbf{G} = \mathbb{R}$) *flow* of an autonomous differentiable IVP, namely the one-parameter family of *diffeomorphisms* (i.e., Φ_t is a bijection and both Φ_t and its inverse are differentiable)
$$\Phi_t : \mathbb{R} \times M \to M : (t, x) \mapsto \Phi_t(x) := \phi(t, x),$$
where $\phi(t, x)$ is the unique solution of the IVP with $\Phi_0(x) = \phi(0, x) = x \in M$. Here x plays the role of the initial value of the IVP.

    (a) Property (2.2) reads
    $$\Phi_{t+s}(x) = (\Phi_t \circ \Phi_s)(x) \qquad \forall t, s \in \mathbb{R},\ x \in M,$$
    which means that the state at time $t + s$ when starting at x is identical to the state at time t when starting at $\Phi_s(x)$.

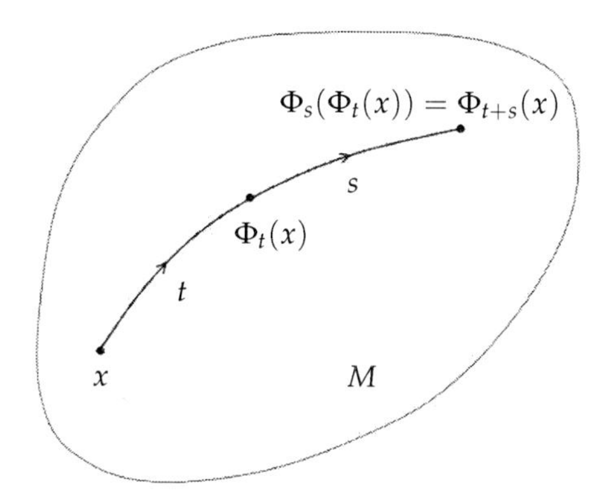

Fig. 2.2. Property (2.2) ([Ku]).

    (b) Property (2.3),
    $$\Phi_0(x) = x \qquad \forall x \in M,$$
    says that x is the prescribed state at time $t = 0$.

    (c) The invertibility property says that
    $$(\Phi_t \circ \Phi_{-t})(x) = x \qquad \forall t \in \mathbb{R},\ x \in M.$$

    In such a situation one says that Φ_t is a *one-parameter global Lie group of diffeomorphisms* on $\mathbb{R}^n$. One says that Φ_t is a *one-parameter local Lie group of diffeomorphisms* on $\mathbb{R}^n$ if instead of $\mathbf{G} = \mathbb{R}$ one has only a subset of $\mathbb{R}$ (containing 0).

- $\{\Phi_k, \mathbb{Z}, M\}$, $M \subset \mathbb{R}^n$, where $\Phi_k \equiv \Phi^k$ is the k-th iterate of a homeomorphism (i.e., Φ is a bijection and both Φ and its inverse are continuous)
$$\Phi : M \to M,$$

namely

$$\Phi^k(x) := (\underbrace{\Phi \circ \cdots \circ \Phi}_{k\,\text{times}})(x), \qquad k \in \mathbb{N},\ x \in M,$$

$$\Phi^k(x) := (\underbrace{\Phi^{-1} \circ \cdots \circ \Phi^{-1}}_{|k|\,\text{times}})(x), \qquad -k \in \mathbb{N},\ x \in M.$$

(a) Property (2.2) reads

$$\Phi^{i+j}(x) = (\Phi^i \circ \Phi^j)(x) \qquad \forall\, i,j \in \mathbb{Z},\ x \in M,$$

which means that the $(i+j)$-th iterate is identical to the composition of the i-th and j-th iterates.

(b) Property (2.3),

$$\Phi^0(x) = x \qquad \forall\, x \in M,$$

says that Φ^0 is the identity map.

► One can consider also dynamical systems whose future states for $t > 0$ (resp. $k > 0$) are completely determined by their initial state at $t = 0$ (resp. $k = 0$), but the history for $t < 0$ (resp. $k < 0$) cannot be reconstructed. Such (noninvertible) dynamical systems are described by semigroup actions defined only for $t \geqslant 0$ (resp. $k \geqslant 0$). In the continuous time case one says that Φ_t is a *semiflow*.

2.2.1 *Orbits, invariant sets, invariant functions and stability*

► Let $\{\Phi_g, \mathbf{G}, M\}$ be a dynamical system. In particular $g \equiv t \in \mathbb{R}$ for a continuous (invertible and globally defined) dynamical system and $g \equiv k \in \mathbb{Z}$ for a discrete (invertible and globally defined) dynamical system.

Definition 2.2

An **orbit** *starting at $x_0 \in M$ is the ordered subset of M defined by*

$$\mathcal{O}(x_0) := \{x \in M \,:\, x = \Phi_g(x_0)\ \forall g \in \mathbf{G}\}.$$

► Remarks:

- Orbits of a continuous dynamical system are curves in M parametrized by the time t and oriented by the direction of time advance.
- Orbits of a discrete dynamical system are sequences of points in M enumerated by increasing integers.

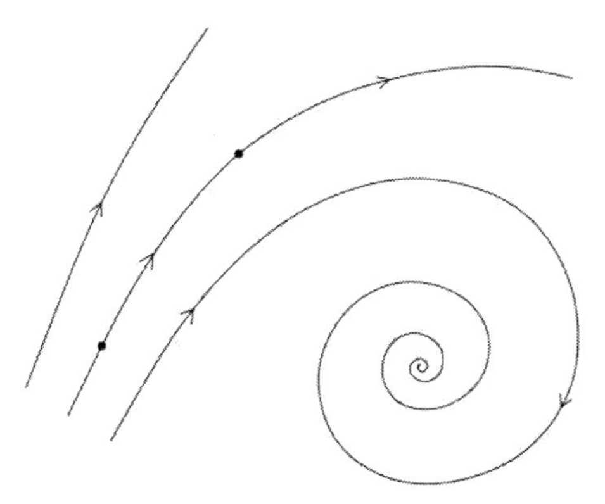

Fig. 2.3. Continuous orbits ([Ku]).

- If $y_0 = \Phi_g(x_0)$ for some $g \in \mathbf{G}$, then the sets $\mathcal{O}(x_0)$ and $\mathcal{O}(y_0)$ coincide. In particular different orbits are disjoint.

► The simplest orbits are *fixed points* and *cycles*.

Definition 2.3

1. *A point $\tilde{x} \in M$ is called a* **fixed point** *(or* **equilibrium point***) if $\tilde{x} = \Phi_g(\tilde{x})$ for all $g \in \mathbf{G}$.*
2. *A* **cycle** *(or* **periodic orbit***) is an orbit $\mathcal{O}(x_0)$ such that each point $x \in \mathcal{O}(x_0)$ satisfies $\Phi_{g+T}(x) = \Phi_g(x)$ with some fixed $T \in \mathbf{G}$, for all $g \in \mathbf{G}$. The minimal T with this property is the* **period** *of the cycle.*

► If a system starts its evolution at x_0 on the cycle, it will return exactly to this point after every T units of time. The system exhibits *periodic oscillations.*

- Cycles of a continuous dynamical system are closed curves in M.
- Cycles of a discrete dynamical system are finite sets of points

$$x_0, \Phi(x_0), \Phi^2(x_0), \dots, \Phi^T(x_0) = x_0, \qquad T \in \mathbb{N}.$$

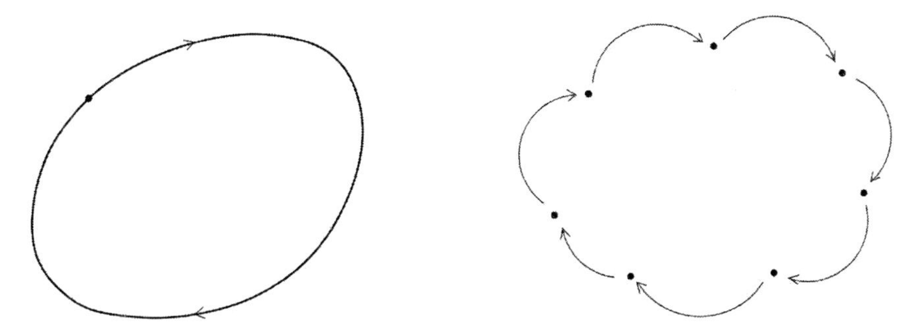

Fig. 2.4. Continuous and discrete periodic orbits ([Ku]).

Definition 2.4

The **phase portrait** *of* $\{\Phi_g, \mathbf{G}, M\}$ *is a partitioning of* M *into orbits.*

► The phase portrait contains a lot of information on the behavior of a dynamical system. By looking at the phase portrait, we can determine number and types of asymptotic states to which the system tends as $g \to +\infty$ (and as $g \to -\infty$ if the system is invertible).

Definition 2.5

1. *An* **invariant set** *of* $\{\Phi_g, \mathbf{G}, M\}$ *is a subset* $S \subset M$ *such that* $x_0 \in S$ *implies* $\Phi_g(x_0) \in S$ *for all* $g \in \mathbf{G}$*, or, equivalently,* $\Phi_g(S) \subseteq S$ *for all* $g \in \mathbf{G}$.
2. *An* **invariant function** *of* $\{\Phi_g, \mathbf{G}, M\}$ *is a function* $F : M \to \mathbb{R}$ *such that*

$$F(\Phi_g(x)) = F(x) \qquad \forall g \in \mathbf{G},\ x \in M.$$

► To represent an observable asymptotic state of a dynamical system, an invariant set must possess some "stability property".

Definition 2.6

Let S *be a closed invariant set.*

1. S *is* **(Lyapunov) stable** *if for any sufficiently small neighborhood* $U \supset S$ *there exists a neighborhood* $V \supset S$ *such that* $\Phi_g(x) \in U$ *for all* $x \in V$ *and all* $g > 0$.
2. S *is* **attracting** *if there exists a neighborhood* $U \supset S$ *such that* $\Phi_g(x) \to S$ *as* $g \to +\infty$*, for all* $x \in U$.

3. *S is* **asymptotically stable** *if it is Lyapunov stable and attracting.*

▶ Remarks:

- If S is a fixed point or a cycle, then Definition 2.6 turns into the definition of stable fixed point or stable cycle.
- There are invariant sets that are Lyapunov stable but not asymptotically stable and invariant sets that are attracting but not asymptotically stable.

▶ From now on we shall consider only continuous dynamical systems, meaning that our time will be a real variable.

2.3 Autonomous IVPs as continuous dynamical systems

▶ We now show how to realize more concretely a continuous dynamical system in terms of the unique solution of an autonomous IVP on $M \subseteq \mathbb{R}^n$:

$$\begin{cases} \dot{x} = f(x), \\ x(t_0) = x_0, \end{cases} \tag{2.5}$$

with $f \in C^k(M, \mathbb{R}^n)$, $k \geqslant 1$.

- Theorems 1.1 (see also Corollary 1.1) and 1.7 assure that (2.5) admits a unique maximal local solution $\phi(t, x_0)$, $x_0 \in M$, defined on a *maximal interval*

$$I_{x_0} := (t_-(x_0), t_+(x_0)) \subseteq \mathbb{R}.$$

- Since (2.5) is autonomous it does not matter at what time t_0 we specify the initial point x_0 (*time translational symmetry*). Indeed, if $\phi(t, x_0)$, $t \in I_{x_0}$, is a solution of (2.5), then so is $\psi(t, x_0) := \phi(t + s, x_0)$ with $t + s \in I_{x_0}$:

$$\dot{\psi}(t, x_0) = \dot{\phi}(t + s, x_0) = f(\phi(t + s, x_0)) = f(\psi(t, x_0)).$$

Therefore we can choose $t_0 = 0$ and I_{x_0} always contains 0.

- Define the set

$$W := \bigcup_{x_0 \in M} I_{x_0} \times \{x_0\} \subseteq \mathbb{R} \times M. \tag{2.6}$$

Then the *flow* (i.e., the continuous time evolution operator (2.4)) of (2.5) is the map

$$\Phi_t : W \to M \; : \; (t, x_0) \mapsto \Phi_t(x_0) := \phi(t, x_0), \tag{2.7}$$

where $\phi(t, x_0)$ is the maximal solution starting at x_0. We have:

$$\Phi_0(x_0) := x_0,$$

which is property (2.3).

- The map Φ_t has the following property (cf. formula (2.2)):

$$\Phi_{s+t}(x_0) = \Phi_s(\Phi_t(x_0)), \qquad x_0 \in M, \tag{2.8}$$

in the sense that if the r.h.s. is defined, so is the l.h.s., and they are equal. Indeed, suppose that $s > t > 0$ and that $\Phi_s(\Phi_t(x_0)$ is defined, i.e., $t \in I_{x_0}$ and $s \in I_{\Phi_t(x_0)}$. Define the map

$$\psi : (t_-(x_0), s+t] \to M,$$

by

$$\psi(r) := \begin{cases} \Phi_r(x_0) & r \in (t_-(x_0), t], \\ \Phi_{r-t}(\Phi_t(x_0)) & r \in [t, t+s]. \end{cases}$$

Then ψ is a solution and $\psi(0) = \Phi_0(x_0) = x_0$. Hence $s + t \in I_{x_0}$. Moreover,

$$\Phi_{t+s}(x_0) = \psi(s+t) = \Phi_s(\Phi_t(x_0)).$$

Property (2.8) expresses the determinism of the system.

- Setting $s = -t$ in (2.8) we see that Φ_t is a local diffeomorphism with inverse Φ_{-t}.

► Let us summarize some of the above results in the following Theorem, whose detailed proof is omitted (note that some of the claims have been justified above).

Theorem 2.1

Consider the IVP (2.5).

1. *For all $x_0 \in M$ there exists a maximal interval $I_{x_0} \subseteq \mathbb{R}$ containing* 0 *and a corresponding unique maximal solution of class C^k w.r.t. t.*
2. *The set W defined by (2.6) is open.*
3. *The map Φ_t defined by (2.7) is a one-parameter local Lie group of diffeomorphisms (of class C^k) on M.*

No Proof.

Example 2.2 (*A one-dimensional continuous dynamical system*)

Consider the continuous dynamical system defined by the scalar IVP

$$\begin{cases} \dot{x} = x^3, \\ x(0) = x_0. \end{cases}$$

The flow is

$$\Phi_t : W \to \mathbb{R} \; : \; (t, x_0) \mapsto \Phi_t(x_0) := \phi(t, x_0) = \frac{x_0}{\sqrt{1 - 2\,x_0^2\,t}},$$

where

$$W := \{(t, x_0) \; : \; 2\,t\,x_0^2 < 1\} \subset \mathbb{R}^2, \qquad I_{x_0} := \left(-\infty, \frac{1}{2\,x_0^2}\right).$$

► Let us now have a closer look at the geometric properties of the IVP (2.5). It turns out that the geometry of the vector field f is closely related to the geometry of the solutions of the IVP when the solutions are viewed as curves in M. One of the main goals of the *geometric method* is to derive qualitative properties of solutions directly from f without solving the IVP.

- The solution $t \mapsto \phi(t, x_0)$ defines for each $x_0 \in M$ two different curves:
  1. A *solution curve* (also called *integral curve* or *trajectory curve*):
  $$\gamma(x_0) := \{(t, x) \; : \; x = \Phi_t(x_0),\, t \in I_{x_0}\} \subset I_{x_0} \times M.$$
  2. An *orbit*, which is the projection of $\gamma(x_0)$ onto M:
  $$\mathcal{O}(x_0) := \{x \; : \; x = \Phi_t(x_0),\, t \in I_{x_0}\} \subset M.$$

  Orbits can be splitted into *forward* (+) and *backward orbits* (-):
  $$\mathcal{O}_+(x_0) := \{x \; : \; x = \Phi_t(x_0),\, t \in (0, t_+(x_0))\},$$
  $$\mathcal{O}_-(x_0) := \{x \; : \; x = \Phi_t(x_0),\, t \in (t_-(x_0), 0)\}.$$
  One can see that $\mathcal{O}(x_0)$ is a periodic orbit (cf. Definition 2.3) if and only if $\mathcal{O}_+(x_0) \cap \mathcal{O}_-(x_0) \neq \emptyset$.

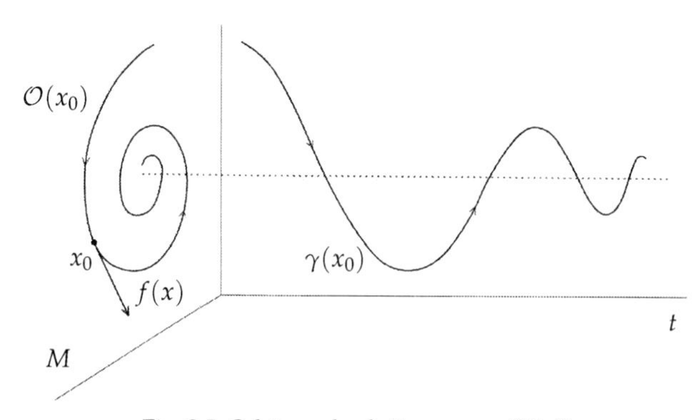

Fig. 2.5. Orbits and solution curves ([Ku]).

Both curves $\gamma(x_0)$ and $\mathcal{O}(x_0)$ are parametrized by time t and oriented by the direction of time advance. The *phase portrait* is the collective graph of orbits in M.

- The map $x \mapsto f(x)$, $x \in M$, defines a *vector field* on M.
  (a) If we replace f by $-f$ in (2.5) then $\Phi_t(x_0) \mapsto \Phi_{-t}(x_0)$.
  (b) The zeros of f define the fixed points of Φ_t. Indeed, if $f(\widetilde{x}) = 0$, then $\Phi_t(\widetilde{x}) = \widetilde{x}$ (cf. Definition 2.3).
  (c) Each orbit $\mathcal{O}(x_0)$ is tangent to f at each point $x \in \mathcal{O}(x_0)$. Note that an orbit cannot cross itself because, in that case, we would have two different tangent vectors of the same vector field at the crossing point. On the other hand, a curve in M as in Fig. 2.6 can exist if the crossing point is a fixed point. Such a curve is the union of the four orbits.

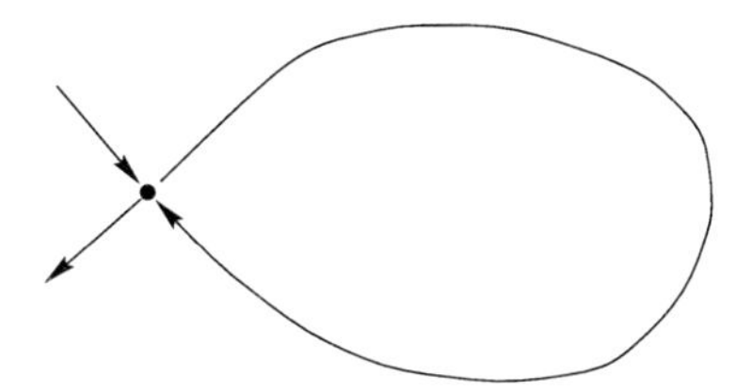

Fig. 2.6. A curve in the phase portrait which consists of four orbits ([Ch]).

► We finally provide a characterization of invariant sets (cf. Definition 2.5) of continuous dynamical systems generated by an IVP (2.5).

- A point $x_0 \in M$ is called $(\pm)$-*complete* if $t_\pm(x_0) = \pm\infty$ and *complete* if $I_{x_0} = \mathbb{R}$.
- As a consequence of Theorem 1.8 one infers that if $\mathcal{O}_+(x_0)$ (resp. $\mathcal{O}_-(x_0)$) lies in a compact subset D of M then x_0 is $(+)$-complete (resp. $(-)$-complete).
- The vector field f and its flow Φ_t are called *complete* if every point $x \in M$ is complete, meaning that Φ_t is globally defined, i.e., $W = \mathbb{R} \times M$. Another way to formulate this is to say that compactly supported vector fields are complete: they generate one-parameter global Lie groups of diffeomorphisms on M.
- A set $S \subset M$ is called $(\pm)$-*invariant* if
$$\mathcal{O}_\pm(x_0) \subseteq S \qquad \forall\, x_0 \in S,$$
and *invariant* if
$$\mathcal{O}(x_0) \subseteq S \qquad \forall\, x_0 \in S.$$
- We conclude saying that if $S \subset M$ is a compact $(\pm)$-invariant set then all points in S are complete.

Example 2.3 (*Phase portrait of a scalar IVP*)

Consider a continuous dynamical system generated by a scalar IVP

$$\begin{cases} \dot{x} = f(x), \\ x(0) = x_0, \end{cases}$$

where $f : \mathbb{R} \to \mathbb{R}$ is a (smooth) function. Looking at the graph of the function $y = f(x)$ in the (x, y)-plane, one concludes that any orbit is either a root of the equation $f(x) = 0$ or an open segment of the x-axis, bounded by such roots or extending to infinity. The orientation of a nontrivial (segment) orbit is determined by the sign of $f(x)$ in the corresponding interval.

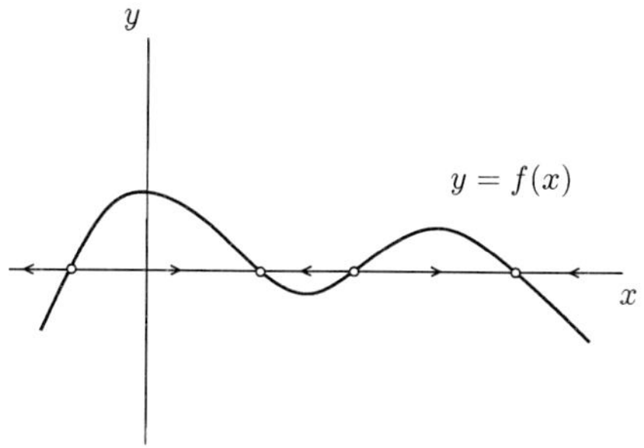

Fig. 2.6. Phase portrait of a scalar ODE ([Ku]).

Example 2.4 (*Phase portrait of a planar IVP*)

Consider a continuous dynamical system generated by the planar IVP

$$\begin{cases} \dot{x}_1 = x_2^2, \\ \dot{x}_2 = x_1, \end{cases} \tag{2.9}$$

with given initial conditions $(x_1(0), x_2(0)) \in \mathbb{R}^2$. The main qualitative properties of the phase portrait can be derived just by looking at the flow lines of the vector field $f(x_1, x_2) = (x_2^2, x_1)$ in the (x_1, x_2)-plane

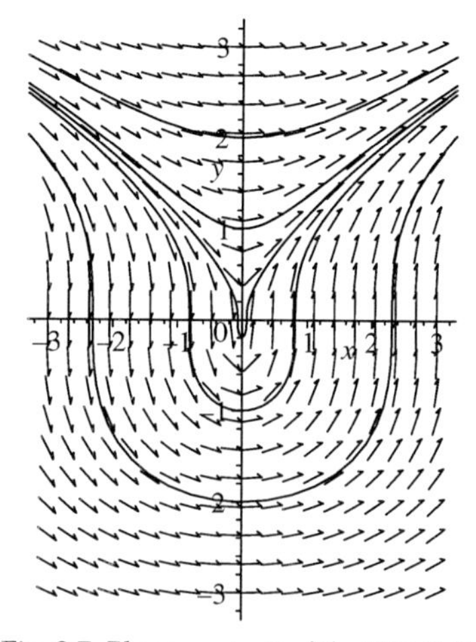

Fig. 2.7. Phase portrait of the IVP (2.9).

2.3.1 Flows, vector fields and invariant functions

▶ In this Subsection we are not interested in the uniqueness problem of solutions of IVPs. Therefore we relax our regularity conditions and we assume that all maps and functions are smooth, i.e., C^∞. Furthermore, we will be mainly interested in properties of flows instead of IVPs. For this reason we slightly change our notation denoting the initial condition x_0 of an IVP by x. Next Theorem 2.2 should clarify any further notational ambiguity.

▶ Consider a one-parameter local Lie group of smooth diffeomorphisms:

$$\Phi_t : I \times M \to M \, : \, (t,x) \mapsto \underline{x}(t,x) := \Phi_t(x), \tag{2.10}$$

where $I \subset \mathbb{R}$ contains 0. In particular we have:

$$\underline{x}(0,x) = \Phi_0(x) = x, \qquad \Phi_{t+s}(x) = \Phi_t(\Phi_s(x)) = \Phi_s(\Phi_t(x)), \tag{2.11}$$

for all $x \in M$ and $t, s \in I$, $t+s \in I$.

▶ The following Theorem establishes a fundamental one-to-one correspondence between one-parameter local Lie groups of smooth diffeomorphisms (i.e., continuous dynamical systems) and solutions of autonomous smooth IVPs.

Theorem 2.2 (*Lie*)

Any one-parameter local Lie group of smooth diffeomorphisms (2.10) is equivalent to the solution of an autonomous IVP

$$\begin{cases} \dot{\underline{x}} = f(\underline{x}), \\ \underline{x}(0,x) = \Phi_0(x) = x, \end{cases} \tag{2.12}$$

where

$$f(\underline{x}) := \left.\frac{\mathrm{d}}{\mathrm{d}t}\right|_{t=0} \Phi_t(\underline{x}).$$

Proof. We proceed by steps.

- Expanding the flow $\Phi_t(x)$ in powers of t around $t = 0$ we get

$$\begin{aligned} \Phi_t(x) &= x + t \left.\frac{\mathrm{d}}{\mathrm{d}t}\right|_{t=0} \Phi_t(x) + O(t^2) \\ &= x + t\, f(x) + O(t^2). \end{aligned}$$

- Consider formula $\Phi_{t+s}(x) = \Phi_s(\Phi_t(x))$. Expanding the l.h.s. in powers of s around $s = 0$ we get

$$\Phi_{t+s}(x) = \Phi_t(x) + s \frac{\mathrm{d}}{\mathrm{d}t}\Phi_t(x) + O(s^2). \tag{2.13}$$

Expanding the r.h.s. in powers of s around $s = 0$ we get

$$\Phi_s(\Phi_t(x)) = \Phi_t(x) + s\, f\,(\Phi_t(x)) + O(s^2). \tag{2.14}$$

- Equating (2.13) and (2.14) we see that $\Phi_t(x)$ satisfies the IVP

$$\begin{cases} \frac{\mathrm{d}}{\mathrm{d}t}\Phi_t(x) = f\,(\Phi_t(x))\,, \\ \Phi_0(x) = x, \end{cases}$$

that is

$$\begin{cases} \dot{\underline{x}} = f\,(\underline{x})\,, \\ \underline{x}(0,x) = x. \end{cases} \tag{2.15}$$

- The IVP (2.15) satisfies the conditions of Theorem 1.1 and therefore it admits a (unique) local solution $\underline{x}(t,x) = \Phi_t(x)$.

The Theorem is proved. ■

► Theorem 2.2 shows that the *infinitesimal transformation*

$$\Phi_t(x) = x + t\, f(x) + O(t^2),$$

contains the essential information to characterize a one-parameter local Lie group of smooth diffeomorphisms. This justifies the next definition.

Definition 2.7

1. *The* **infinitesimal generator** *of Φ_t is the linear differential operator*

$$\mathbf{v} := \sum_{i=1}^{n} f_i(x)\frac{\partial}{\partial x_i}, \tag{2.16}$$

where $f_i(x)$ is the i-th component of

$$f(x) := \left.\frac{\mathrm{d}}{\mathrm{d}t}\right|_{t=0} \Phi_t(x).$$

2. *The action of* $\mathbf{v}$ *on a function $F \in \mathscr{F}(M,\mathbb{R})$ defines the* **Lie derivative of** F **along v**, *denoted by $\mathfrak{L}_{\mathbf{v}}F$:*

$$(\mathfrak{L}_{\mathbf{v}}F)(x) := \mathbf{v}[F(x)] = \sum_{i=1}^{n} f_i(x)\frac{\partial F}{\partial x_i} = \langle\, f(x), \,\mathrm{grad}_x F(x)\,\rangle\,. \tag{2.17}$$

Here $\langle\,\cdot,\cdot\,\rangle$ is the scalar product in $\mathbb{R}^n$ and

$$\mathrm{grad}_x F(x) := \left(\frac{\partial F}{\partial x_1}, \ldots, \frac{\partial F}{\partial x_n}\right)$$

is the gradient of F w.r.t. x.

▶ Remarks:

- $\mathbf{v}$ is the tangent vector to the orbit of $\Phi_t(x)$ at each point $x \in M$. In Differential Geometry $\mathbf{v}$ takes the name of (smooth) *tangent vector field*, or simply (smooth) *vector field*. Technically, it would be more precise to use the notation $\mathbf{v}|_x$ (or $\mathbf{v}_x$) instead of $\mathbf{v}$ and $\partial/\partial x_i|_x$ (or $(\partial/\partial x_i)_x$) instead of $\partial/\partial x_i$. It may happen that we call both $\mathbf{v}$ and f vector field.

- Smooth vector fields $\mathbf{v}$ over M form a real vector space, denoted by $\mathfrak{X}(M)$, w.r.t. the operations

$$(\mathbf{v}+\mathbf{w})|_x = \mathbf{v}|_x + \mathbf{w}|_x \qquad (\lambda\,\mathbf{v})|_x = \lambda\,\mathbf{v}|_x,$$

  for all $\mathbf{v},\mathbf{w} \in \mathfrak{X}(M)$ and $\lambda \in \mathbb{R}$. We can multiply $\mathbf{v}$ by smooth functions as well, by the rule

$$(F\,\mathbf{v})|_x := F(x)\mathbf{v}|_x.$$

- From (2.16) we see that $\mathbf{v}$ is regarded as a differential operator which naturally acts on smooth functions on M. Indeed, each smooth vector field $\mathbf{v}$ becomes a *derivation* on the algebra of smooth functions $\mathscr{F}(M,\mathbb{R})$ when we define $\mathbf{v}[F]$ to be the element of $\mathscr{F}(M,\mathbb{R})$ whose value at a point $x \in M$ is the *directional derivative* (2.17) of F at x in the direction $\mathbf{v}|_x$.

- Saying that $\mathbf{v}$ is a *derivation* on $\mathscr{F}(M,\mathbb{R})$ means that

$$\mathbf{v}[F\,G] = \mathbf{v}[F]G + F\,\mathbf{v}[G] \qquad \forall\, F,G \in \mathscr{F}(M,\mathbb{R}),$$

  which is called *Leibniz rule*. It can be proved any derivation on $\mathscr{F}(M,\mathbb{R})$ arises in this fashion from a uniquely determined smooth vector field $\mathbf{v}$.

▶ The next Theorem (claim *2.*) shows how to use the infinitesimal generator $\mathbf{v}$ to find the explicit solution of the associated IVP (2.12).

Theorem 2.3

Let Φ_t be the one-parameter local Lie group of smooth diffeomorphisms (2.10). Let $\mathbf{v} \in \mathfrak{X}(M)$ be its infinitesimal generator.

1. *There holds*

$$(\mathfrak{L}_{\mathbf{v}}F)(x) = \mathbf{v}[F(x)] = \left.\frac{\mathrm{d}}{\mathrm{d}t}\right|_{t=0} (F\circ\Phi_t)(x),$$

for any $F \in \mathscr{F}(M,\mathbb{R})$. In particular,

$$(\mathfrak{L}_{\mathbf{v}} x_i)(x) = \mathbf{v}[x_i] = \left.\frac{\mathsf{d}}{\mathsf{d}t}\right|_{t=0} (\Phi_t(x))_i, \qquad i = 1,\dots,n.$$

2. *There holds (**Lie series**):*

$$(\Phi_t(x))_i = \exp(t\,\mathbf{v})\, x_i := \sum_{k=0}^{\infty} \frac{t^k}{k!} \mathbf{v}^k[x_i], \qquad i = 1,\dots,n.$$

where $\mathbf{v}^k := \mathbf{v}\,\mathbf{v}^{k-1}$, $\mathbf{v}^0$ being the identity.

Proof. We prove both claims.

1. From Theorem 2.2 we know that $\underline{x}(t,x) = \Phi_t(x)$ is the solution of the IVP (2.12). Hence by the chain rule we have:

$$\begin{aligned} \left.\frac{\mathsf{d}}{\mathsf{d}t}\right|_{t=0} (F \circ \Phi_t)(x) &= \left\langle \mathsf{grad}_x F(x), \left.\frac{\mathsf{d}}{\mathsf{d}t}\right|_{t=0} \Phi_t(x) \right\rangle \\ &= \langle \mathsf{grad}_x F(x), f(x) \rangle = (\mathfrak{L}_{\mathbf{v}} F)(x). \end{aligned}$$

In particular, setting $F(x) = x_i$ we get

$$\begin{aligned} (\mathfrak{L}_{\mathbf{v}} x_i)(x) &= \mathbf{v}[x_i] = \left.\frac{\mathsf{d}}{\mathsf{d}t}\right|_{t=0} (x_i \circ \Phi_t)(x) \\ &= \left.\frac{\mathsf{d}}{\mathsf{d}t}\right|_{t=0} (\Phi_t(x))_i = f_i(x). \end{aligned}$$

2. Expanding component-wise the flow $\Phi_t(x)$ in powers of t around $t = 0$ we get

$$\begin{aligned} (\Phi_t(x))_i &= \sum_{k=0}^{\infty} \frac{t^k}{k!} \left.\frac{\mathsf{d}^k}{\mathsf{d}t^k}\right|_{t=0} (\Phi_t(x))_i \\ &= x_i + t \left.\frac{\mathsf{d}}{\mathsf{d}t}\right|_{t=0} (\Phi_t(x))_i + \frac{t^2}{2} \left.\frac{\mathsf{d}^2}{\mathsf{d}t^2}\right|_{t=0} (\Phi_t(x))_i + \dots \end{aligned}$$

Now we have:

$$\left.\frac{\mathsf{d}}{\mathsf{d}t}\right|_{t=0} (\Phi_t(x))_i = \mathbf{v}[x_i],$$

and

$$\left.\frac{\mathsf{d}^2}{\mathsf{d}t^2}\right|_{t=0} (\Phi_t(x))_i = \mathbf{v}\left[\left.\frac{\mathsf{d}}{\mathsf{d}t}\right|_{t=0} (\Phi_t(x))_i \right] = \mathbf{v}[\mathbf{v}[x_i]] = \mathbf{v}^2[x_i],$$

and, in general,

$$\left.\frac{\mathsf{d}^k}{\mathsf{d}t^k}\right|_{t=0} (\Phi_t(x))_i = \mathbf{v}^k[x_i].$$

Both claims are proved. ■

► In summary, there are, in principle, two ways to construct explicitly Φ_t:

1. Integrate the IVP (2.12).
2. Express Φ_t in terms of its Lie series.

Example 2.5 (*Linear IVPs*)

1. On $\mathbb{R}^n$ consider the infinitesimal generator

$$\mathbf{v} := \sum_{i=1}^{n} a_i \frac{\partial}{\partial x_i}, \qquad a_i \in \mathbb{R}.$$

Then, for $x \in \mathbb{R}^n$, the corresponding flow,

$$\underline{x}(t,x) = \Phi_t(x) = \exp(t\,\mathbf{v})x = \left(1 + t\sum_{i=1}^{n} a_i \frac{\partial}{\partial x_i} + O(t^2)\right) x = x + t\,a,$$

with $a := (a_1, \dots, a_n)$, is the flow of the IVP

$$\begin{cases} \dot{\underline{x}} = a, \\ \underline{x}(0,x) = \Phi_0(x) = x. \end{cases}$$

2. On $\mathbb{R}^n$ consider the infinitesimal generator

$$\mathbf{v} := \sum_{i=1}^{n} \left(\sum_{j=1}^{n} A_{ij}x_j\right) \frac{\partial}{\partial x_i}, \qquad A_{ij} \in \mathbb{R}.$$

Then, introducing the $n \times n$ real matrix $A := (A_{ij})_{1 \leqslant i,j \leqslant n}$, the corresponding flow,

$$\underline{x}(t,x) = \Phi_t(x) = \exp(t\,\mathbf{v})x = \mathrm{e}^{t\,A}x,$$

is the flow of the IVP

$$\begin{cases} \dot{\underline{x}} = A\,x, \\ \underline{x}(0,x) = \Phi_0(x) = x. \end{cases}$$

► A consequence of Theorem 2.3 is the following claim.

Corollary 2.1

Let $F \in \mathscr{F}(M,\mathbb{R})$. Then there holds

$$F(\Phi_t(x)) = \exp(t\,\mathbf{v})F(x) \qquad \forall\, t \in I,\ x \in M.$$

No Proof.

► According to Definition 2.5, an *invariant function* of Φ_t is a function $F \in \mathscr{F}(M,\mathbb{R})$ such that

$$F(\Phi_t(x)) = F(x) \qquad \forall\, t \in I,\ x \in M. \tag{2.18}$$

- Condition (2.18) means that F is constant on every orbit of Φ_t. The hypersurface
$$S_h := \{x \in M \,:\, F(x) = h\},$$
where $h \in \mathbb{R}$ is a fixed constant (depending on the initial condition), is called *level set* of F. Therefore, if F is an invariant function under the action of Φ_t, then clearly every level set of F is an invariant set of Φ_t. However, it is not true that if the set of zeros of a function, $\{x \in M \,:\, F(x) = 0\}$, is an invariant set then the function itself is an invariant function. Nevertheless it can be proved that if every level set of F is an invariant set, then F is an invariant function.

- It may happen that a given flow Φ_t admits more than one invariant function, say m functions $F_1, \ldots, F_m \in \mathscr{F}(M, \mathbb{R})$. These functions define m distinct invariant functions if they satisfy condition (2.18) and they are *functionally independent* on M, that is
$$\mathsf{rank}\,(\,\mathsf{grad}_x F_1(x), \ldots, \mathsf{grad}_x F_m(x)) = m \qquad \forall\, x \in M.$$

- The knowledge of invariant functions provides useful information about solutions of the corresponding IVP. In particular, if the number m of invariant functions is $m = n - 1$ then the IVP is *solvable*. Indeed, for any initial point the orbit through that point lies in the intersection of the $n - 1$ level sets
$$S_{h_i} := \{x \in M \,:\, F_i(x) = h_i\}, \qquad i = 1, \ldots, n-1,$$
with $h_i \in \mathbb{R}$ being fixed by the initial condition. The IVP is then reduced to a one-dimensional problem which is solvable by separation of variables. The mechanism behind such solvability is called *reduction to quadratures*.

- In the context of dynamical systems a function satisfying condition (2.18) is called *integral of motion* or *conserved quantity*.

- The existence of one or more integrals of motion is not guaranteed a priori. Given a dynamical system, both continuous and discrete, it is not easy to find them (if exist). Often, one falls back on either physical intuition and guesswork. A deeper fact, due to E. Noether, is that integrals of motion are the result of underlying symmetry properties of the dynamical system (see Theorem 3.8).

Example 2.6 (*Invariant sets, invariant functions*)

1. On $\mathbb{R}^2$ consider the infinitesimal generator
$$\mathbf{v} := a\,\frac{\partial}{\partial x_1} + \frac{\partial}{\partial x_2}, \qquad a \in \mathbb{R}.$$
This generates the flow
$$\Phi_t : (t, (x_1, x_2)) \mapsto (x_1 + a\,t, x_2 + t).$$
   - The set
$$S_h := \{(x_1, x_2) \in \mathbb{R}^2 \,:\, F(x_1, x_2) := x_1 - a\,x_2 = h\},$$

where $h \in \mathbb{R}$ is a constant, is evidently an invariant set of Φ_t being precisely the orbits of Φ_t.

- The function $F(x_1, x_2) := x_1 - a\,x_2$ is an invariant function since

$$F(\Phi_t(x_1, x_2)) = x_1 + t\,a - a\,(x_2 + t) = x_1 - a\,x_2 = F(x_1, x_2),$$

for all $t \in \mathbb{R}$ and $(x_1, x_2) \in \mathbb{R}^2$. Indeed, every function $F(x_1 - a\,x_2)$ is invariant under Φ_t.

2. On $\mathbb{R}^2$ consider the infinitesimal generator

$$\mathbf{v} := x_1 \frac{\partial}{\partial x_1} + x_2 \frac{\partial}{\partial x_2}, \qquad a \in \mathbb{R}.$$

This generates the flow

$$\Phi_t : (t, (x_1, x_2)) \mapsto (\mathrm{e}^t\, x_1, \mathrm{e}^t\, x_2)\,.$$

- The set

$$S := \{(x_1, x_2) \in \mathbb{R}^2 \;:\; F(x_1, x_2) := x_1 x_2 = 0\}$$

is evidently an invariant set of Φ_t, but the function $F(x_1, x_2) := x_1 x_2$ is not an invariant function.

- The function

$$F_1(x_1, x_2) := \frac{x_1}{x_2}$$

is an invariant function defined on $\mathbb{R}^2 \setminus \{x_2 = 0\}$. Another invariant function is

$$F_2(x_1, x_2) := \frac{x_1 x_2}{x_1^2 + x_2^2}$$

defined on $\mathbb{R}^2 \setminus \{(0,0)\}$. Note however that F_1 and F_2 are functionally dependent, $\mathsf{rank}\,(\mathsf{grad}_x F_1(x_1, x_2), \mathsf{grad}_x F_2(x_1, x_2)) = 1$. Indeed, $F_2 = F_1/(1 + F_1^2)$.

► Here is an important characterization of integral of motions.

Theorem 2.4

A function $F \in \mathscr{F}(M, \mathbb{R})$ is an integral of motion of Φ_t if and only if

$$(\mathfrak{L}_{\mathbf{v}} F)(x) = 0 \qquad \forall\, x \in M. \tag{2.19}$$

Proof. From Corollary 2.1 we have:

$$\begin{aligned} F(\Phi_t(x)) &= \exp(t\,\mathbf{v})\, F(x) = \sum_{k=0}^{\infty} \frac{t^k}{k!} \mathbf{v}^k\,[F(x)] \\ &= F(x) + t\,\mathbf{v}[F(x)] + \frac{t^2}{2} \mathbf{v}^2[F(x)] + \ldots \end{aligned}$$

If condition (2.18) holds true then

$$\mathbf{v}[F(x)] = (\mathfrak{L}_{\mathbf{v}} F)(x) = 0$$

for all $x \in M$.

Conversely, if $(\mathfrak{L}_{\mathbf{v}}F)(x) = 0$ for all $x \in M$ then $\mathbf{v}^k\left[F(x)\right] = 0$ for all $k = 2,\ldots,\infty$, which implies condition (2.18). ■

► Condition (2.19) can be explicitly written as

$$\sum_{i=1}^{n} f_i(x)\frac{\partial F}{\partial x_i} = 0 \qquad \forall\, x \in M. \tag{2.20}$$

Therefore $F \in \mathscr{F}(M,\mathbb{R})$ is an invariant function of Φ_t if and only if F is a solution to the homogeneous linear first-order partial differential equation (2.20).

► We give an illustration of the above notions with a couple of examples.

Example 2.7 (*Rotations in* $\mathbb{R}^2$)

Consider the one-parameter Lie group of smooth diffeomorphisms on $\mathbb{R}^2$:

$$\begin{aligned} \Phi_t: \quad [0,2\pi)\times\mathbb{R}^2 &\to \mathbb{R}^2, \\ (t,(x_1,x_2)) &\mapsto (\underline{x}_1,\underline{x}_2) := (x_1\cos t + x_2\sin t, -x_1\sin t + x_2\cos t). \end{aligned}$$

Note that Φ_t satisfies properties (2.11) and it defines planar rotations of angle t.

- The planar IVP which has Φ_t as flow is easily constructed. We have:

$$\begin{cases} \dot{\underline{x}}_1 = f_1(\underline{x}_1,\underline{x}_2), \\ \dot{\underline{x}}_2 = f_2(\underline{x}_1,\underline{x}_2), \\ (\underline{x}_1(0,(x_1,x_2)),\underline{x}_2(0,(x_1,x_2)) = (x_1,x_2), \end{cases} \tag{2.21}$$

with

$$\begin{aligned} (f_1(\underline{x}_1,\underline{x}_2),f_2(\underline{x}_1,\underline{x}_2)) &:= \left.\frac{\mathrm{d}}{\mathrm{d}t}\right|_{t=0} (\underline{x}_1\cos t + \underline{x}_2\sin t, -\underline{x}_1\sin t + \underline{x}_2\cos t) \\ &= (\underline{x}_2,-\underline{x}_1). \end{aligned}$$

- The infinitesimal generator of the flow Φ_t is

$$\mathbf{v} := x_2\frac{\partial}{\partial x_1} - x_1\frac{\partial}{\partial x_2}.$$

- The Lie series of Φ_t reproduces the solution of the original IVP. Note that

$$\begin{aligned} \mathbf{v}^0[x_1] &= x_1, \\ \mathbf{v}[x_1] &= x_2, \\ \mathbf{v}[\mathbf{v}[x_1]] &= \mathbf{v}^2[x_1] = -x_1, \\ \mathbf{v}[\mathbf{v}[\mathbf{v}[x_1]]] &= \mathbf{v}^3[x_1] = -x_2, \\ \mathbf{v}[\mathbf{v}[\mathbf{v}[\mathbf{v}[x_1]]]] &= \mathbf{v}^4[x_1] = x_1, \end{aligned}$$

and, in general,

$$\mathbf{v}^{2k+1}[x_1] = (-1)^k x_2, \qquad \mathbf{v}^{2k}[x_1] = (-1)^k x_1, \qquad k \in \mathbb{N}_0.$$

This leads to

$$\begin{aligned}(\Phi_t(x))_1 &= \exp(t\,\mathbf{v})\,x_1 = \sum_{k=0}^{\infty} \frac{t^k}{k!}\mathbf{v}^k[x_1] \\ &= x_1 \sum_{k=0}^{\infty} \frac{(-1)^k t^{2k}}{(2k)!} + x_2 \sum_{k=0}^{\infty} \frac{(-1)^k t^{2k+1}}{(2k+1)!} \\ &= x_1 \cos t + x_2 \sin t.\end{aligned}$$

Similarly for $(\Phi_t(x))_2 = -x_1 \sin t + x_2 \cos t$.

- An integral of motion F of Φ_t can be found by solving the linear homogeneous partial differential equation

$$(\mathfrak{L}_{\mathbf{v}} F)(x_1, x_2) = x_2 \frac{\partial F}{\partial x_1} - x_1 \frac{\partial F}{\partial x_2} = 0,$$

whose general solution (as one may expect from some geometric intuition) is

$$F(x_1, x_2) = F\left(x_1^2 + x_2^2\right).$$

- It is evident that

$$F_1(x_1, x_2) := x_1^2 + x_2^2, \qquad F_2(x_1, x_2) := \log\left(x_1^2 + x_2^2\right),$$

are two functionally dependent integrals of motion.

Example 2.8 (*Construction of integrals of motion*)

Let Φ_t be a flow acting on $\mathbb{R}^3$ with infinitesimal generator

$$\mathbf{v} := -x_2 \frac{\partial}{\partial x_1} + x_1 \frac{\partial}{\partial x_1} + \left(1 + x_3^2\right) \frac{\partial}{\partial x_3}.$$

Note that this corresponds to the IVP

$$\begin{cases} \dot{\underline{x}}_1 = -\underline{x}_2, \\ \dot{\underline{x}}_2 = \underline{x}_1, \\ \dot{\underline{x}}_3 = 1 + \underline{x}_3^2, \\ \underline{x}_i(0, (x_1, x_2, x_3)) = x_i, \ i = 1, 2, 3. \end{cases}$$

- From condition (2.19) we have that $F = F(x_1, x_2, x_3)$ is an integral of motion of Φ_t if and only if

$$-x_2 \frac{\partial F}{\partial x_1} + x_1 \frac{\partial F}{\partial x_2} + \left(1 + x_3^2\right) \frac{\partial F}{\partial x_3} = 0, \qquad \forall\, x \in \mathbb{R}^3.$$

This linear homogeneous partial differential equation has a corresponding *characteristic system of ODEs* given by

$$-\frac{\mathrm{d}x_1}{x_2} = \frac{\mathrm{d}x_2}{x_1} = \frac{\mathrm{d}x_3}{1 + x_3^2}.$$

- The first of these two ODEs,

$$-\frac{\mathrm{d}x_1}{x_2} = \frac{\mathrm{d}x_2}{x_1},$$

is easily solved to give

$$x_1^2 + x_2^2 - c_1^2 = 0, \qquad c_1 \in \mathbb{R}.$$

Therefore the first invariant function is

$$F_1(x_1, x_2) := x_1^2 + x_2^2.$$

- Now set $x_1^2 = c_1^2 - x_2^2$ and solve the second of the two ODEs,

$$\frac{\mathrm{d}x_2}{\sqrt{c_1^2 - x_2^2}} = \frac{\mathrm{d}x_3}{1 + x_3^2},$$

to find

$$\arcsin \frac{x_2}{c_1} = \arctan x_3 + c_2, \qquad c_2 \in \mathbb{R}.$$

A straightforward manipulation gives a second invariant function:

$$F_2(x_1, x_2, x_3) := \frac{x_1 x_3 - x_2}{x_2 x_3 + x_1},$$

which is functionally independent on $F_1(x_1, x_2)$.

► Let Φ_t and Ψ_s, $t, s \in I$, be two distinct one-parameter local Lie groups of smooth diffeomorphisms on M whose infinitesimal generators are respectively given by

$$\mathbf{v} := \sum_{i=1}^{n} f_i(x) \frac{\partial}{\partial x_i}, \qquad \mathbf{w} := \sum_{i=1}^{n} g_i(x) \frac{\partial}{\partial x_i}, \tag{2.22}$$

with

$$f(x) := \left.\frac{\mathrm{d}}{\mathrm{d}t}\right|_{t=0} \Phi_t(x), \qquad g(x) := \left.\frac{\mathrm{d}}{\mathrm{d}s}\right|_{s=0} \Psi_s(x).$$

In general the flows Φ_t and Ψ_s *do not commute*, namely

$$(\Phi_t \circ \Psi_s)(x) \neq (\Psi_s \circ \Phi_t)(x).$$

Example 2.9 (***Rotations and translations in*** $\mathbb{R}^2$)

Consider the following one-parameter Lie groups of smooth diffeomorphisms on $\mathbb{R}^2$:

$$\Phi_t : (t, (x_1, x_2)) \mapsto (x_1 \cos t + x_2 \sin t, -x_1 \sin t + x_2 \cos t),$$

and

$$\Psi_s : (s, (x_1, x_2)) \mapsto (x_1 + s, x_2).$$

Note that Φ_t corresponds to planar rotations of angle t, while Ψ_s corresponds to translations in the x_1-direction.

- The corresponding infinitesimal generators are:

$$\mathbf{v} := x_2 \frac{\partial}{\partial x_1} - x_1 \frac{\partial}{\partial x_2}, \qquad \mathbf{w} := \frac{\partial}{\partial x_1}.$$

- As it is clear from Fig. 2.8 the flows do not commute:

$$(\Phi_t \circ \Psi_s)(x_1, x_2) \neq (\Psi_s \circ \Phi_t)(x_1, x_2).$$

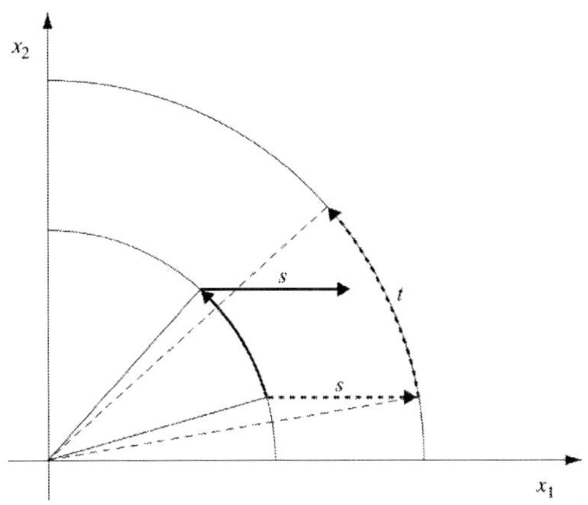

Fig. 2.8. Non-commutativity of translations and rotations in $\mathbb{R}^2$ ([FaMa]).

► We introduce a new object which will give a measure of the lack of commutativity of two flows (see Theorem 2.5).

Definition 2.8

The **Lie bracket** *of two vector fields* $\mathbf{v}, \mathbf{w} \in \mathfrak{X}(M)$*, defined in (2.22), is the vector field denoted by* $[\mathbf{v}, \mathbf{w}]$ *and defined by*

$$\begin{aligned} [\mathbf{v}, \mathbf{w}] &:= \sum_{i,j=1}^{n} \left(f_j(x) \frac{\partial g_i}{\partial x_j} - g_j(x) \frac{\partial f_i}{\partial x_j} \right) \frac{\partial}{\partial x_i} \\ &= \sum_{i=1}^{n} \left(\mathfrak{L}_{\mathbf{v}} g_i - \mathfrak{L}_{\mathbf{w}} f_i \right)(x) \frac{\partial}{\partial x_i}. \end{aligned} \tag{2.23}$$

► Remarks:

- It is not difficult to check that the Lie bracket (2.23) satisfies the following properties:

  1. *(Bi)linearity*: $[\lambda_1 \mathbf{v} + \lambda_2 \mathbf{w}, \mathbf{r}] = \lambda_1 [\mathbf{v}, \mathbf{r}] + \lambda_2 [\mathbf{w}, \mathbf{r}]$,
  2. *Skew-symmetry*: $[\mathbf{v}, \mathbf{w}] = -[\mathbf{w}, \mathbf{v}]$,
  3. *Jacobi identity*: $[\mathbf{v}, [\mathbf{w}, \mathbf{r}]] + \circlearrowleft (\mathbf{v}, \mathbf{w}, \mathbf{r}) = 0$.

  for all $\mathbf{v}, \mathbf{w}, \mathbf{r} \in \mathfrak{X}(M)$ and $\lambda_1, \lambda_2 \in \mathbb{R}$. Here $\circlearrowleft (\mathbf{v}, \mathbf{w}, \mathbf{r})$ means cyclic permutation of $(\mathbf{v}, \mathbf{w}, \mathbf{r})$.

- In general, a vector space V equipped with a bilinear mapping $[\cdot,\cdot] : V \times V \to V$ satisfying properties 1., 2. and 3. defines a *Lie algebra* $(V, [\cdot,\cdot])$ (see Chapter 4, Subsection 4.2.7, for further details on Lie algebras). Therefore we call $(\mathfrak{X}(M), [\cdot,\cdot])$, with $[\cdot,\cdot]$ defined by (2.23), the *Lie algebra of smooth vector fields*. Indeed, $(\mathfrak{X}(M), [\cdot,\cdot])$ is the Lie algebra of the one-parameter local Lie group Φ_t.

► Let us give the following useful result.

Lemma 2.1

Let $\mathbf{v}, \mathbf{w} \in \mathfrak{X}(M)$. *Then*

$$\left(\mathfrak{L}_{[\mathbf{v},\mathbf{w}]}F\right)(x) = ((\mathfrak{L}_{\mathbf{v}}\mathfrak{L}_{\mathbf{w}} - \mathfrak{L}_{\mathbf{w}}\mathfrak{L}_{\mathbf{v}})F)(x) \qquad \forall F \in \mathscr{F}(M, \mathbb{R}).$$

Proof. Let $F \in \mathscr{F}(M, \mathbb{R})$. We have:

$$\begin{aligned}
((\mathfrak{L}_{\mathbf{v}}\mathfrak{L}_{\mathbf{w}} - \mathfrak{L}_{\mathbf{w}}\mathfrak{L}_{\mathbf{v}})F)(x) &= (\mathfrak{L}_{\mathbf{v}}(\mathfrak{L}_{\mathbf{w}}F))(x) - (\mathfrak{L}_{\mathbf{w}}(\mathfrak{L}_{\mathbf{v}}F))(x) \\
&= \sum_{i,j=1}^{n}\left(f_i(x)\frac{\partial}{\partial x_i}\left(g_j(x)\frac{\partial F}{\partial x_j}\right) - g_i(x)\frac{\partial}{\partial x_i}\left(f_j(x)\frac{\partial F}{\partial x_j}\right)\right) \\
&= \sum_{i,j=1}^{n}\left(f_i(x)g_j(x)\frac{\partial^2 F}{\partial x_i \partial x_j} + f_i(x)\frac{\partial g_j}{\partial x_i}\frac{\partial F}{\partial x_j} - g_i(x)f_j(x)\frac{\partial^2 F}{\partial x_i \partial x_j} - g_i\frac{\partial f_j}{\partial x_i}\frac{\partial F}{\partial x_j}\right) \\
&= \sum_{i,j=1}^{n}\left(f_i(x)\frac{\partial g_j}{\partial x_i} - g_i(x)\frac{\partial f_j}{\partial x_i}\right)\frac{\partial F}{\partial x_j} \\
&= \sum_{j=1}^{n}[\mathbf{v},\mathbf{w}]_j\frac{\partial F}{\partial x_j} = \left(\mathfrak{L}_{[\mathbf{v},\mathbf{w}]}F\right)(x).
\end{aligned}$$

The Theorem is proved. ∎

► We are now in the position to show that the Lie bracket of two vector fields $\mathbf{v}, \mathbf{w} \in \mathfrak{X}(M)$ gives a measure of the degree of non-commutativity of the corresponding flows Φ_t, Ψ_s.

Theorem 2.5

Let Φ_t *and* Ψ_s, $t, s \in I$, *be two distinct one-parameter groups of smooth diffeomorphisms on* M *whose infinitesimal generators are* $\mathbf{v}, \mathbf{w} \in \mathfrak{X}(M)$. *Then*

$$(\Phi_t \circ \Psi_s)(x) = (\Psi_s \circ \Phi_t)(x) \qquad \forall t, s \in I,\ x \in M,$$

if and only if

$$[\mathbf{v}, \mathbf{w}] = 0 \qquad \forall\, x \in M.$$

Proof. We prove only that $[\mathbf{v}, \mathbf{w}] = 0$ is a necessary condition for the commutativity of the flows Φ_t and Ψ_s.

- Let $F \in \mathscr{F}(M, \mathbb{R})$. Define the smooth function

$$\Delta \equiv \Delta(t, s, x) := F\left((\Psi_s \circ \Phi_t)(x)\right) - F\left((\Phi_t \circ \Psi_s)(x)\right).$$

  Note that $\Delta(0, 0, x) = 0$ and if $\Phi_t \circ \Psi_s = \Psi_s \circ \Phi_t$ then $\Delta \equiv 0$ for all $t, s \in I$.

- Consider the Taylor expansion of Δ around $(t, s) = (0, 0)$:

$$\begin{aligned}
\Delta &= t\left.\frac{\partial \Delta}{\partial t}\right|_{(t,s)=(0,0)} + s\left.\frac{\partial \Delta}{\partial s}\right|_{(t,s)=(0,0)} \\
&\quad + \frac{1}{2}\left(t^2\left.\frac{\partial^2 \Delta}{\partial t^2}\right|_{(t,s)=(0,0)} + s^2\left.\frac{\partial^2 \Delta}{\partial s^2}\right|_{(0,0)} + 2\,s\,t\left.\frac{\partial^2 \Delta}{\partial s \partial t}\right|_{(t,s)=(0,0)}\right) \\
&\quad + O(s^2 t, s\, t^2) \\
&= st\left.\frac{\partial^2 \Delta}{\partial s \partial t}\right|_{(t,s)=(0,0)} + O(s^2 t, s\, t^2),
\end{aligned}$$

  because $\Delta(t, 0, x) = \Delta(0, s, x) = 0$.

- By Theorem 2.3 (claim *1.*) we know that

$$\left.\frac{\partial}{\partial t}\right|_{t=0} F(\Phi_t \circ \Psi_s(x)) = (\mathfrak{L}_{\mathbf{v}} F)(\Psi_s(x))$$

  so that

$$\left.\frac{\partial}{\partial t}\right|_{s=0} \left.\frac{\partial}{\partial t}\right|_{t=0} F(\Phi_t \circ \Psi_s(x)) = (\mathfrak{L}_{\mathbf{w}}(\mathfrak{L}_{\mathbf{v}} F))(x),$$

  and similarly

$$\left.\frac{\partial}{\partial t}\right|_{t=0} \left.\frac{\partial}{\partial s}\right|_{s=0} F(\Psi_s \circ \Phi_t(x)) = (\mathfrak{L}_{\mathbf{v}}(\mathfrak{L}_{\mathbf{w}} F))(x).$$

- Subtracting the last two equations we have:

$$\left.\frac{\partial^2 \Delta}{\partial s \partial t}\right|_{(0,0)} = \left(\left(\mathfrak{L}_{\mathbf{v}} \mathfrak{L}_{\mathbf{w}} - \mathfrak{L}_{\mathbf{w}} \mathfrak{L}_{\mathbf{v}}\right) F\right)(x) = \left(\mathfrak{L}_{[\mathbf{v}, \mathbf{w}]} F\right)(x),$$

  where we used Lemma 2.1.

- Therefore,
$$\Delta = s\,t\left(\mathfrak{L}_{[\mathbf{v},\mathbf{w}]}F\right)(x) + O(s^2t, s\,t^2) = O(s^2t, s\,t^2)$$
if $[\mathbf{v},\mathbf{w}] = 0$.

The claim is proved. ∎

Example 2.10 (*Rotations and translations in* $\mathbb{R}^2$)

Consider the non-commuting flows on $\mathbb{R}^2$ of Example 2.9:
$$\Phi_t : (t,(x_1,x_2)) \mapsto (x_1\cos t + x_2\sin t, -x_1\sin t + x_2\cos t),$$
and
$$\Psi_s : (s,(x_1,x_2)) \mapsto (x_1 + s, x_2).$$
Their infinitesimal generators are given respectively by
$$\mathbf{v} := x_2\frac{\partial}{\partial x_1} - x_1\frac{\partial}{\partial x_2}, \qquad \mathbf{w} := \frac{\partial}{\partial x_1}.$$
It is easy to verify that $[\mathbf{v},\mathbf{w}] \neq 0$. Take $F \in \mathscr{F}(\mathbb{R}^2,\mathbb{R})$ and compute
$$\begin{aligned}[\mathbf{v},\mathbf{w}][F(x)] &= \left(x_2\frac{\partial}{\partial x_1} - x_1\frac{\partial}{\partial x_2}\right)\frac{\partial F}{\partial x_1} - \frac{\partial}{\partial x_1}\left(x_2\frac{\partial F}{\partial x_1} - x_1\frac{\partial F}{\partial x_2}\right)\\ &= x_2\frac{\partial^2 F}{\partial x_1^2} - x_1\frac{\partial^2 F}{\partial x_1\partial x_2} - x_2\frac{\partial^2 F}{\partial x_1^2} + \frac{\partial F}{\partial x_2} + x_1\frac{\partial^2 F}{\partial x_1\partial x_2}\\ &= \frac{\partial F}{\partial x_2},\end{aligned}$$
that is
$$[\mathbf{v},\mathbf{w}] = \frac{\partial}{\partial x_2}.$$

2.3.2 *Evolution of phase space volume*

► Let $D \subset M$ be a compact set. Then the *volume* of D is defined by the Riemann integral
$$\mathrm{Vol}(D) := \int_D \mathrm{d}x.$$
Here $\mathrm{d}x := \mathrm{d}x_1 \wedge \cdots \wedge \mathrm{d}x_n$ is the *volume n-form* on M.

► Consider a one-parameter global Lie group of smooth diffeomorphisms: $\Phi_t : \mathbb{R} \times D \to D : (t,x) \mapsto \underline{x}(t,x) := \Phi_t(x)$, whose infinitesimal generator is
$$\mathbf{v} := \sum_{i=1}^{n} f_i(x)\frac{\partial}{\partial x_i},$$
with
$$f(x) := \left.\frac{\mathrm{d}}{\mathrm{d}t}\right|_{t=0}\Phi_t(x).$$

Then $D(t) := \Phi_t(D)$ and the volume of D at time t is

$$\mathsf{Vol}(D)(t) := \int_{D(t)} \mathrm{d}\underline{x}, \qquad \mathrm{d}\underline{x} := \mathrm{d}\underline{x}_1 \wedge \cdots \wedge \mathrm{d}\underline{x}_n.$$

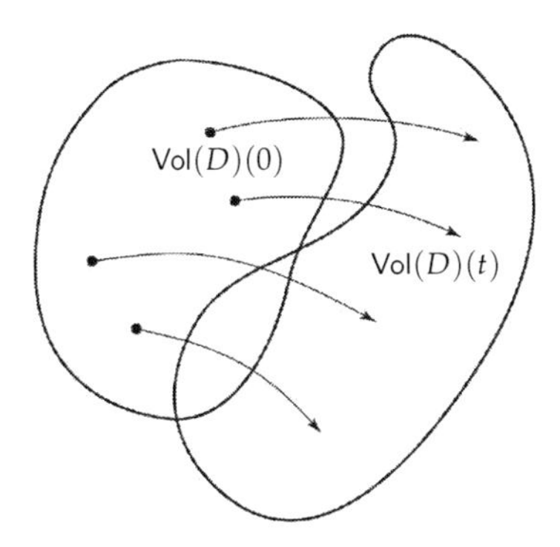

Fig. 2.9. Time evolution of the volume $\mathsf{Vol}(D)(t)$ ([MaRaAb]).

Theorem 2.6 (*Liouville*)

If

$$\mathrm{div}_x f(x) := \sum_{i=1}^{n} \frac{\partial f}{\partial x_i} = 0 \qquad \forall\, x \in M,$$

then Φ_t preserves $\mathsf{Vol}(D)$, that is

$$\mathsf{Vol}(D)(t) = \mathsf{Vol}(D) \qquad \forall\, t \in \mathbb{R}.$$

Proof. We proceed by steps.

- We have:

$$\mathsf{Vol}(D)(t) := \int_{D(t)} \mathrm{d}\underline{x} = \int_D \delta(t,x)\, \mathrm{d}x,$$

  where

$$\delta(t,x) := \det\left(\frac{\partial \Phi_t(x)}{\partial x}\right)$$

  is the determinant of the $n \times n$ Jacobian matrix of Φ_t. Note that $\delta(0,x) = 1$.

- Therefore,

$$\left.\frac{\mathrm{d}}{\mathrm{d}t}\right|_{t=s} \mathsf{Vol}(D)(t) = \int_D \left.\frac{\mathrm{d}}{\mathrm{d}t}\right|_{t=s} \delta(t,x)\, \mathrm{d}x.$$

- Note that

$$\left.\frac{\mathsf{d}}{\mathsf{d}t}\right|_{t=s} \delta(t,x) = \left.\frac{\mathsf{d}}{\mathsf{d}t}\right|_{t=s} \det\left(\frac{\partial \Phi_t(x)}{\partial \Phi_s(x)}\right) \det\left(\frac{\partial \Phi_s(x)}{\partial x}\right)$$
$$= \left.\frac{\mathsf{d}}{\mathsf{d}t}\right|_{t=s} \det\left(\frac{\partial \Phi_t(x)}{\partial \Phi_s(x)}\right) \delta(s,x),$$

where, using $\Phi_{t+s}(x) = \Phi_t(\Phi_s(x))$,

$$\left.\frac{\mathsf{d}}{\mathsf{d}t}\right|_{t=s} \det\left(\frac{\partial \Phi_t(x)}{\partial \Phi_s(x)}\right) = \left.\frac{\mathsf{d}}{\mathsf{d}t}\right|_{t=s} \det\left(\frac{\partial \Phi_{t-s}(\Phi_s(x))}{\partial \Phi_s(x)}\right)$$
$$= \left.\frac{\mathsf{d}}{\mathsf{d}t}\right|_{t=s} \delta(t-s,\Phi_s(x))$$
$$= \left.\frac{\mathsf{d}}{\mathsf{d}t}\right|_{t=0} \delta(t,\Phi_s(x)).$$

- Therefore the claim follows if we prove that

$$\left.\frac{\mathsf{d}}{\mathsf{d}t}\right|_{t=0} \delta(t,x) = 0$$

for all $x \in M$.

- Recalling that

$$\Phi_t(x) = x + t\, f(x) + O(t^2),$$

we have

$$\frac{\partial \Phi_t(x)}{\partial x} = \mathbb{1}_n + t\, A + O(t^2), \qquad A := (A_{ij})_{1\leqslant i,j\leqslant n}, \quad A_{ij} := \frac{\partial f_i}{\partial x_j}.$$

- For any $n \times n$ matrix A we have

$$\det\left(\mathbb{1}_n + t\, A\right) = 1 + t\, \mathsf{Trace}\, A + O(t^2).$$

- Therefore,

$$\delta(t,x) = 1 + t\, \mathsf{Trace}\, A + O(t^2) = 1 + t\, \mathsf{div}_x f(x) + O(t^2).$$

- Hence we get

$$\left.\frac{\mathsf{d}}{\mathsf{d}t}\right|_{t=0} \delta(t,x) = \mathsf{div}_x f(x).$$

The Theorem is proved. ■

► We now give the following fundamental claim.

Theorem 2.7 (***Poincaré***)

Let $D \subset M$ be a compact set. Consider a one-parameter global Lie group of smooth diffeomorphisms $\Phi_t : \mathbb{R} \times D \to D$ which is volume preserving, i.e.,

$$\mathsf{Vol}(D)(t) = \mathsf{Vol}(D) \qquad \forall t \in \mathbb{R}.$$

Then for each open set $U \subset D$ and for each fixed time $s > 0$ there exists $x \in U$ and $t > s$ such that $\Phi_t(x) \in U$.

Proof. We proceed by steps.

- Fix a time $s > 0$ and consider the set $\Phi_s(U)$. If $\Phi_s(U) \cap U \neq \emptyset$ then the claim follows since, by continuity, $\Phi_s(U)$ intersects U for $t > s$ sufficiently close to s.

- Suppose $\Phi_s(U) \cap U = \emptyset$. Define $U_n := \Phi_{ns}(U)$, $n \in \mathbb{N}_0$, $U_0 = U$. Suppose that
$$U_n \cap U = \emptyset \qquad \forall n \in \mathbb{N}. \tag{2.24}$$
This implies
$$U_n \cap U_m = \emptyset \qquad \forall n > m \in \mathbb{N}. \tag{2.25}$$

- If (2.25) is not satisfied for some $n > m > 0$ then this would imply $U_n \cap U_m \neq \emptyset$. But this implies $U_{n-1} \cap U_{m-1} \neq \emptyset$. Indeed, if there exists $\underline{x} \in U_n \cap U_m$ and $U_{n-1} \cap U_{m-1} = \emptyset$ then one would have $\underline{x} = \Phi_s(x_n)$ and $\underline{x} = \Phi_t(x_m)$ with $x_n \in U_{n-1}$ and $x_m \in U_{m-1}$. But this violates the unicity of solutions.

- By iteration of this argument we would end up with $U_{n-m} \cap U \neq \emptyset$, which contradicts our hypothesis (2.24).

- Condition (2.24) implies that
$$\sum_{n\in\mathbb{N}_0} \mathsf{Vol}(U_n) = \mathsf{Vol}\left(\bigcup_{n\in\mathbb{N}_0} U_n\right).$$
Our map Φ_t is, by assumption, volume preserving, meaning that
$$\mathsf{Vol}(U_n) = \mathsf{Vol}(U) \qquad \forall n \in \mathbb{N}.$$
Therefore we find:
$$\infty = \sum_{n\in\mathbb{N}_0} \mathsf{Vol}(U_n) = \mathsf{Vol}\left(\bigcup_{n\in\mathbb{N}_0} U_n\right) \leqslant \mathsf{Vol}(D) < \infty,$$
where the last inequality follows from the compactness of D. We obtained a contradiction.

The Theorem is proved. ■

▶ Remarks:

- Theorem 2.7 shows that any arbitrary neighborhood U (even a single point x) in a compact space, will evolve, under the action of a volume preserving flow, in such a way that U will return arbitrarily close to any point it starts from. The time it takes the map to return depends on the map, and on the size of the neighborhood U. In general, this time may be extremely large.

Fig. 2.10. A Poincaré recurrence ([Ge]).

- Reformulation of Theorem 2.7: *Let $D \subset M$ be a compact set. Consider a one-parameter global Lie group of smooth diffeomorphisms $\Phi_t : \mathbb{R} \times D \to D$ which is volume preserving. Then, for any $x \in D$ there exists $\varepsilon > 0$ and $\widetilde{x} \in D$ such that if $\|x - \widetilde{x}\| < \varepsilon$ then $\|x - \Phi_t(\widetilde{x})\| < \varepsilon$ for some $t > 0$.*

Example 2.11 (*Maxwell Gedankenexperiment*)

Suppose to have a box partitioned into two chambers, say A and B. Suppose that A contains a gas, while B is empty. We make a hole in the wall separating A and B. Then it is reasonable to expect that after some time the gas will be uniformly distributed in A and B.

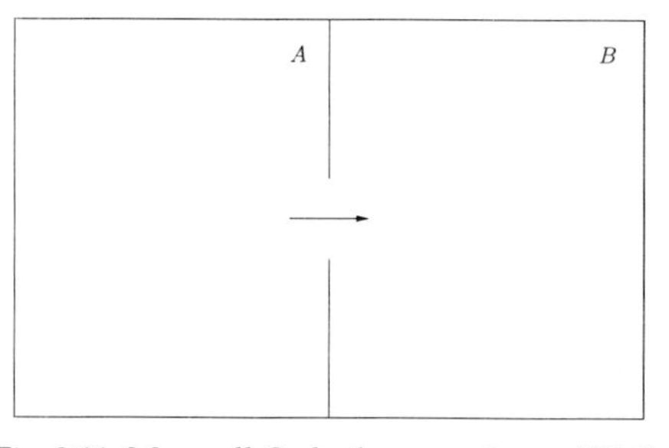

Fig. 2.11. Maxwell Gedankenexperiment ([Ge]).

- We are under the conditions of Theorem 2.7. Indeed, the volume of the box is finite and, assuming elastic and regular interactions between molecules, the conservation of the global energy assures that molecular velocities are limited. Then the phase space is a compact space and Theorem 2.7 says that there exists a time s for which all molecules constituting the gas will come back to a configuration which is close to the initial configuration, that is all molecules in the chamber A!
- The resolution of this paradoxical situation lies in the fact that s is longer than the duration of the solar system's existence (...billions of years). Furthermore, one of the assumption we used in this ideal experiment is that the system is isolated (i.e., no external perturbations are admitted). This assumption is not realistic, especially on long times.

2.3.3 *Stability of fixed points and Lyapunov functions*

▶ One of the major tasks of dynamical systems theory is to analyze the long-time behavior of a dynamical system. Of course, one might try to solve this problem by brute force, merely computing many orbits numerically (by simulations). However, the most useful aspect of the theory is that one can predict some features of the phase portrait without actually solving the IVP. The simplest example of such information is the number and the positions of fixed points.

▶ Let Φ_t be a one-parameter global Lie group of smooth diffeomorphisms, $\Phi_t : \mathbb{R} \times M \to M$. Let $\mathbf{v} \in \mathfrak{X}(M)$ be the infinitesimal generator of Φ_t. We know from Definition 2.3 that a point $\widetilde{x} \in M$ is a *fixed point* of the dynamical system $\{\Phi_t, \mathbb{R}, M\}$ if $\widetilde{x} = \Phi_t(\widetilde{x})$ for all $t \in \mathbb{R}$. Such a point is uniquely defined by the vanishing of the function f in the corresponding IVP, i.e., $f(\widetilde{x}) = 0$.

▶ We now adapt Definition 2.6 to the case of a dynamical system $\{\Phi_t, \mathbb{R}, M\}$.

Definition 2.9

Let $\widetilde{x} \in M$ be a fixed point of Φ_t.

1. *$\widetilde{x}$ is* **(Lyapunov) stable** *if for any $\varepsilon > 0$ there exists $\delta(\varepsilon) > 0$ such that whenever $\|x - \widetilde{x}\| < \delta(\varepsilon)$ then $\|\Phi_t(x) - \widetilde{x}\| < \varepsilon$ for all $t \geqslant 0$.*

2. *$\widetilde{x}$ is* **attracting** *if there exists a neighborhood of $\widetilde{x}$, say $\widetilde{M}$, such that*

$$\lim_{t \to +\infty} \|\Phi_t(x) - \widetilde{x}\| = 0 \qquad \forall\, x \in \widetilde{M}. \tag{2.26}$$

The maximal open set $\widetilde{M}$ where (2.26) holds true is called **basin of attraction** *of $\widetilde{x}$.*

3. *$\widetilde{x}$ is* **asymptotically stable** *if it is stable and attracting. Equivalently, if it is stable and there exists $\delta > 0$ such that whenever $\|x - \widetilde{x}\| < \delta$ then $\lim_{t \to +\infty} \Phi_t(x) = \widetilde{x}$.*

4. $\widetilde{x}$ *is* **unstable** *if it is not stable.*

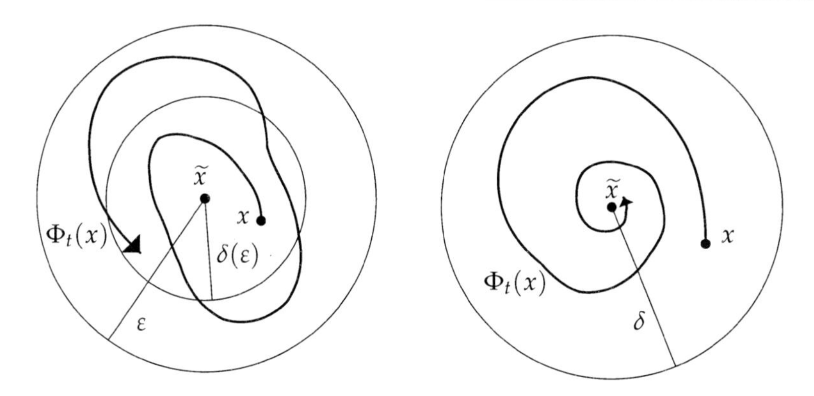

Fig. 2.12. Stable and asymptotically stable fixed point ([Ol1]).

► Remarks:

- In Definition 2.9, item *1.*, we assumed that $\Phi_t(x)$ with initial point $x \in \mathbb{R}^n$ such that $\|x - \widetilde{x}\| < \delta(\varepsilon)$, is globally defined for $t \geqslant 0$. In fact, $\Phi_t(x)$ admits a limited maximal interval of existence I_x only if the solution of the corresponding IVP leaves every compact $D \subset M$ in a finite time. Therefore, to require $\|\Phi_t(x) - \widetilde{x}\| < \varepsilon$ for all $t \in I_x \cap \mathbb{R}_+$ automatically implies $I_x \cap \mathbb{R}_+ = \mathbb{R}_+$.
- A fixed point $\widetilde{x}$ can be attractive but not asymptotically stable. It is sufficient that $\widetilde{x}$ is unstable.

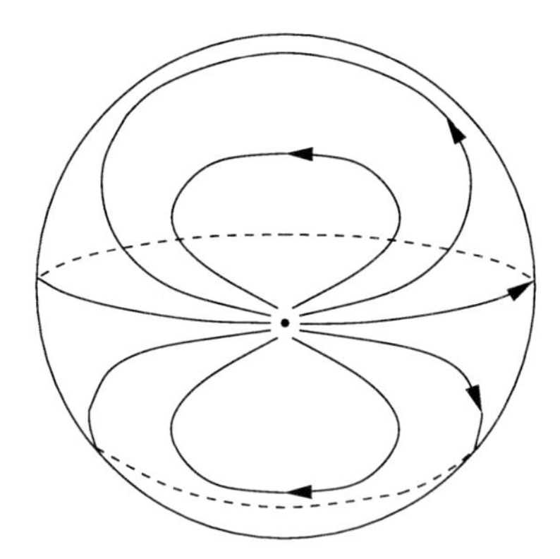

Fig. 2.13. An attracting fixed point which is unstable ([BuNe]).

- An important consequence of Theorem 2.7 is that if M is compact and Φ_t is a volume-preserving (global) flow, then Φ_t cannot possess asymptotically stable fixed points.

Example 2.12 (*Stability analysis of a scalar IVP*)

Consider the dynamical system defined by the flow Φ_t of the scalar IVP

$$\begin{cases} \dot{x} = x\left(1 - x^2\right), \\ x(0) = x_0. \end{cases} \tag{2.27}$$

- The fixed points are given by the solutions of the algebraic equation $f(x) := x\left(1 - x^2\right) = 0$:
$$\widetilde{x}_1 = -1, \qquad \widetilde{x}_2 = 0, \qquad \widetilde{x}_3 = 1.$$
- Let $x < -1$. As x increases, the graph of the function $f(x)$ switches from positive to negative at $\widetilde{x}_1 = -1$ which proves its stability. Any solution with $x_0 < 0$ will end up, asymptotically, at $\widetilde{x}_1 = -1$.
- Similarly, the graph goes from negative to positive at $\widetilde{x}_2 = 0$ establishing its instability.
- The last fixed point $\widetilde{x}_3 = 1$ is stable because $f(x)$ again changes from positive to negative there. Any solution with $x_0 > 0$ will end up, asymptotically, at $\widetilde{x}_3 = 1$.

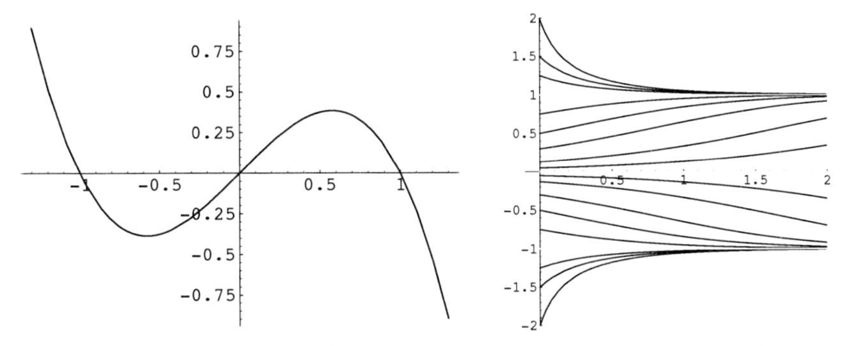

Fig. 2.14. Plot of the function $f(x) := x(1 - x^2)$ and solution curves of (2.27) ([Ol1]).

- The only solution that does not end up at ± 1 is the unstable stationary solution $\phi(t) \equiv 0$ for all t. Any perturbation of it will force the solutions to choose one of the stable fixed points.

▶ Definition 2.9 does not provide a practical tool to detect the stability of a fixed point. The most practical way to investigate the stability of fixed points is based on the use of the so called *Lyapunov functions*. The *direct method of Lyapunov functions* finds its motivation in a physical observation. Systems which have damping, viscosity and/or frictional effects do not typically possess integrals of motion. From a physical standpoint, these effects imply that the *total energy* of the system is a decreasing function of time. In other words, asymptotically, the system returns to a (stable) configuration, where its energy has a minimum and the extra energy has been dissipated away.

▶ Here is the mathematical definition of a Lyapunov function.

Definition 2.10

Let $\widetilde{x} \in M$ be a fixed point of Φ_t and $\widetilde{M}$ be a neighborhood of $\widetilde{x}$. A function

$F \in C^1(\widetilde{M}, \mathbb{R})$ is called **Lyapunov function** *of Φ_t if the following conditions hold:*

1. *$F(\widetilde{x}) = 0$ and $F(x) > 0$ for all $x \in \widetilde{M} \setminus \{\widetilde{x}\}$.*
2. *$(\mathfrak{L}_{\mathbf{v}} F)(x) \leqslant 0$ for all $x \in \widetilde{M}$.*

If $(\mathfrak{L}_{\mathbf{v}} F)(x) < 0$ for all $x \in \widetilde{M} \setminus \{\widetilde{x}\}$ then F is called **strict Lyapunov function**.

► The following Theorem gives sufficient conditions for stability and asymptotic stability of fixed points of Φ_t. It is customary to call this way of detecting stability *Lyapunov direct method*.

Theorem 2.8 (*Lyapunov*)

Let $\widetilde{x} \in M$ be a fixed point of Φ_t and $\widetilde{M}$ be a neighborhood of $\widetilde{x}$.

1. *If there exists a Lyapunov function $F \in C^1(\widetilde{M}, \mathbb{R})$ then $\widetilde{x}$ is stable.*
2. *If there exists a strict Lyapunov function $F \in C^1(\widetilde{M}, \mathbb{R})$ then $\widetilde{x}$ is asymptotically stable.*

Proof. We prove only the first claim.

- Let $\widetilde{M}_\varepsilon \subset \widetilde{M}$ a neighborhood of radius $\varepsilon > 0$ and center $\widetilde{x}$. Define

$$\alpha(\varepsilon) := \min_{x \in \partial \widetilde{M}_\varepsilon} F(x),$$

so that $\alpha(\varepsilon) > 0$ because $F(x) > 0$ for all $x \in \widetilde{M} \setminus \{\widetilde{x}\}$.

- Define the set

$$U := \left\{ x \in \widetilde{M}_\varepsilon \, : \, F(x) < \frac{1}{2}\alpha(\varepsilon) \right\},$$

which is open due to the fact that $F \in C^1(\widetilde{M}, \mathbb{R})$. Therefore there exists a neighborhood $\widetilde{M}_\delta \subset U$ with $\delta < \varepsilon$. Note that the open set U is not necessarily connected, but there exists always a connected component which contains $\widetilde{x}$ and $\widetilde{M}_\delta$.

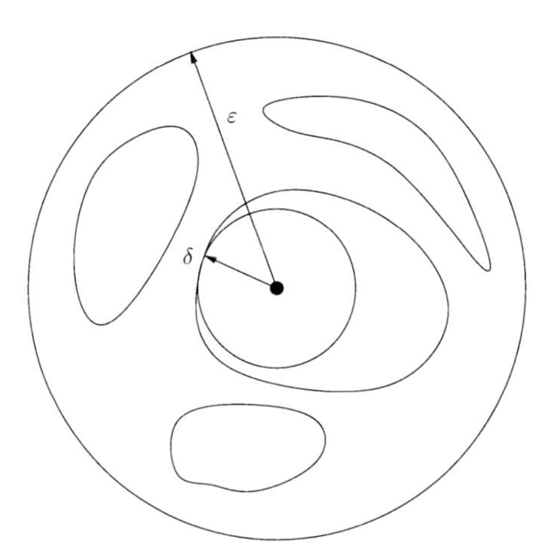

Fig. 2.15. The set U ([Ge]).

- To prove the stability of $\widetilde{x}$ it suffices to show that given an initial condition $x \in \widetilde{M}_\delta$ then $\Phi_t(x) \in \widetilde{M}_\delta$ for all $t \geqslant 0$.
- Assume that there exists a time s such that for $x \in \widetilde{M}_\delta$ then $\Phi_s(x) \in \partial\widetilde{M}_\delta$. Then we would have:

$$0 < \frac{1}{2}\alpha(\varepsilon) < F(\Phi_s(x)) - F(x) = \int_0^s \frac{\mathrm{d}F(\Phi_t(x))}{\mathrm{d}t}\mathrm{d}t \leqslant 0,$$

  which is a contradiction. Therefore $\widetilde{x}$ must be stable.

The first claim is proved. ■

▶ Remarks:

- The idea of the proof of claim 2. in Theorem 2.8 is the following: if there exists a Lyapunov function $F \in C^1(\widetilde{M}, \mathbb{R})$ then the condition $(\mathfrak{L}_{\mathbf{v}}F)(x) \leqslant 0$ for all $x \in \widetilde{M}$ implies that all orbits which start from points close to $\widetilde{x}$ remain close to the same points: such orbits do not go towards those points where F grows. Condition $(\mathfrak{L}_{\mathbf{v}}F)(x) < 0$ for all $x \in \widetilde{M} \setminus \{\widetilde{x}\}$, defining a strict Lyapunov function, implies that all orbits go towards those points where the value of F is smaller, and therefore towards $\widetilde{x}$. This means that $\widetilde{x}$ is stable and attracting, namely asymptotically stable.
- The advantage of Theorem 2.8 is that one does not need to solve any IVP to understand the stability properties of its fixed points. Nevertheless, the drawback is that the method is not constructive and the form of the Lyapunov function has to be guessed case by case.

Example 2.13 (*Gradient systems in* $\mathbb{R}^n$)

A *gradient system* is a dynamical system defined through the flow Φ_t (with infinitesimal generator $\mathbf{v}$) of the IVP

$$\begin{cases} \dot{x} = -\,\mathsf{grad}_x G(x), \\ x(0) = x_0, \end{cases} \tag{2.28}$$

where $G \in C^2(M, \mathbb{R})$, with $M \subseteq \mathbb{R}^n$. The main features of these systems are:

1. $(\mathfrak{L}_{\mathbf{v}} G)(x) \leqslant 0$ for all $x \in M$ and $(\mathfrak{L}_{\mathbf{v}} G)(x) = 0$ if and only if x is a fixed point.
2. If $\tilde{x} \in M$ is an isolated local minimum of G then $\tilde{x}$ is an asymptotically stable fixed point.
3. If $\tilde{x}$ is a regular point of (2.28), i.e., $\mathsf{grad}_x G(\tilde{x}) \neq 0$, then the vector field $\mathbf{v}$ generating the flow Φ_t is orthogonal to each level set $S_h := \{x \in M \,:\, G(x) = h\}$, with $h \in \mathbb{R}$, at the point $\tilde{x}$.

Let us prove explicitly the first two claims.

1. The infinitesimal generator of Φ_t is

$$\mathbf{v} := -\sum_{i=1}^{n} \frac{\partial G}{\partial x_i} \frac{\partial}{\partial x_i}.$$

We have:

$$(\mathfrak{L}_{\mathbf{v}} G)(x) = -\sum_{i,j=1}^{n} \frac{\partial G}{\partial x_i} \frac{\partial G}{\partial x_j} = -\left(\sum_{i=1}^{n} \frac{\partial G}{\partial x_i}\right)^2 = -(\,\mathsf{grad}_x G(x))^2 \leqslant 0.$$

The above Lie derivative vanishes if and only if $\mathsf{grad}_x G(x) = 0$, that means that x is a fixed point.

2. The asymptotic stability of an isolated minimum $\tilde{x} \in M$ of G can be proven by using Theorem 2.8 with the strict Lyapunov function

$$F(x) := G(x) - G(\tilde{x}).$$

Example 2.14 (*Newton equation of motion in* $\mathbb{R}$)

Consider a unit mass particle moving on the real line $\mathbb{R}$ under the influence of a smooth *potential energy* $U : \mathbb{R} \to \mathbb{R}$. Then Newton equation of motion takes the form

$$\ddot{x} = -\frac{\mathrm{d}U}{\mathrm{d}x} =: f(x), \tag{2.29}$$

where $f : \mathbb{R} \to \mathbb{R}$ is the force field acting on the particle.

- The scalar second-order ODE (2.29) is equivalent to the following planar system of first-order ODEs:

$$\begin{cases} \dot{x} = v, \\ \dot{v} = f(x), \end{cases} \tag{2.30}$$

which is an IVP once we fix some initial condition $(x(0), v(0))$. Theorem 1.1 assures the existence and uniqueness of its solution

$$(t, (x(0), v(0))) \mapsto (x, v) = (x(t, (x(0), v(0))), v(t, (x(0), v(0)))).$$

- The *kinetic energy* of the system is by definition

$$T(v) := \frac{1}{2} v^2,$$

while the *potential energy* reads

$$U(x) := -\int f(x)\, \mathrm{d}x,$$

and it is determined up to a constant (of integration).

- The *total energy* of the system,

$$E(x,v) := T(v) + U(x) = \frac{1}{2}v^2 + U(x), \tag{2.31}$$

is an integral of motion of (2.30). Denote by $\mathbf{v}$ the infinitesimal generator of the flow of (2.30). Then,

$$(\mathfrak{L}_{\mathbf{v}}E)(x,v) = v\frac{\partial E}{\partial x} + f(x)\frac{\partial E}{\partial v} = v\frac{\partial U}{\partial x} + f(x)v = 0 \qquad \forall\,(x,v) \in \mathbb{R}^2,$$

thanks to (2.29).

- Fixing the value of the constant energy equal to $h \in \mathbb{R}$, we can solve (2.31) for v to get

$$v = \dot{x} = \pm\sqrt{2(h - U(x))},$$

that gives the *quadrature* of (2.30):

$$t = \pm \int_{x(0)}^{x} \frac{\mathrm{d}\xi}{\sqrt{2(h - U(\xi))}}. \tag{2.32}$$

Equation (2.32) defines implicitly the solution $(t,(x(0)) \mapsto x = x(t,x(0))$ of (2.30). This gives the time t as a function of x. Inverting such a relation we obtain the solution $x = x(t,x(0))$ satisfying the given initial condition.

- The fixed points of (2.30) in the (x,v)-plane are determined by $(v, f(\tilde{x})) = (0,0)$. Hence they correspond to extremal points of the potential energy.

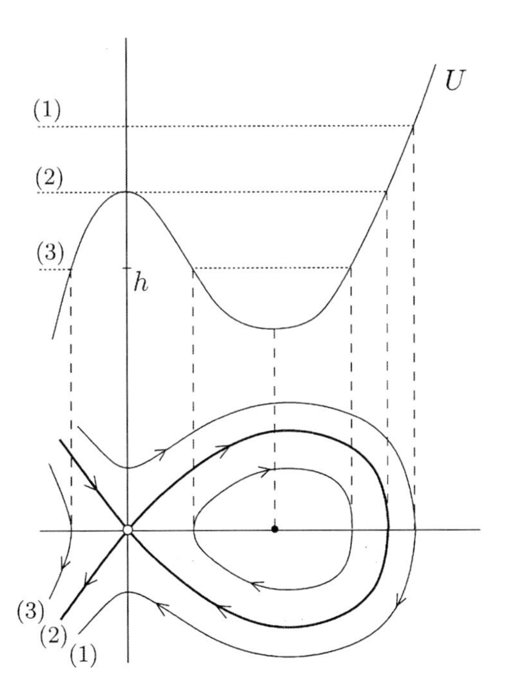

Fig. 2.16. Plot of the potential energy U and corresponding orbits of (2.30).

- One can prove that if U has a local minimum at $\tilde{x}$ then the function $\tilde{x}$ is a stable fixed point, called *center* (see Theorem 3.5). If U has a local maximum at $\tilde{x}$ then the function $\tilde{x}$ is an unstable fixed point, called *saddle*.

- One can prove that a connected subset of any level set

$$S_h := \{(x,v) \in \mathbb{R}^2 \, : \, E(x,v) = h\},$$

where $h \in \mathbb{R}$ is a regular value, is diffeomorphic to either a circle (closed orbit, see orbit (2) in Fig. 2.16) or a line (open orbit, see orbits (1) and (3) in Fig. 2.16). Any closed orbit defines a periodic solution.

Example 2.15 (*Planar pendulum and harmonic oscillator*)

1. The planar pendulum described in Example 2.1 is a one-dimensional mechanical system obtained from Newton equation of motion (2.29) with

$$f(x) := -\sin x, \qquad x \in (-\pi, \pi]. \tag{2.33}$$

Here x describes the displacement angle from the position at rest ($x = 0$) of a mass $m = 1$ suspended from a pivot. It can swing freely moving on a plane which is parallel to the gravitational field. The potential energy is chosen to be

$$U(x) := 1 - \cos x.$$

Then the IVP (2.30) reads

$$\begin{cases} \dot{x} = v, \\ \dot{v} = -\sin x, \end{cases} \tag{2.34}$$

where some initial conditions $(x(0), v(0)$ are fixed.

- The fixed points of (2.34) in the (x,v)-plane are

$$(\widetilde{x}_1, \widetilde{v}) = (0,0), \qquad (\widetilde{x}_2, \widetilde{v}) = (\pi, 0).$$

- Note that U has a local minimum at $\widetilde{x} = 0$. Therefore $(\widetilde{x}_1, \widetilde{v}) = (0,0)$ is a stable fixed point. On the other hand, U has a local maximum at $\widetilde{x} = \pi$. Therefore $(\widetilde{x}_2, \widetilde{v}) = (\pi, 0)$ is an unstable fixed point. The stability structure of the system is (2π)-periodic.
- The total energy of the system,

$$E(x,v) := \frac{1}{2}v^2 + 1 - \cos x,$$

is an integral of motion and every level set

$$S_h := \{(x,v) \in \mathbb{R}^2 \, : \, E(x,v) = h\}$$

is an invariant set. In particular:

(a) For $h = 0$ we have $S_0 = (0,0)$.

(b) For $0 < h < 2$ the level set S_h is homeomorphic to a circle. Since this circle contains no fixed points, it is a regular periodic orbit. We have an oscillatory motion.

(c) For $h = 2$ the level set consists of the fixed point $(\pi, 0)$ and two non-closed orbits connecting two fixed points $(-\pi, 0)$ and $(\pi, 0)$ (such an orbit is called *separatrix*).

(d) For $h > 2$ the level sets are open orbits. We have a rotational motion.

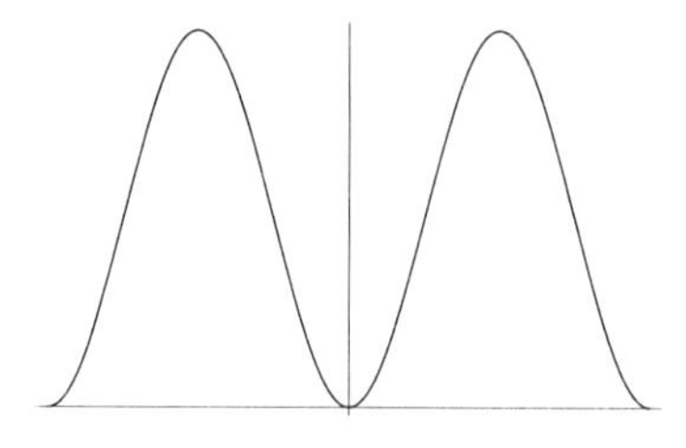

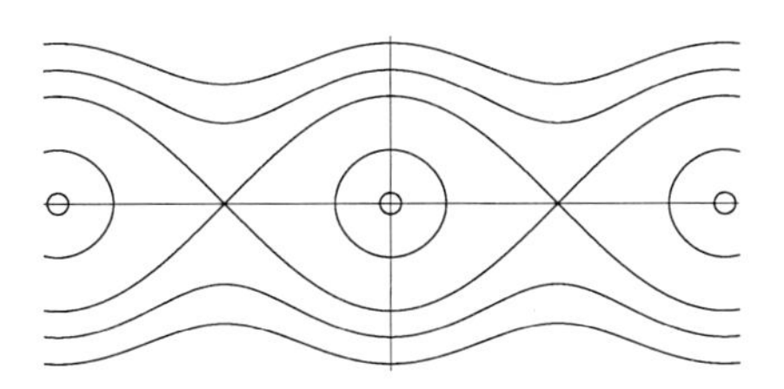

Fig. 2.17. Plot of the potential energy U and corresponding orbits of (2.34).

- It turns out that the exact solution of the IVP (2.34) is highly non-trivial. Indeed, x depends on t in terms of elliptic functions.

2. A simpler one-dimensional system is obtained assuming small oscillations around the fixed point $(0,0)$ in the Newton equation of motion, so that

$$f(x) := -x.$$

The corresponding system is called *harmonic oscillator* and the IVP reads

$$\begin{cases} \dot{x} = v, \\ \dot{v} = -x, \end{cases}$$

where some initial conditions $(x(0), v(0))$ are fixed. Note that this is exactly the same system consider in Example 2.7. In particular, the phase portrait consists of circular orbits centered at $(0,0)$, which is a stable fixed point.

2.3.4 Stability properties of linear homogeneous IVPs

► We consider a continuous dynamical system $\{\Phi_t, \mathbb{R}, \mathbb{R}^n\}$, generated by the flow of a *linear autonomous and homogeneous IVP*:

$$\begin{cases} \dot{x} = A\,x, \\ x(0) = x_0, \end{cases} \tag{2.35}$$

with $A \in \mathfrak{gl}(n,\mathbb{R})$, where $\mathfrak{gl}(n,\mathbb{R})$ is the vector space of all linear maps from $\mathbb{R}^n$ to $\mathbb{R}^n$ (i.e., the *general linear Lie algebra on* $\mathbb{R}^n$). In other words, $A \in \mathfrak{gl}(n,\mathbb{R})$ can be represented as a $n \times n$ matrix, not necessarily invertible, with real entries (see Chapter 4, Subsection 4.2.7, for further details on Lie algebras).

Theorem 2.9

The flow of the IVP (2.35) is a global Lie group of smooth diffeomorphisms given by

$$\Phi_t : (t, x_0) \mapsto \Phi_t(x_0) := e^{A\,t} x_0, \tag{2.36}$$

where

$$e^{A\,t} := \sum_{k=0}^{\infty} \frac{t^k}{k!} A^k$$

is the **exponential of the matrix** $t\,A$.

Proof. First of all, note that the IVP (2.35) satisfies the condition of Theorem 1.1. Therefore its solution exists and is unique. Moreover it is globally defined over $\mathbb{R}$ thanks to Theorem 1.9.

To construct explicitly the flow we can use the following Picard iteration:

$$\begin{aligned} x_0(t,x_0) &= x_0, \\ x_1(t,x_0) &= x_0 + \int_0^t A\,x_0(s,x_0)\,\mathrm{d}s = x_0 + t\,A\,x_0, \\ x_2(t,x_0) &= x_0 + \int_0^t A\,x_1(s,x_0)\,\mathrm{d}s = x_0 + t\,A\,x_0 + \frac{t^2}{2}A^2\,x_0, \end{aligned}$$

and hence, by induction,

$$x_m(t,x_0) = \sum_{k=0}^{m} \frac{t^k}{k!} A^k\,x_0.$$

The limit as $m \to \infty$ gives the unique solution of (2.35):

$$\phi(t,x_0) = \lim_{m\to\infty} x_m(t,x_0) = \left(\sum_{k=0}^{\infty} \frac{t^k}{k!} A^k\right) x_0 = \mathrm{e}^{t\,A} x_0.$$

The Picard iteration ensures convergence of $\mathrm{e}^{tA}x_0$ for all $x_0 \in \mathbb{R}^n$. Note that:

$$\begin{aligned} \frac{\mathrm{d}}{\mathrm{d}t}\mathrm{e}^{t\,A} &= \sum_{k=0}^{\infty} \frac{\mathrm{d}}{\mathrm{d}t}\left(\frac{t^k}{k!}A^k\right) = \sum_{k=1}^{\infty} \frac{k\,t^{k-1}}{k!}A^k \\ &= A\sum_{k=1}^{\infty} \frac{t^{k-1}}{(k-1)!}A^{k-1} = A\,\mathrm{e}^{t\,A}, \end{aligned}$$

where we used the absolute and uniform convergence of the series to derive under the sum. The fact that the flow $\Phi_t : (t,x_0) \mapsto \phi(t,x_0)$ is a global Lie group of smooth diffeomorphisms is obvious. The Theorem is proved. ■

► To investigate the behavior of the flow (2.36) it is essential to understand and compute the exponential of A. To do so we will use $\mathfrak{gl}(n,\mathbb{C})$ (i.e., the *general linear Lie algebra on* $\mathbb{C}^n$), instead of $\mathfrak{gl}(n,\mathbb{R})$, since $\mathbb{C}$ is algebraically closed (which will be important later on, when we compute the matrix exponential with the help of the *Jordan canonical form*).

► Let us recall some definitions from Linear Algebra. Let $A := (A_{ij})_{1\leqslant i,j\leqslant n} \in \mathfrak{gl}(n,\mathbb{C})$ be a matrix acting on $\mathbb{C}^n$.

- The *matrix norm* of A is defined by

$$\|A\|_* := \sup_{\|x\| \leqslant 1} \|A\,x\|, \qquad (A\,x)_i = \sum_{j=1}^{n} A_{ij}\,x_j.$$

  The space $(\mathfrak{gl}(n,\mathbb{C}), \|\cdot\|_*)$ is a Banach space.

- A vector $v_j \in \mathbb{C}^n$ is an *eigenvector* of A associated with the *eigenvalue* $\lambda_j \in \mathbb{C}$ if $A\,v_j = \lambda_j\,v_j$. The *eigenspace* associated with λ_j is the linear subspace

$$G_{\lambda_j} := \{x \in \mathbb{C}^n : (A - \lambda_j\,\mathbb{1}_n)x = 0\},$$

  whose dimension, say g_j, is called *geometric multiplicity*. Thus $v_j \in G_{\lambda_j}$.

- The *characteristic polynomial* of A is

$$P_A(\lambda) := \det\,(A - \lambda\,\mathbb{1}_n) = (\lambda_1 - \lambda)^{m_1} \cdots (\lambda_p - \lambda)^{m_p},$$

  where the complex λ_j's are p distinct eigenvalues of A and m_j is the *algebraic multiplicity* of λ_j, with $m_1 + \cdots + m_p = n$. In general $1 \leqslant g_j \leqslant m_j$.

► We now distinguish between two cases according to the fact that A is or is not diagonalizable.

- A is diagonalizable. This happens if $g_j = m_j$ for all $j = 1, \ldots, p$. Then there exists a basis of $\mathbb{C}^n$ consisting of eigenvectors of A, namely $\mathbb{C}^n = \bigoplus G_{\lambda_j}$. Moreover, the eigenspaces G_{λ_j} are invariant under A: if $x \in E_{\lambda_j}$ then $A\,x \in E_{\lambda_j}$. In such a case there exists an invertible matrix T, whose columns are the eigenvectors, such that the matrix $T^{-1}A\,T$ is diagonal. Thus, after the *diagonalization* of A, it is easy to compute $\mathrm{e}^{t\,A}$.

- A is not diagonalizable. This happens if there exists at least one λ_j such that $g_j < m_j$. Then it is not true that $\mathbb{C}^n = \bigoplus G_{\lambda_j}$. For such matrices one defines the *generalized eigenspace* associated with λ_j:

$$E_{\lambda_j} := \left\{x \in \mathbb{C}^n : (A - \lambda_j\,\mathbb{1}_n)^{\ell_j}x = 0, \text{ for some } \ell_j \geqslant 1\right\},$$

  which has dimension m_j. Elements of E_{λ_j} are called *generalized eigenvectors*. Note that, for $\ell_j = 1$, $E_{\lambda_j} = G_{\lambda_j}$. It turns out that the generalized eigenvectors form a basis of $\mathbb{C}^n$, namely $\mathbb{C}^n = \bigoplus E_{\lambda_i}$, and they are invariant under A: if $x \in E_{\lambda_j}$ then $A\,x \in E_{\lambda_j}$.

► Here is the procedure to compute $\mathrm{e}^{t\,A}$ in the case that A is not diagonalizable. We start with $A \in \mathfrak{gl}(n,\mathbb{C})$ but we then consider the real reduction of the procedure since our original problem is defined for $A \in \mathfrak{gl}(n,\mathbb{R})$.

1. *Jordanization of A.* There exists an invertible matrix T, whose columns are the generalized eigenvectors, such that

$$\widetilde{A} := T^{-1} A\, T = \mathsf{diag}(J_1, \ldots, J_p),$$

where each block J_j, corresponding to an eigenvalue λ_j, is a $m_j \times m_j$ matrix

$$J_j := \lambda_j \mathbb{1}_{m_j} + N_j = \begin{pmatrix} \lambda_j & 1 & & & \\ & \lambda_j & 1 & & \\ & & \lambda_j & \ddots & \\ & & & \ddots & 1 \\ & & & & \lambda_j \end{pmatrix}, \tag{2.37}$$

where N_j is a matrix with ones in the first diagonal above the main diagonal and zeros elsewhere. In particular we have:

- N_j is a *nilpotent* matrix of order ℓ_j, i.e., $N_j^{\ell_j} = 0$.
- $[\lambda_j \mathbb{1}_{m_j}, N_j] = 0$, hence $e^{J_j} = e^{\lambda_j \mathbb{1}_{m_j}} e^{N_j}$, where e^{N_j} terminates after ℓ_j terms.

2. *Exponentiation of Jordan blocks.* If A is in its Jordan form $\widetilde{A}$ then

$$e^{\widetilde{A}} = \mathsf{diag}\left(e^{J_1}, \ldots, e^{J_p}\right).$$

Thus the problem is reduced to the computation of the exponential of a single block e^{J_j}. Since N_j is nilpotent of order ℓ_j we have:

$$e^{J_j} = e^{\lambda_j \mathbb{1}_{m_j}} e^{N_j} = e^{\lambda_j} \begin{pmatrix} 1 & 1 & 1/2! & \ldots & 1/(\ell_j - 1)! \\ 0 & 1 & 1 & \ldots & 1/(\ell_j - 2)! \\ \ldots & \ldots & \ldots & \ldots & \ldots \\ \ldots & \ldots & \ldots & \ldots & 1 \\ \ldots & \ldots & \ldots & \ldots & 1 \end{pmatrix}. \tag{2.38}$$

3. *Real Jordanization of A.* If $A \in \mathfrak{gl}(n, \mathbb{R})$ the eigenvalues of A are either real or complex conjugate, $\lambda_j := \alpha_j + \mathrm{i}\,\beta_j$, $\lambda_j^* := \alpha_j - \mathrm{i}\,\beta_j$, $\alpha_j, \beta_j \in \mathbb{R}$. Then the real Jordan form of A is

$$\widetilde{A} := T^{-1} A\, T = \mathsf{diag}(J_1, \ldots, J_p),$$

with the following prescription:

- If $\lambda_j \in \mathbb{R}$ then the corresponding block is of the form (2.37).

- If $\lambda_j \in \mathbb{C}$, we have two complex conjugate blocks $J_j := \lambda_j \mathbb{1}_{m_j} + N_j$ and $J_j^* := \lambda_j^* \mathbb{1}_{m_j} + N_j$. The double block $\mathrm{diag}(J_j, J_j^*)$ can be written as the matrix

$$\begin{pmatrix} R_j & \mathbb{1}_2 & & & \\ & R_j & \mathbb{1}_2 & & \\ & & R_j & \ddots & \\ & & & \ddots & \mathbb{1}_2 \\ & & & & R_j \end{pmatrix}, \qquad R_j := \begin{pmatrix} \alpha_j & \beta_j \\ -\beta_j & \alpha_j \end{pmatrix}.$$

4. *Exponentiation of real Jordan blocks.*

- If $\lambda_j \in \mathbb{R}$ then the corresponding block is of the form (2.37). Its exponential is given by (2.38).
- If $\lambda_j \in \mathbb{C}$, the exponential of $\mathrm{diag}(J_j, J_j^*)$ is of the form

$$\begin{pmatrix} e^{R_j} & e^{R_j} & e^{R_j}/2! & \dots & e^{R_j}/(n_j-1)! \\ 0 & e^{R_j} & e^{R_j} & \dots & e^{R_j}/(n_j-2)! \\ \dots & \dots & \dots & \dots & \dots \\ \dots & \dots & \dots & \dots & e^{R_j} \\ \dots & \dots & \dots & \dots & e^{R_j} \end{pmatrix},$$

for some $n_j \leqslant n$. Here

$$e^{R_j} = e^{\alpha_j} \begin{pmatrix} \cos\beta_j & \sin\beta_j \\ -\sin\beta_j & \cos\beta_j \end{pmatrix}.$$

▶ Remarks:

- It is now clear that the solution of the linear IVP (2.35), with $A \in \mathfrak{gl}(n, \mathbb{R})$, is completely determined by the eigenvalues and the generalized eigenvectors of A. In fact, also its stability is completely fixed by the eigenstructure of A (see Theorem 2.13).

- The point $x = 0$ is is always a fixed point of the linear IVP (2.35). Furthermore, if A is invertible, that is

$$A \in \mathbf{GL}(n, \mathbb{R}) := \{A \in \mathfrak{gl}(n, \mathbb{R}) \ : \ \det A \neq 0\},$$

the point $x = 0$ is the only fixed point. Indeed, if $\det A = 0$, then A has at least one eigenvalue equal to 0 and any point x in the corresponding eigenspace is a fixed point. The space $\mathbf{GL}(n, \mathbb{R})$ here introduced is called *general linear Lie group on* $\mathbb{R}^n$: it is a smooth manifold equipped with a group structure and not a linear vector space as $\mathfrak{gl}(n, \mathbb{R})$ (see Chapter 4, Subsection 4.2.7, for further details on Lie groups).

► The next Theorem summarizes our results concerning the structure of the solution of the IVP (2.35)

Theorem 2.10

Consider the IVP (2.35). Assume that $A \in \mathfrak{gl}(n, \mathbb{R})$ has p distinct real eigenvalues $\lambda_1, \dots, \lambda_p$ with algebraic multiplicity $m_1, \dots, m_p$, and q distinct complex conjugates eigenvalues $\alpha_1 \pm \mathrm{i}\,\beta_1, \dots, \alpha_q \pm \mathrm{i}\,\beta_q$, with algebraic multiplicity $r_1, \dots, r_q$ ($m_1 + \dots + m_p + r_1 + \dots + r_q = n$). Then the flow of the IVP (2.35) is a global Lie group of smooth diffeomorphisms given by

$$\Phi_t : (t, x_0) \mapsto \Phi_t(x_0) := \mathrm{e}^{A\,t} x_0,$$

where the entries of the matrix $\mathrm{e}^{A\,t} \in \mathbf{GL}(n, \mathbb{R})$ are linear combinations of terms of the form

$$\mathrm{e}^{\lambda_j\,t} P_j(t), \qquad \mathrm{e}^{\alpha_k\,t} \cos(\beta_k\,t) Q_k(t), \qquad \mathrm{e}^{\alpha_k\,t} \sin(\beta_k\,t) R_k(t), \tag{2.39}$$

with $j = 1, \dots, p$ and $k = 1, \dots, q$. Here P_j and Q_k, R_k are polynomials of t of degree less than m_j and r_k respectively.

► The next definition is useful.

Definition 2.11

1. *The linear subspaces*

$$E^+ := \bigoplus_{\lambda_j : \alpha_j < 0} E_{\lambda_j}, \qquad E^- := \bigoplus_{\lambda_j : \alpha_j > 0} E_{\lambda_j}, \qquad E^0 := \bigoplus_{\lambda_j : \alpha_j = 0} E_{\lambda_j},$$

   are called resp. **stable, unstable, center subspaces** *of the IVP (2.35).*

2. *If $E^0 = \emptyset$ then the flow Φ_t of the IVP (2.35) is called* **hyperbolic flow**. *Correspondingly, the fixed point $\widetilde{x} = 0$ is called* **hyperbolic fixed point**. *In addition:*

   (a) *If $E^+ = \emptyset$ then the flow Φ_t of the IVP (2.35) is called* **expansion** *and $\widetilde{x} = 0$ is called* **source**.

   (b) *If $E^- = \emptyset$ then the flow Φ_t of the IVP (2.35) is called* **contraction** *and $\widetilde{x} = 0$ is called* **sink**.

► The next claim shows that, if Φ_t is an expansion, then all non-stationary solutions of the IVP (2.35) escape exponentially fast from the source $\widetilde{x} = 0$. Viceversa, if Φ_t is a contraction, then all non-stationary solutions of the IVP (2.35) converge exponentially fast towards the sink $\widetilde{x} = 0$.

Theorem 2.11

Consider the IVP (2.35).

1. *If Φ_t is an expansion (i.e., $E^+ = E^0 = \emptyset$) then there exists constants $C, \gamma > 0$ such that*
$$\left\| \mathrm{e}^{-A\,t} x \right\| \leqslant C\,\mathrm{e}^{-\gamma\,t} \|x\| \qquad \forall\, x \in \mathbb{R}^n,\ t \geqslant 0.$$

2. *If Φ_t is a contraction (i.e., $E^- = E^0 = \emptyset$) then there exists constants $C, \gamma > 0$ such that*
$$\left\| \mathrm{e}^{A\,t} x \right\| \leqslant C\,\mathrm{e}^{-\gamma\,t} \|x\| \qquad \forall\, x \in \mathbb{R}^n,\ t \geqslant 0.$$

No Proof.

▶ The next claim provides a characterization of the stability of the fixed point $\widetilde{x} = 0$ for hyperbolic flows.

Theorem 2.12

Consider the IVP (2.35). Let the fixed point $\widetilde{x} = 0$ be hyperbolic.

1. *The phase space $\mathbb{R}^n$ admits a unique decomposition*
$$\mathbb{R}^n = E^+ \oplus E^-, \qquad E^{\pm} = \pi_{\pm} \mathbb{R}^n,$$
where $\pi_{\pm}$, $\pi_+ + \pi_- = \mathbb{1}_n$ are projection operators. The invariant subspaces $E^{\pm}$ are such that restriction of the flow Φ_t of (2.35) on E^+ is a contraction and the restriction of the flow Φ_t of (2.35) on E^- is an expansion.

2. *There exists constants $C, \gamma_{\pm} > 0$ such that*
$$\left\| \mathrm{e}^{\pm A\,t} \pi_{\pm} x \right\| \leqslant C\,\mathrm{e}^{-\gamma_{\pm} t} \| \pi_{\pm} x \| \qquad \forall\, x \in \mathbb{R}^n,\ t \geqslant 0.$$

3. *The invariant stable and unstable subspaces admit the following characterization:*
$$E^+ = \left\{ x \in \mathbb{R}^n \ : \ \sup_{t \geqslant 0} \left\| \mathrm{e}^{A\,t} x \right\| < +\infty \right\},$$
$$E^- = \left\{ x \in \mathbb{R}^n \ : \ \sup_{t \leqslant 0} \left\| \mathrm{e}^{A\,t} x \right\| < +\infty \right\}.$$

No Proof.

► The next claim provides a characterization of the stability of the fixed point $\widetilde{x} = 0$ according to the eigenvalues of the matrix A.

Theorem 2.13

Consider the IVP (2.35). Assume that $A \in \mathfrak{gl}(n, \mathbb{R})$ has p distinct eigenvalues $\lambda_1, \ldots, \lambda_p$ (either real or complex conjugate) with algebraic multiplicity $m_1, \ldots, m_p$ ($m_1 + \cdots + m_p = n$) and geometric multiplicity $g_1, \ldots, g_p$.

1. *If $\mathfrak{Re}\,\lambda_i < 0$ for all $i = 1, \ldots, p$ (i.e., $E^- = E^0 = \emptyset$) then $\widetilde{x} = 0$ is asymptotically stable.*
2. *If $\mathfrak{Re}\,\lambda_i \leqslant 0$ for all $i = 1, \ldots, p$ (i.e., $E^- = \emptyset$) and for those eigenvalues λ_k such that $\mathfrak{Re}\,\lambda_k = 0$ there holds $m_k = g_k$ then $\widetilde{x} = 0$ is stable.*
3. *If there exists at least one λ_k with $\mathfrak{Re}\,\lambda_k > 0$ (i.e., $E^- \neq \emptyset$) then $\widetilde{x} = 0$ is unstable.*

Proof. We prove all claims.

1. If $\mathfrak{Re}\,\lambda_i < 0$ for all $i = 1, \ldots, p$, then the solution of the IVP is a linear combination of terms (see (2.39))
$$e^{\lambda_i t} P_i(t),$$
when λ_i is real, and
$$e^{\mathfrak{Re}\,\lambda_i t} \cos(\mathfrak{Im}\,\lambda_i t) Q_i(t), \qquad e^{\mathfrak{Re}\,\lambda_i t} \sin(\mathfrak{Im}\,\lambda_i t) R_i(t),$$
when λ_i is complex. All of these terms converge to 0 as $t \to \infty$.
2. If there exists λ_k such that $\mathfrak{Re}\,\lambda_k = 0$ then the divergent terms are given by
$$e^{\mathfrak{Re}\,\lambda_k t} \cos(\mathfrak{Im}\,\lambda_k t) Q_k(t), \qquad e^{\mathfrak{Re}\,\lambda_k t} \sin(\mathfrak{Im}\,\lambda_k t) R_k(t).$$
The polynomials Q_k, R_k are generated by exponentials of nilpotent matrices. Therefore they vanish identically if and only if $m_k = g_k$.
3. If there exists at least one λ_k with $\mathfrak{Re}\,\lambda_k > 0$ then one of the terms
$$e^{\lambda_k t} P_k(t),$$
if λ_k is real, or
$$e^{\mathfrak{Re}\,\lambda_k t} \cos(\mathfrak{Im}\,\lambda_k t) Q_k(t), \qquad e^{\mathfrak{Re}\,\lambda_k t} \sin(\mathfrak{Im}\,\lambda_k t) R_k(t),$$
if λ_k is complex, is divergent.

The Theorem is proved. ■

Example 2.16 (*Linear autonomous and homogeneous IVPs in* $\mathbb{R}^2$)

Consider the following IVP in $\mathbb{R}^2$:

$$\begin{cases} \dot{x} = A\,x, \\ x(0) = (x_1(0), x_2(0)), \end{cases}$$

with $A \in \mathbf{GL}(2,\mathbb{R})$. The point $(0,0)$ is the only fixed point. There exists always a basis such that A takes the four forms A_1, A_2, A_3, A_4 described below.

1. $\lambda_1 \neq \lambda_2$.

   (i) $\lambda_1, \lambda_2 \in \mathbb{R}$. Then

   $$A_1 := \begin{pmatrix} \lambda_1 & 0 \\ 0 & \lambda_2 \end{pmatrix}.$$

   The flow of the IVP is

   $$\Phi_t(x(0)) = \left(e^{\lambda_1 t} x_1(0), e^{\lambda_2 t} x_2(0)\right).$$

   Orbits through $x(0)$ are

   $$\mathcal{O}(x(0)) := \left\{(x_1, x_2) \in \mathbb{R}^2 \,:\, x_2 = c\, x_1^{\lambda_2/\lambda_1}\right\},$$

   where $c \in \mathbb{R}$ depends on $x(0)$. The fixed point $(0,0)$ is a *node* if $\lambda_1\lambda_2 > 0$ (Fig. 2.18 (a)), a *saddle* if $\lambda_1\lambda_2 < 0$ (Fig. 2.18 (b)).

   (ii) $\lambda_1 := \alpha + i\,\beta = \lambda_2^* \in \mathbb{C}$. Then

   $$A_2 := \begin{pmatrix} \alpha & -\beta \\ \beta & \alpha \end{pmatrix}.$$

   The flow of the IVP is

   $$\Phi_t(x(0)) = e^{\alpha t}(\cos(\beta t)\, x_1(0) - \sin(\beta t)\, x_2(0), \sin(\beta t)\, x_1(0) + \cos(\beta t)\, x_2(0)).$$

   Orbits through $x(0)$ are spirals if $\alpha \neq 0$ and $(0,0)$ is a *focus* (Fig. 2.18 (c)) or ellipses if $\alpha = 0$ and $(0,0)$ is a *center* (Fig. 2.18 (d)).

2. $\lambda_1 = \lambda_2 =: \lambda$.

   (i) If the geometric multiplicity of λ is 2, then

   $$A_3 := \begin{pmatrix} \lambda & 0 \\ 0 & \lambda \end{pmatrix}.$$

   The flow of the IVP is

   $$\Phi_t(x(0)) = e^{\lambda t}(x_1(0), x_2(0)).$$

   Orbits through $x(0)$ are lines and $(0,0)$ is a *degenerate node* (Fig. 2.18 (e)).

   (ii) If the geometric multiplicity of λ is 1, then

   $$A_4 := \begin{pmatrix} \lambda & 1 \\ 0 & \lambda \end{pmatrix}.$$

   The flow of the IVP is

   $$\Phi_t(x(0)) = e^{\lambda t}(x_1(0) + t\, x_2(0), x_2(0)).$$

   Orbits through $x(0)$ are curves and $(0,0)$ is an *improper node* (Fig. 2.18 (f)).

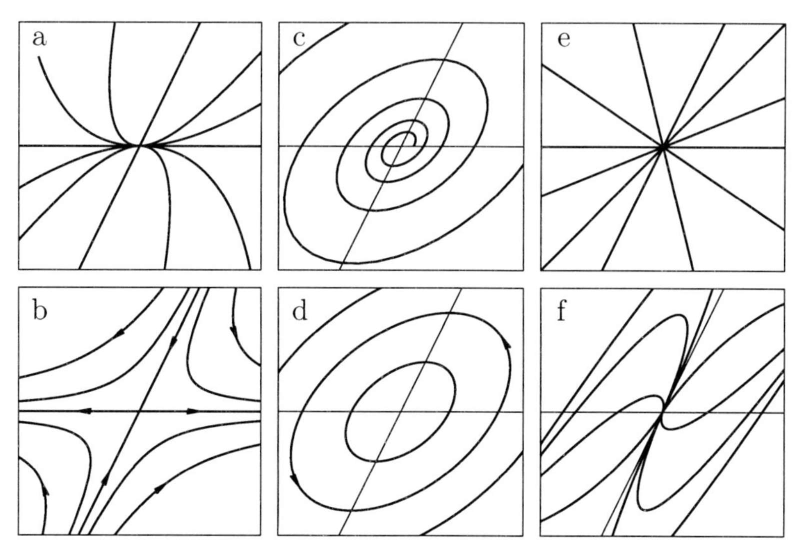

Fig. 2.18. Phase portraits of linear 2 × 2 systems ([Be]).

► The next Theorem provides a characterization of a Lyapunov function for the linear IVP (2.35). Such a result will be used later to prove Theorem 2.15.

Theorem 2.14

Consider the IVP (2.35). The fixed point $\tilde{x} = 0$ *is asymptotically stable, i.e.,* $\mathfrak{Re}\,\lambda_i < 0$ *for all* $i = 1, \ldots, p$*, if and only if for any positive definite symmetric matrix* $Q \in \mathbf{GL}(n, \mathbb{R})$ *there exists a positive definite symmetric matrix* $P \in \mathbf{GL}(n, \mathbb{R})$ *that satisfies the matrix equation*

$$P\,A + A^\top P = -Q. \tag{2.40}$$

Proof. We prove the Theorem in both directions.

- The sufficiency follows from Theorem 2.8 with Lyapunov function

$$F(x) := \langle\, x, P\,x \,\rangle\,,$$

where $P \in \mathbf{GL}(n, \mathbb{R})$ is a positive definite symmetric matrix. It is clear that $F(x) > 0$ for all $x \in \mathbb{R}^n$ and $F(x) = 0$ if and only if $x = 0$. According to Definition 2.10 we need to prove that $(\mathfrak{L}_\mathbf{v} F)(x) < 0$ for all $x \in \mathbb{R}^n \setminus \{0\}$. Here $\mathbf{v}$ is the infinitesimal generator of the flow $\Phi_t : (t, x_0) \mapsto \Phi_t(x_0) := \mathrm{e}^{A\,t} x_0$, that is

$$\mathbf{v} := \sum_{i=1}^{n} (A\,x)_i \frac{\partial}{\partial x_i}.$$

Note that $\text{grad}_x F(x) = 2\,P\,x$. Therefore we get:

$$\begin{aligned}
(\mathfrak{L}_{\mathbf{v}}F)(x) &= 2\,\langle\,A\,x, P\,x\,\rangle \\
&= \left\langle\,x, A^\top P\,x\,\right\rangle + \langle\,x, P\,A\,x\,\rangle \\
&= \left\langle\,x, \left(A^\top P + P\,A\right)\,x\,\right\rangle \\
&= -\,\langle\,x, Q\,x\,\rangle\,,
\end{aligned}$$

which is negative if Q is positive definite as assumed.

- To prove necessity we assume that $\mathfrak{Re}\,\lambda_i < 0$ for all $i = 1, \ldots, p$, and we define the matrix
$$P := \int_0^\infty \mathrm{e}^{A^\top t}\,Q\,\mathrm{e}^{A\,t}\,\mathrm{d}t.$$
Note that P is positive definite and the integral converges because A is the matrix corresponding to an asymptotically stable IVP. Substituting this integral into $-A^\top P - PA$ we get:
$$\begin{aligned}
-\,A^\top\left(\int_0^\infty \mathrm{e}^{A^\top t}\,Q\,\mathrm{e}^{A\,t}\,\mathrm{d}t\right) &- \left(\int_0^\infty \mathrm{e}^{A^\top t}\,Q\,\mathrm{e}^{A\,t}\,\mathrm{d}t\right)\,A \\
&= -\int_0^\infty \left(A^\top \mathrm{e}^{A^\top t} Q\,\mathrm{e}^{At} + \mathrm{e}^{A^\top t} Q\,\mathrm{e}^{A\,t}\,A\right)\mathrm{d}t \\
&= -\int_0^\infty \frac{\mathrm{d}}{\mathrm{d}t}\left(\mathrm{e}^{A^\top t} Q\,\mathrm{e}^{A\,t}\right)\mathrm{d}t \\
&= -\mathrm{e}^{A^\top t} Q\,\mathrm{e}^{A\,t}\Big|_0^\infty \\
&= Q,
\end{aligned}$$
which shows that P is indeed a solution of (2.40)

The Theorem is proved. ■

2.3.5 *Stability properties of linearized IVPs*

▶ Consider the following autonomous IVP in $M \subseteq \mathbb{R}^n$:

$$\begin{cases} \dot{x} = f(x), \\ x(0) = x_0, \end{cases} \tag{2.41}$$

with $f \in C^k(M, \mathbb{R}^n)$, $k \geqslant 1$. Let $\widetilde{x} \in M$ be a fixed point. We can always assume that $\widetilde{x} = 0$ by defining a new variable $x - \widetilde{x}$. Hence without loss of generality we fix $\widetilde{x} = 0$.

Definition 2.12

1. *The* **linearization** *of the IVP (2.41) around the fixed point* $\tilde{x} = 0$ *is the linear IVP*
$$\begin{cases} \dot{x} = A\,x, \\ x(0) = x_0, \end{cases} \tag{2.42}$$
where
$$A := \left.\frac{\partial f}{\partial x}\right|_{x=0} \in \mathfrak{gl}(n, \mathbb{R}).$$

2. *The point* $\tilde{x} = 0$ *is a* **hyperbolic fixed point** *of the IVP (2.41) if it is a hyperbolic fixed point of the linearized system (2.42), i.e., if* $E^0 = \emptyset$.

3. *The point* $\tilde{x} = 0$ *is a* **linearly asymptotically stable fixed point** *of the IVP (2.41) if it is a asymptotically stable stable fixed point of the linearized system (2.42). Similarly for the stable and unstable fixed points.*

► Remarks:

- Definition 2.12 is simply based on a Taylor expansion of the function f around $\tilde{x} = 0$:
$$\dot{x} = f(x) = f(0) + A\,x + g(x) = A\,x + g(x),$$
where g is a (nonlinear) function such that $g(0) = 0$ and $\|g(x)\| = O(\|x\|)$ as $\|x\| \to 0$.

- Nonlinear dynamical systems generated by IVPs (2.41) can be always approximated by linear dynamical systems generated by IVPs (2.42). Nevertheless it is not guaranteed, a priori, that the dynamics of the linearized system reflects the dynamics of the original nonlinear system.

► A useful method to determine the relation between linear and nonlinear stability, in the case of hyperbolic fixed points, is given by the *principle of linearized stability*, also known as "Poincaré-Lyapunov Theorem". This Theorem shows that if $\tilde{x} = 0$ is a hyperbolic fixed point of the linearized system (2.42) then the nonlinear system (2.41) exhibits at $\tilde{x} = 0$ the same stability of the linearized system.

Theorem 2.15 (*Poincaré-Lyapunov*)

Consider the IVP (2.41) and let (2.42) be its linearization around the fixed point $\tilde{x} = 0$.

1. *If* $\tilde{x} = 0$ *is linearly asymptotically stable then it is asymptotically stable.*

2. *If $\widetilde{x} = 0$ is linearly unstable then it is unstable.*

Proof. We prove only the first claim.

- The fixed point $\widetilde{x} = 0$ is linearly asymptotically stable, i.e., $\Re\mathfrak{e}\,\lambda_i < 0$ for all $i = 1, \ldots, p$. From Theorem 2.14 we know that there exists a positive definite symmetric matrix $P \in \mathbf{GL}(n, \mathbb{R})$ such that the function

$$F(x) := \langle\, x, P\,x \,\rangle$$

  is a strict Lyapunov function if $-P\,A - A^\top P = Q$ is a positive definite symmetric matrix.

- Denote by $q_i > 0$ the eigenvalues of Q and by $p_i > 0$ the eigenvalues of P. We set

$$p_0 := \max p_i > 0, \qquad q_0 := \min q_i > 0.$$

  In particular we have the inequalities

$$\|P\,x\| \leqslant p_0\|x\|, \qquad \langle\, x, Q\,x \,\rangle \geqslant q_0\|x\|^2.$$

- Consider F as a strict Lyapunov function candidate also for the nonlinear system (2.41).We have $F(x) > 0$ and $F(x) = 0$ if and only if $x = 0$. Recall that the nonlinear function g in (2.43) is such that for any $\varepsilon > 0$ there exists $\delta > 0$ such that $\|g(x)\| < \varepsilon\|x\|$ for $\|x\| < \delta$. We can now estimate the Lie derivative of the function F:

$$\begin{aligned}
(\mathfrak{L}_{\mathbf{v}}F)(x) &= 2\,\langle\, f(x), P\,x \,\rangle \\
&= \langle\, f(x), P\,x \,\rangle + \langle\, x, P\,f(x) \,\rangle \\
&= \langle\, A\,x + g(x), P\,x \,\rangle + \langle\, x, P\,(A\,x + g(x)) \,\rangle \\
&= -\langle\, x, Q\,x \,\rangle + 2\,\langle\, g(x), P\,x \,\rangle \\
&\leqslant -q_0\|x\|^2 + 2\,\|g(x)\|\,\|P\,x\| \\
&\leqslant -q_0\|x\|^2 + 2\,\varepsilon\,p_0\|x\|^2 \\
&\leqslant (-q_0 + 2\,\varepsilon\,p_0)\,\|x\|^2,
\end{aligned}$$

  which is negative if we choose ε such that $-q_0 + 2\,\varepsilon\,p_0 < 0$.

The first claim is proved. ■

► For completeness we provide an alternative formulation of Theorem 2.15.

Theorem 2.16 (*Poincaré-Lyapunov*)

Consider the IVP (2.41) and let (2.42) be its linearization around the fixed point

$\widetilde{x} = 0$. Let $\lambda_1, \ldots, \lambda_p$ be the p distinct eigenvalues of the matrix A.

1. *If $\mathfrak{Re}\,\lambda_i < -\gamma$ for all $i = 1, \ldots, p$, with $\gamma > 0$, then there exists a neighborhood $\widetilde{M}$ of $\widetilde{x} = 0$ such that:*
   (a) *The flow Φ_t of (2.41) is such that $\Phi_t(x) \in \widetilde{M}$ for all $x \in \widetilde{M}$ and for all $t \geqslant 0$.*
   (b) *There exists $C > 0$ such that*
   $$\|\Phi_t(x)\| \leqslant C\,\mathrm{e}^{-\gamma t/2}\|x\| \qquad \forall\, x \in \widetilde{M},\, t \geqslant 0.$$
   In particular, $\widetilde{x} = 0$ is asymptotically stable.
2. *If there exists at least one λ_k with $\mathfrak{Re}\,\lambda_k > 0$ then $\widetilde{x} = 0$ is an unstable fixed point of (2.41).*

No Proof.

► From Theorem 2.15 (and its equivalent formulation 2.16) we see that the borderline situation occurs when there exists at least one λ_k with $\mathfrak{Re}\,\lambda_k = 0$, i.e., $\widetilde{x} = 0$ is not hyperbolic. In this case Theorems 2.15 and 2.16 are inconclusive.

Example 2.17 (*Van der Pol system*)

Consider the following IVP in $\mathbb{R}^2$:

$$\begin{cases} \dot{x}_1 = x_2, \\ \dot{x}_2 = x_2\left(1 - x_1^2\right) - x_1, \\ (x_1(0), x_2(0)) \in \mathbb{R}^2. \end{cases}$$

- The only fixed point is $(0,0)$. The linearized system has matrix
  $$A := \begin{pmatrix} 0 & 1 \\ -1 & 1 \end{pmatrix},$$
  which has eigenvalues $\lambda_{1,2} = (1 \pm \mathrm{i}\sqrt{3})/2$.
- The point $(0,0)$ is an unstable focus of the linearized system and by Theorem 2.15 we conclude that $(0,0)$ is an unstable fixed point also for the nonlinear system. All non-stationary solutions starting out near $(0,0)$ eventually spiral away.

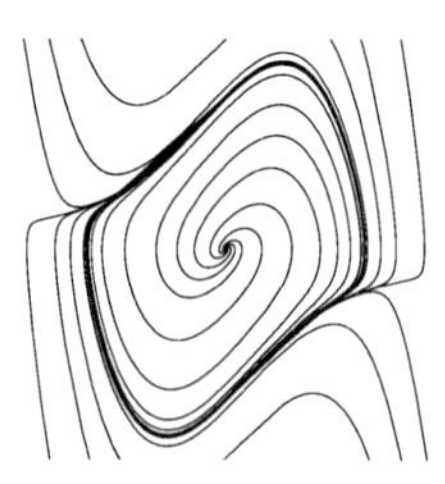

Fig. 2.19. Phase portrait of the Van der Pol system ([Ol1]).

- A detailed analysis shows the existence of a stable limit cycle centered at $(0,0)$: all non-stationary solutions spiral towards it.

Example 2.18 (*Lotka-Volterra system*)

Consider the following IVP in $\mathbb{R}^2$:

$$\begin{cases} \dot{x}_1 = x_1(1-x_2), \\ \dot{x}_2 = \alpha\, x_2(x_1-1), \\ (x_1(0), x_2(0)) \in \mathbb{R}^2. \end{cases} \tag{2.43}$$

with $\alpha > 0$. This system describes the competition between two populations, one prey species x_1 and one predator species x_2.

- There are two fixed points: $(0,0)$ and $(1,1)$. The lines $x_1 = 0$ and $x_2 = 0$ are invariant under the flow:
$$\Phi_t(0, x_2(0)) = (0, x_2(0)\mathrm{e}^{-\alpha t}), \qquad \Phi_t(x_1(0), 0) = (x_1(0)\mathrm{e}^{t}, 0).$$
In particular, since no other solution can cross these lines, the first quadrant $Q := \{(x_1, x_2) \in \mathbb{R}^2 : x_1 > 0, x_2 > 0\}$ is invariant. This is the region we are interested in and one can consider $(x_1(0), x_2(0)) \in Q$. In particular, we see that $(0,0)$ is unstable.
- Let us consider the fixed point $(1,1)$. The eigenvalues of the matrix of the linearized system are purely imaginary and thus Theorem 2.15 does not apply. To understand the stability of $(1,1)$ we use the method of Lyapunov functions. Note that
$$\frac{\mathrm{d}x_2}{\mathrm{d}x_1} = \alpha\,\frac{x_2(x_1-1)}{x_1(1-x_2)}.$$
This equation can be solved implicitly by
$$F(x_1,x_2) := \alpha(x_1 - \log x_1) + x_2 - \log x_2 = c,$$
where $c \in \mathbb{R}$ is a constant. Indeed, the function F is an integral of motion:
$$(\mathfrak{L}_{\mathbf{v}}F)(x_1,x_2) = \alpha\, x_1(1-x_2)\left(1 - \frac{1}{x_1}\right) + \alpha\, x_2(x_1-1)\left(1-\frac{1}{x_2}\right) = 0,$$
for all $(x_1, x_2) \in Q$.
- The two-dimensional surface $x_3 = F(x_1, x_2)$ is convex and it exhibits an isolated minimum at $(1,1)$. The orbits of the IVP (2.43) are obtained by the intersection of the surface $x_3 = F(x_1,x_2)$ with the planes $x_3 = c$.

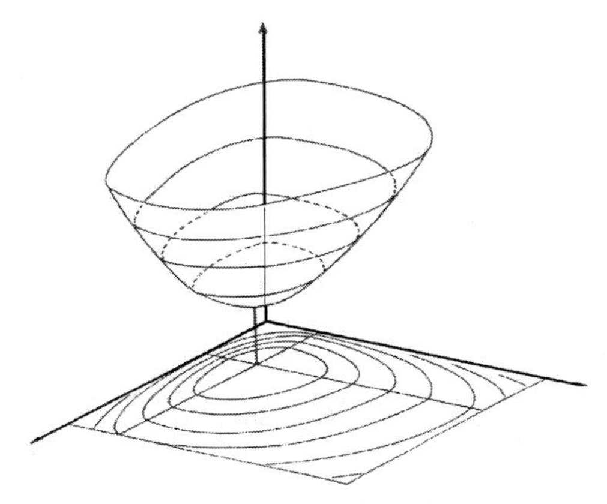

Fig. 2.20. A plot of the surface $x_3 = F(x_1, x_2)$ and some representative orbits.

- The function F serves as Lyapunov function. Theorem 2.8 shows that $(1,1)$ is a stable fixed point. Moreover, due to the convexity of the surface $x_3 = F(x_1, x_2)$ we conclude that all orbits of the IVP (2.43) starting in Q are closed and they encircle the fixed point $(1,1)$.

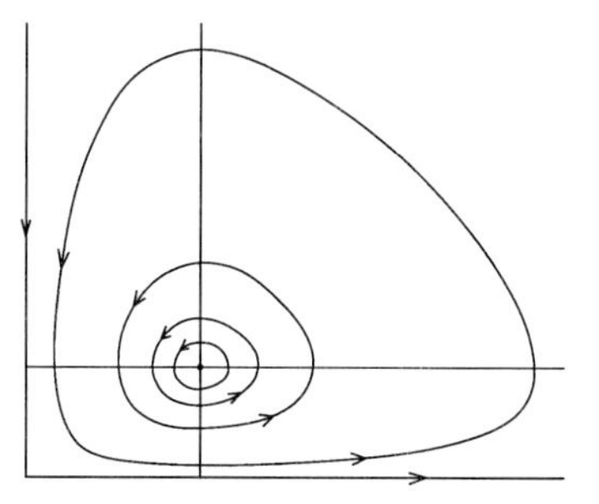

Fig. 2.21. Phase portrait of the Lotka-Volterra system ([Ku]).

2.4 Topological equivalence of dynamical systems

► The (ambitious) aim of qualitative theory of dynamical systems is to provide a catalogue of *equivalence classes* between dynamical systems.

- The comparison of dynamical systems is based on *equivalence relations*, allowing us to define classes of equivalent "objects" and to study transitions between these classes.
- Thus we have to specify when we define two dynamical systems as being qualitatively similar or equivalent. Such a definition must meet some general intuitive criteria.
- It is natural to expect that two equivalent systems have the same number of fixed points and cycles of the same stability types. The relative position of these invariant sets and the shape of their regions of attraction should also be similar for equivalent systems. In other words, we consider two dynamical systems as equivalent if their phase portraits are qualitatively similar, namely, if one portrait can be obtained from another by a continuous transformation.

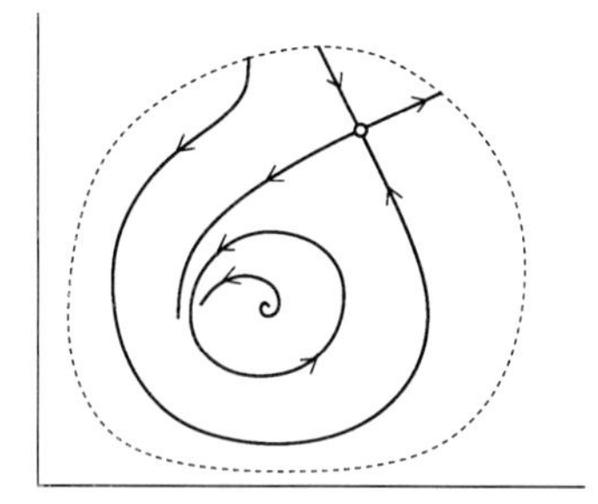

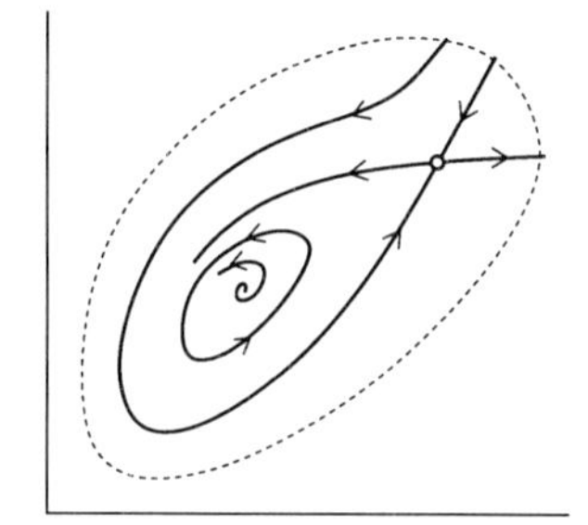

Fig. 2.22. Two topologically equivalent phase portraits ([Ku]).

► The "Poincaré-Lypaunov Theorem" 2.15 indicates one way in which the flow near a fixed point of an autonomous IVP resembles the flow of its linearization. We will give a Theorem (see Theorem 2.17) which provides further insight into the extent of the resemblance.

► Let Φ_t and Ψ_t, $t \in \mathbb{R}$, be two distinct one-parameter groups of smooth diffeomorphisms on M_1 and M_2 respectively, with $M_1, M_2 \subseteq \mathbb{R}^n$. Their infinitesimal generators are vector fields on M_1, M_2 respectively:

$$\mathbf{v} := \sum_{i=1}^{n} f_i(x)\frac{\partial}{\partial x_i}, \qquad \mathbf{w} := \sum_{i=1}^{n} g_i(x)\frac{\partial}{\partial y_i},$$

with

$$f(x) := \left.\frac{\mathrm{d}}{\mathrm{d}t}\right|_{t=0} \Phi_t(x), \qquad g(y) := \left.\frac{\mathrm{d}}{\mathrm{d}t}\right|_{t=0} \Psi_t(y).$$

Therefore we have two distinct dynamical systems $\{\Phi_t, \mathbb{R}, M_1\}$ and $\{\Psi_t, \mathbb{R}, M_2\}$.

Definition 2.13

1. $\{\Phi_t, \mathbb{R}, M_1\}$ *and* $\{\Psi_t, \mathbb{R}, M_2\}$ *are* **topologically equivalent** *if there exists a homeomorphism* $\Lambda : M_1 \longrightarrow M_2$ *(i.e., both* Λ *and* Λ^{-1} *are continuous) which maps the orbits of* $\{\Phi_t, \mathbb{R}, M_1\}$ *onto the orbits of* $\{\Psi_t, \mathbb{R}, M_2\}$ *preserving the direction of time, that is*

$$(\Lambda \circ \Phi_t)(x) = (\Psi_\tau \circ \Lambda)(x), \qquad \forall\, x \in M_1,\ t \in \mathbb{R}$$

   where $\tau : \mathbb{R} \to \mathbb{R} : t \mapsto \tau(t)$ *is a homeomorphism. If* Λ *is a diffeomorphism of class* C^k *(i.e., both* Λ *and* Λ^{-1} *are* C^k*-differentiable) then* $\{\Phi_t, \mathbb{R}, M_1\}$ *and* $\{\Psi_t, \mathbb{R}, M_2\}$ *are* **C^k-diffeomorphic**.

2. $\{\Phi_t, \mathbb{R}, M_1\}$ *and* $\{\Psi_t, \mathbb{R}, M_2\}$ *are* **topologically conjugate** *if they are topologically equivalent with* $\tau(t) = t$ *for all* $t \in \mathbb{R}$.

3. *Let* $M_1 = M_2 = M \subseteq \mathbb{R}^n$. $\{\Phi_t, \mathbb{R}, M\}$ *and* $\{\Psi_t, \mathbb{R}, M\}$ *are* **orbitally equivalent** *if there exists a smooth scalar positive definite function* $\mu : \mathbb{R} \to (0, \infty)$ *such that*

$$g(x) = \mu(x)\, f(x) \qquad \forall\, x \in M.$$

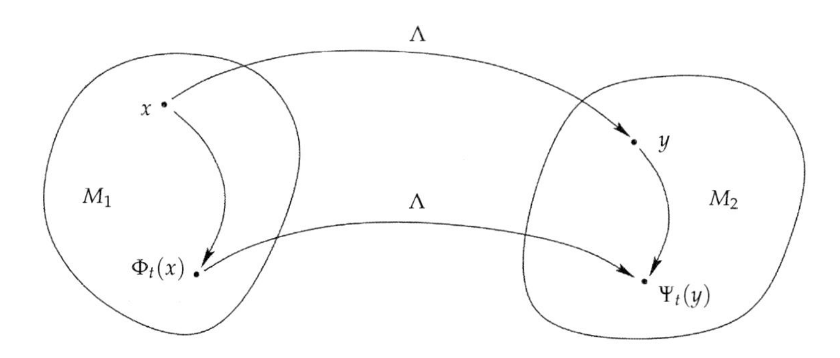

Fig. 2.23. Topological conjugacy ([Ku]).

► Remarks:

- Definition 2.13 is given in a global formulation. Clearly, one can define the same properties only locally in a neighborhood of a fixed point.
- Assume that $\{\Phi_t, \mathbb{R}, M\}$ and $\{\Psi_t, \mathbb{R}, M\}$ are C^k-diffeomorphic with $\tau(t) = t$ for all $t \in \mathbb{R}$. Then there exists a diffeomorphism $\Lambda \in C^k(M, M)$ such that
$$\Lambda(\Phi_t(x)) = \Psi_t(\Lambda(x)),$$
for all $x \in M$. Differentiating w.r.t. t at $t = 0$ we get:
$$\frac{\partial \Lambda}{\partial x} f(x) = g(\Lambda(x)),$$
namely
$$f(x) = \left(\frac{\partial \Lambda}{\partial x}\right)^{-1} g(\Lambda(x)).$$
Two diffeomorphic systems are practically identical and can be viewed as the same system written using different coordinates.
- For two orbitally equivalent systems the map Λ is just the identity map and the systems $\{\Phi_t, \mathbb{R}, M\}$ and $\{\Psi_t, \mathbb{R}, M\}$ differ only by a time parametrization along the orbits. Note that such systems can be non-diffeomorphic, having cycles that look like the same closed curve in M but have different periods.

Example 2.19 (*Topologically conjugate phase portraits*)

Consider the following planar linear systems of ODEs:

$$\begin{cases} \dot{x}_1 = -2\,x_1, \\ \dot{x}_2 = -2\,x_2, \end{cases} \qquad \begin{cases} \dot{x}_1 = -2\,x_1 + x_2, \\ \dot{x}_2 = -x_1 - 2\,x_2, \end{cases} \qquad \begin{cases} \dot{x}_1 = -2\,x_1, \\ \dot{x}_2 = x_1 - 2\,x_2. \end{cases} \tag{2.44}$$

- The eigenvalues of the corresponding matrices are:

$$\lambda_{1,2} = -2, \qquad \lambda_{1,2} = -2 \pm \mathrm{i}, \qquad \lambda_{1,2} = -2,$$

hence the fixed point $(0,0)$ is always asymptotically stable.

- The phase portraits are:

Fig. 2.24. Phase portraits of systems (2.44) ([McMe]).

- Writing the first two systems in polar coordinates, say $(r,\theta) \in (0,\infty) \times [0, 2\pi)$ and $(s,\varphi) \in (0,\infty) \times [0, 2\pi)$, we get:

$$\begin{cases} \dot{r} = -2r, \\ \dot{\theta} = 0, \end{cases} \qquad \begin{cases} \dot{s} = -2s, \\ \dot{\varphi} = -1. \end{cases}$$

The homeomorphism

$$\Lambda : (s,\varphi) \mapsto (r,\theta) = \left(s, \varphi - \frac{1}{2}\log s\right)$$

gives the topological conjugacy between the first two systems.

- The first and the third system are topologically conjugate. The homeomorphism relating the two systems is given by

$$\Lambda : (x_1, x_2) \mapsto \left(x_1, x_2 - \frac{x_1}{2}\log|x_1|\right).$$

► The following Theorem is an important result about the local behavior of continuous dynamical systems in the neighborhood of a hyperbolic fixed point $\widetilde{x} \in M$.

Theorem 2.17 (*Hartman-Grobman*)

Any hyperbolic linear IVP in $M \subseteq \mathbb{R}^n$,

$$\begin{cases} \dot{x} = A\,x, \\ x(0) = x_0, \end{cases}$$

with $A \in \mathfrak{gl}(n,\mathbb{R})$, is locally topologically conjugate in a neighborhood $\widetilde{M}$ of $\widetilde{x} = 0$ to any nonlinear IVP in $\widetilde{M}$,

$$\begin{cases} \dot{y} = A\,y + f(y), \\ y(0) = x_0, \end{cases} \tag{2.45}$$

where $f \in C^1(\widetilde{M}, \mathbb{R}^n)$ *with*

$$f(0) = 0, \qquad \left.\frac{\partial f}{\partial y}\right|_{y=0} = 0.$$

Precisely, if Φ_t *is the flow of (2.45), then there exists a homeomorphism* $x \mapsto y := \Lambda(x)$ *for which*

$$\left(\Lambda \circ e^{t\,A}\right)(x) = (\Phi_t \circ \Lambda)(x) \qquad \forall x \in \widetilde{M}.$$

No Proof.

► Theorem 2.17 gives us conditions under which a map Λ between certain flows may exist. Some limitations are:

- The conditions it gives are sufficient, but certainly not necessary, for Λ to exist.
- It does not give a simple way to construct Λ (in closed form, at least).
- It does not indicate how smooth Λ might be.

► Topological conjugacy is not a very strong property because Λ need not be differentiable. In fact, one can show that all linear systems with the same number of eigenvalues with positive a negative real parts are topologically equivalent. On the other hand, topological conjugacy of class C^k with $k > 0$ is harder to achieve.

Example 2.20 (*Topological equivalence*)

Consider the following IVP in $\mathbb{R}^2$:

$$\begin{cases} \dot{x}_1 = 2\,x_1 + x_2^2, \\ \dot{x}_2 = x_2, \\ (x_1(0), x_2(0)) \in \mathbb{R}^2. \end{cases} \tag{2.46}$$

- The orbits are given by the curves
$$x_1 = x_2^2(\alpha + \log|x_2|), \tag{2.47}$$
where $\alpha \in \mathbb{R}$ is a constant depending on the initial conditions.
- The linearization of (2.46) around the fixed point $(0,0)$ reads
$$\begin{cases} \dot{x}_1 = 2\,x_1, \\ \dot{x}_2 = x_2, \end{cases} \tag{2.48}$$
and $(0,0)$ is a hyperbolic fixed point.
- The orbits of the linearized system are given by the curves
$$x_1 = \alpha\,x_2^2,$$
where $\alpha \in \mathbb{R}$ is a constant depending on the initial conditions.

- The nonlinear system (2.46) is topologically conjugate to its linearization (2.48) in a sufficiently small neighborhood of $(0,0)$, but the two flows are C^1- but not C^2-conjugate because of the logarithm in (2.47).

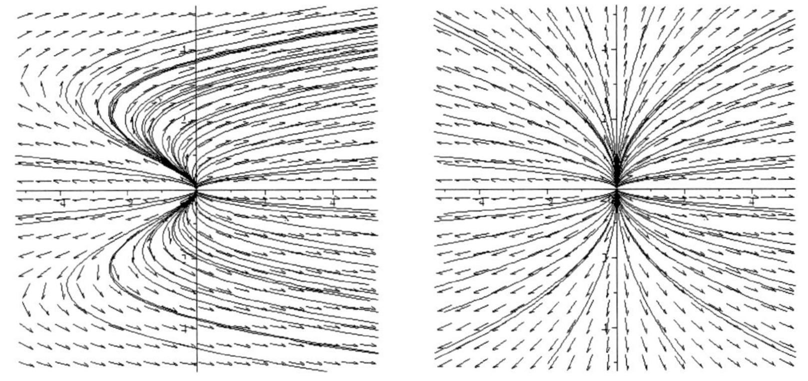

Fig. 2.25. Phase portraits of systems (2.46) and (2.48).

2.5 Stability properties of nonlinear IVPs

▶ Consider the following autonomous IVP in $M \subseteq \mathbb{R}^n$:

$$\begin{cases} \dot{x} = f(x), \\ x(0) = x_0, \end{cases} \tag{2.49}$$

with $f \in C^k(M, \mathbb{R}^n)$, $k \geqslant 1$. Let $\widetilde{x} \in M$ be a fixed point. Without loss of generality we fix $\widetilde{x} = 0$.

▶ According to Definition 2.12 the *linearization* of (2.49) around the fixed point $\widetilde{x} = 0$ is the linear IVP

$$\begin{cases} \dot{x} = A\,x, \\ x(0) = x_0, \end{cases} \tag{2.50}$$

where

$$A := \left.\frac{\partial f}{\partial x}\right|_{x=0} \in \mathfrak{gl}(n, \mathbb{R}).$$

Therefore the system of ODEs in (2.49) is written as

$$\dot{x} = A\,x + g(x),$$

where g is a (nonlinear) function such that $g(0) = 0$ and $\|g(x)\| = O(\|x\|)$ as $\|x\| \to 0$.

2.5.1 Existence of invariant stable and unstable manifolds

► We assume that $\tilde{x} = 0$ is a hyperbolic fixed point, i.e., $E^0 = \emptyset$. Let m be the number of eigenvalues of A with negative real part and $n - m$ be the number of eigenvalues of A with positive real part (counting multiplicities). Consider the linearized IVP (2.50).

- From Theorem 2.12 we know that

$$\mathbb{R}^n = E^+ \oplus E^-, \qquad \mathsf{dim}(E^+) = m, \qquad \mathsf{dim}(E^-) = n - m. \tag{2.51}$$

- After a change of basis in $\mathbb{R}^n$ we can always assume that $E^\pm$ have the form

$$E^+ := \{x \in \mathbb{R}^n \,:\, x = (x_+, 0) \in \mathbb{R}^m \times \mathbb{R}^{n-m}\},$$

$$E^- := \{x \in \mathbb{R}^n \,:\, x = (0, x_-) \in \mathbb{R}^m \times \mathbb{R}^{n-m}\}.$$

In particular, A is a $\mathfrak{gl}(n, \mathbb{R})$ matrix with block-form

$$A := \begin{pmatrix} A_+ & 0 \\ 0 & A_- \end{pmatrix},$$

where $A_+ \in \mathfrak{gl}(m, \mathbb{R})$ defines a contraction and $A_- \in \mathfrak{gl}(n-m, \mathbb{R})$ defines an expansion. In other words, after a linear change of coordinates, all linear systems of ODEs $\dot{x} = A\,x$ satisfying (2.51) are topologically conjugate to the *standard saddle system*

$$\begin{cases} \dot{\xi}_1 = -\xi_1, \\ \dot{\xi}_2 = \xi_2, \end{cases}$$

where $(\xi_1, \xi_2) \in \mathbb{R}^m \times \mathbb{R}^{n-m}$.

- Define the projectors $\pi_\pm : \mathbb{R}^n \to E^\pm$,

$$\pi_+ x := (x_+, 0), \qquad \pi_- x := (0, x_-) \qquad \forall\, x = (x_+, x_-) \in \mathbb{R}^n.$$

- Then we have:

$$\begin{aligned} A\,x &= \pi_+(A\,x) + \pi_-(A\,x) = A\,\pi_+ x + A\,\pi_- x \\ &= (A_+ x_+, A_- x_-), \end{aligned}$$

and

$$\begin{aligned} \mathrm{e}^{A\,t} &= \pi_+\left(\mathrm{e}^{A\,t} x\right) + \pi_-\left(\mathrm{e}^{A\,t} x\right) = \mathrm{e}^{A\,t}\pi_+ x + \mathrm{e}^{A\,t}\pi_- x \\ &= \left(\mathrm{e}^{A_+ t} x_+, \mathrm{e}^{A_- t} x_-\right). \end{aligned}$$

- We conclude that there exists constants $C, \gamma_\pm \geqslant 0$ such that

$$\left\|e^{At}\pi_+ x\right\| = \left\|e^{A_+ t}x_+\right\| \leqslant C e^{-\gamma_+ t}\|\pi_+ x\| = C e^{-\gamma_+ t}\|x_+\|,$$

$$\left\|e^{-At}\pi_- x\right\| = \left\|e^{-A_- t}x_-\right\| \leqslant C e^{-\gamma_- t}\|\pi_- x\| = C e^{-\gamma_- t}\|x_-\|,$$

for all $x = (x_+, x_-) \in \mathbb{R}^n$, $t \geqslant 0$.

► In a neighborhood $\widetilde{M} \subset M$ of the hyperbolic fixed point $\widetilde{x} = 0$, we have a similar situation also for the nonlinear IVP (2.49). We give the following definitions.

Definition 2.14

Consider the IVP (2.49) and denote by Φ_t its flow. Let $\widetilde{x} = 0$ be a hyperbolic fixed point of Φ_t and define the neighborhood

$$\widetilde{M} := \{x \in \mathbb{R}^n \,:\, \|\pi_\pm x\| < \delta\},$$

where $\delta > 0$.

1. *The* **local stable manifold** *of $\widetilde{x} = 0$ is the invariant set*

$$W^+ := \left\{x \in \widetilde{M} \,:\, \Phi_t(x) \in \widetilde{M}\ \forall t \geqslant 0,\ \lim_{t\to+\infty} \Phi_t(x) = 0\right\}.$$

2. *The* **local unstable manifold** *of $\widetilde{x} = 0$ is the invariant set*

$$W^+ := \left\{x \in \widetilde{M} \,:\, \Phi_{-t}(x) \in \widetilde{M}\ \forall t \geqslant 0,\ \lim_{t\to+\infty} \Phi_{-t}(x) = 0\right\}.$$

► The next Theorem establishes the existence and uniqueness of the local stable and unstable manifolds of (2.49).

Theorem 2.18

Consider the IVP (2.49) and denote by Φ_t its flow. Let $\widetilde{x} = 0$ be a hyperbolic fixed point of Φ_t. Assume that $\dim(E^+) = m$, $\dim(E^-) = n - m$.

1. *There exists a unique non-empty m-dimensional local stable manifold W^+ of class C^k. Its tangent space at $\widetilde{x} = 0$ is E^+.*

2. *There exists a unique non-empty $(n-m)$-dimensional local unstable manifold W^- of class C^k. Its tangent space at $\widetilde{x} = 0$ is E^-.*

3. There holds:

$$W^+ = \left\{ x = (x_+, x_-) \in \widetilde{M} \, : \, x_- = h(x_+) \right\},$$

where h is a function of class C^k such that $h(0) = 0$ and $\partial h / \partial x_+|_0 = 0$. Similarly for W^-.

No Proof.

Example 2.21 (***Stable and unstable manifolds for a planar IVP***)

Consider following IVP in $\mathbb{R}^2$:

$$\begin{cases} \dot{x}_1 = -x_1 + x_2 + 3\,x_2^2, \\ \dot{x}_2 = x_2, \\ (x_1(0), x_2(0)) \in \mathbb{R}^2. \end{cases} \tag{2.52}$$

- The flow is given by
$$\Phi_t(x_1(0), x_2(0)) = \left(x_1(0)\mathrm{e}^{-t} + x_2(0)\sinh t + x_2^2(0)(\mathrm{e}^{2t} - \mathrm{e}^{-t}), x_2(0)\mathrm{e}^t \right).$$
- There is only one fixed point, $(0,0)$, which is hyperbolic. It turns out that the stable and unstable manifolds are:
$$W^+ = \{(x_1, x_2) \in \mathbb{R}^2 : x_2 = 0\},$$
and
$$W^- = \left\{ (x_1, x_2) \in \mathbb{R}^2 : x_1 = \frac{1}{2}x_2 + x_2^2 \right\}.$$
- The tangent spaces of $W^\pm$ at $(0,0)$ give the linear stable and unstable subspaces for the flow of the linearized system:
$$\Phi_t(x_1(0), x_2(0)) = \left(x_1(0)\mathrm{e}^{-t} + x_2(0)\sinh t, x_2(0)\mathrm{e}^t \right).$$
One finds
$$E^+ = \{(x_1, x_2) \in \mathbb{R}^2 : x_2 = 0\} = W^+,$$
and
$$E^- = \left\{ (x_1, x_2) \in \mathbb{R}^2 : x_1 = \frac{1}{2}x_2 \right\}.$$

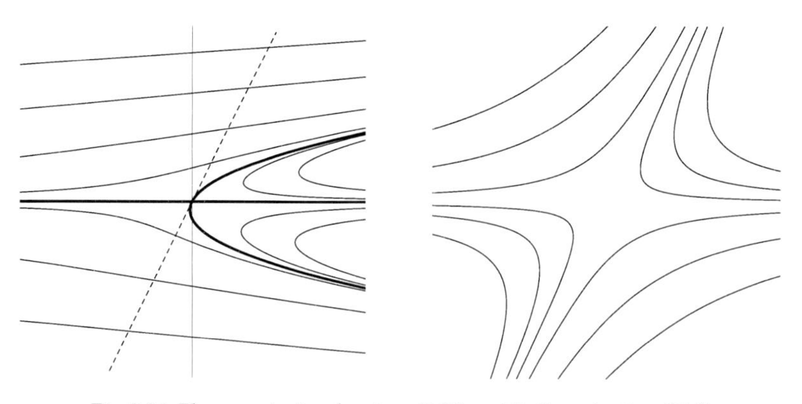

Fig. 2.26. Phase portraits of system (2.52) and its linearization ([Te]).

2.5.2 Existence of invariant center manifolds

▶ We assume that $\tilde{x} = 0$ is a non-hyperbolic fixed point, i.e., $E^0 \neq \emptyset$. Let m be the number of eigenvalues of A with negative real part and ℓ be the number of eigenvalues of A with positive real part and $n - m - \ell$ be the number of eigenvalues of A with vanishing real part (counting multiplicities). Then we have:

$$\mathbb{R}^n = E^+ \oplus E^- \oplus E^0,$$

with

$$\mathsf{dim}(E^+) = m, \qquad \mathsf{dim}(E^-) = \ell, \qquad \mathsf{dim}(E^0) = n - m - \ell.$$

Theorem 2.19

Consider the IVP (2.49) and denote by Φ_t its flow. Let $\tilde{x} = 0$ be a non-hyperbolic fixed point of Φ_t. Assume that $\mathsf{dim}(E^+) = m$, $\mathsf{dim}(E^-) = \ell$, $\mathsf{dim}(E^0) = n - m - \ell$.

1. *There exists a unique non-empty m-dimensional local stable manifold W^+ of class C^k. Its tangent space at $\tilde{x} = 0$ is E^+.*
2. *There exists a unique non-empty ℓ-dimensional local unstable manifold W^- of class C^k. Its tangent space at $\tilde{x} = 0$ is E^-.*
3. *There exists a non-empty $(n - m - \ell)$-dimensional local invariant manifold W^0 of class C^k whose tangent space at $\tilde{x} = 0$ is E^0. The set W^0 is called* **local center manifold**.

No Proof.

▶ The following example shows the non-uniqueness of center manifolds.

Example 2.22 (*Center manifolds for a planar IVP*)

Consider the following IVP in $\mathbb{R}^2$:

$$\begin{cases} \dot{x}_1 = x_1^2, \\ \dot{x}_2 = -x_2, \\ (x_1(0), x_2(0)) \in \mathbb{R}^2. \end{cases} \tag{2.53}$$

- Its flow is given by
$$\Phi_t(x_1(0), x_2(0)) = \left(\frac{1}{t - x_1(0)}, x_2(0)\mathrm{e}^{-t} \right).$$
- The fixed point is $(0, 0)$, which is not hyperbolic. The linearized system has a center subspace
$$E^0 = \{(x_1, x_2) \in \mathbb{R}^2 \, : \, x_2 = 0\}.$$
- One can prove that there exists a one-parameter family of center manifolds given by $W^0_\alpha \cup \widetilde{W}^0$, with
$$W^0_\alpha := \left\{ (x_1, x_2) \in \mathbb{R}^2 : \, x_1 < 0, \, x_2 = \alpha \, \mathrm{e}^{1/x_1} \right\},$$

where $\alpha \in \mathbb{R}$, and

$$\widetilde{W}^0 := \{(x_1, x_2) \in \mathbb{R}^2 : x_1 \geqslant 0, x_2 = 0\}.$$

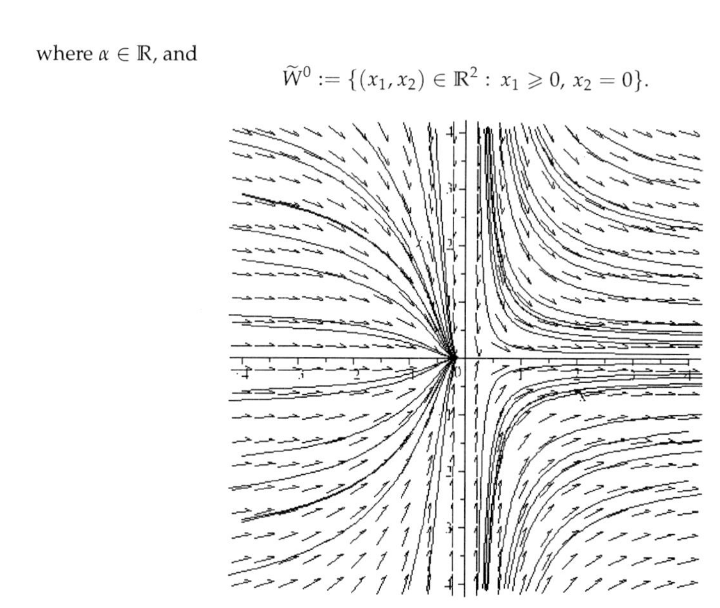

Fig. 2.27. Phase portrait of system (2.53).

► While Theorem 2.18, valid for hyperbolic fixed points, provides a complete description of the local dynamics of (2.49) in a neighborhood $\widetilde{M}$ of $\widetilde{x} = 0$, Theorem 2.19 gives such description provided we determine the behavior of solutions on the center manifold W^0. We illustrate how this can be done in the special case when $E^- = \emptyset$, i.e., $\ell = 0$, so that for the linearized IVP there holds

$$\mathbb{R}^n = E^+ \oplus E^0,$$

with

$$\mathsf{dim}(E^+) = m, \qquad \mathsf{dim}(E^0) = n - m.$$

- By a linear change of coordinates the system of ODEs (2.49) can be written in the form

$$\begin{cases} \dot{\xi} = B\,\xi + u(\xi, \eta), \\ \dot{\eta} = C\,\eta + v(\xi, \eta), \end{cases} \tag{2.54}$$

  where $(\xi, \eta) \in \mathbb{R}^{n-m} \times \mathbb{R}^m$, $B \in \mathfrak{gl}(n-m, \mathbb{R})$ has $n - m$ eigenvalues with vanishing real part, $C \in \mathfrak{gl}(m, \mathbb{R})$ has m eigenvalues with negative real part, and u, v are (nonlinear) functions such that $u(0,0) = v(0,0) = 0$.

- The tangent space at $\widetilde{x} = 0$ of a local center manifold is the linear invariant center subspace E^0. It is given by the graph of a function $q \in C^k(\mathbb{R}^{n-m}, \mathbb{R}^m)$:

$$W^0 := \left\{(\xi, \eta) \in \mathbb{R}^{n-m} \times \mathbb{R}^m : \eta = q(\xi), \|\xi\| \leqslant \varepsilon\right\},$$

  where $\varepsilon > 0$ is sufficiently small.

- If Φ_t is the the flow of (2.49), then the restriction of Φ_t on W^0 obeys the system of ODEs
$$\dot{\xi} = B\,\xi + u(\xi, q(\xi)).$$

- Since $\eta = q(\xi)$ we have:
$$\dot{\eta} = \frac{\partial q}{\partial \xi}\,\dot{\xi},$$
so that
$$C\,q(\xi) + v(\xi, q(\xi)) = \frac{\partial q}{\partial \xi}\,(B\,\xi + u(\xi, q(\xi))), \qquad (2.55)$$
that is a first-order nonlinear system of partial differential equations for $q(\xi)$.

- The idea is to expand the function $q(\xi)$ in a power series of ξ and determine the (approximate) form of $q(\xi)$ by using (2.55).

Example 2.23 (*Approximation of the center manifold for a planar IVP*)

Assume that the system of ODEs of an IVP in $\mathbb{R}^2$ can be written, after a change of coordinates, in the form
$$\begin{cases} \dot{\xi} = \xi^2\eta - \xi^5, \\ \dot{\eta} = -\eta + \xi^2. \end{cases} \qquad (2.56)$$

- Comparing (2.54) with (2.56) we see that:
$$B = 0, \qquad C = -1, \qquad u(\xi,\eta) = \xi^2\eta - \xi^5, \qquad v(\xi,\eta) = \xi^2.$$

- The point $(0,0)$ is a non-hyperbolic fixed point. The linear center subspace is
$$E^0 := \Big\{(\xi,\eta) \in \mathbb{R}^2 : \eta = 0\Big\}.$$

- The local center manifold is given by the graph of a differentiable function $q = q(\xi)$ whose power expansion is
$$\eta = q(\xi) = a\,\xi^2 + b\,\xi^3 + c\,\xi^4 + \cdots,$$
where $a, b, c \in \mathbb{R}$ are unknown coefficients to be determined. We get
$$\frac{\mathrm{d}q}{\mathrm{d}\xi} = 2\,a\,\xi + 3\,b\,\xi^2 + 4\,c\,\xi^3 + \cdots,$$

- We substitute these expansions in (2.55), collect the powers of ξ and find:
$$a = 1, \qquad b = 0, \qquad c = 0.$$
Therefore:
$$W^0 := \Big\{(\xi,\eta) \in \mathbb{R}^2 \,:\, \eta = \xi^2 + O(\xi^5),\ |\xi| \leqslant \varepsilon\Big\}.$$

- The restriction of equations (2.56) on W^0 is
$$\begin{aligned} \dot{\xi} &= B\,\xi + u(\xi, q(\xi)) = \xi^2\eta + O(\xi^5) \\ &= \xi^2(\xi^2 + O(\xi^5)) = \xi^4 + O(\xi^5), \end{aligned}$$
from which we find that $(0,0)$ is an unstable fixed point.

► The following Theorem describes the general situation and illustrates the importance in stability theory of local center manifolds.

Theorem 2.20

Consider the IVP (2.49) and denote by Φ_t its flow. Let $\tilde{x} = 0$ be a non-hyperbolic fixed point of Φ_t. Assume that $\dim(E^+) = m$, $\dim(E^-) = \ell$, $\dim(E^0) = n - m - \ell$.

1. *By a linear change of coordinates the system of ODEs of (2.49) can be written in the form*

$$\begin{cases} \dot{\xi} = B\,\xi + u(\xi,\eta), \\ \dot{\eta} = C\,\eta + v(\xi,\eta), \end{cases} \tag{2.57}$$

where $(\xi,\eta) \in \mathbb{R}^{n-m-\ell} \times \mathbb{R}^{m+\ell}$, $B \in \mathfrak{gl}(n-m-\ell,\mathbb{R})$ has $n-m-\ell$ eigenvalues with vanishing real part, $C \in \mathfrak{gl}(m+\ell,\mathbb{R})$ has m eigenvalues with negative real part and ℓ eigenvalues with positive real part. Here u, v are (nonlinear) functions such that $u(0,0) = v(0,0) = 0$.

2. *System (2.57) has a local center manifold around $(\tilde{\xi},\tilde{\eta}) = (0,0)$ given by*

$$W^0 := \left\{ (\xi,\eta) \in \mathbb{R}^{n-m-\ell} \times \mathbb{R}^{m+\ell} \,:\, \eta = q(\xi),\ \|\xi\| \leqslant \varepsilon \right\},$$

where $\varepsilon > 0$ is sufficiently small and q is a function of class C^k such that

$$q(0) = 0, \qquad \left.\frac{\partial q}{\partial \xi}\right|_0 = 0.$$

3. *In a neighborhood of $(\tilde{\xi},\tilde{\eta}) = (0,0)$ system (2.57) is topologically conjugate to the decoupled system of ODEs*

$$\begin{cases} \dot{\xi} = B\,\xi + u(\xi,q(\xi)), \\ \dot{\eta} = C\,\eta. \end{cases}$$

4. *If $\ell = 0$, i.e., $E^- = \emptyset$, and $\xi = 0$ is a stable fixed point of the restriction of (2.57) to W^0,*

$$\dot{\xi} = B\,\xi + u(\xi,q(\xi)),$$

then $(\tilde{\xi},\tilde{\eta}) = (0,0)$ is a stable fixed point of (2.57).

No Proof.

2.6 Basic facts on non-autonomous linear IVPs

▶ We consider the following *non-autonomous homogeneous linear IVP* in $\mathbb{R}^n$:

$$\begin{cases} \dot{x} = A(t)\,x, \\ x(t_0) = x_0, \end{cases} \tag{2.58}$$

with $(t_0, x_0) \in I \times \mathbb{R}^n$ and $A \in C(I, \mathfrak{gl}(n, \mathbb{R}))$, where $I \subseteq \mathbb{R}$.

Theorem 2.21

Consider the IVP (2.58).

1. *Every solution admits an extension to I.*
2. *The set of all solutions is a linear subspace of $C^1(I, \mathbb{R}^n)$ isomorphic to $\mathbb{R}^n$.*
3. *There exists a $\mathfrak{gl}(n, \mathbb{R})$-valued function $\Pi(t, t_0)$ defined by*

$$\Pi(t, t_0) := \mathbb{1}_n + \int_{t_0}^{t} A(s)\,\Pi(s, t_0)\,\mathrm{d}s, \qquad t \in I, \tag{2.59}$$

and called **principal matrix solution** *such that*

$$\phi(t, t_0, x_0) = \Pi(t, t_0)\,x_0$$

is the unique solution of (2.58).

4. *The function $\Pi(t, t_0)$ satisfies the following properties:*

$$\Pi(t_0, t_0) = \mathbb{1}_n, \quad \Pi(t, s)\,\Pi(s, t_0) = \Pi(t, t_0), \quad (\Pi(t, t_0))^{-1} = \Pi(t_0, t),$$

for all $t_0, t, s \in I$.

Proof. We prove all claims.

1. It is a consequence of Theorem 1.9 where $S_2(T) := \max_{t \in [0,T]} \|A(t)\|_*$ for every T in I.
2. We observe that linear combinations of solutions of $\dot{x} = A(t)\,x$ are again solutions. Indeed, let $\mathcal{S}$ be the set of all solutions of $\dot{x} = A(t)\,x$. Let $\phi_1, \phi_2 \in \mathcal{S}$ be two solutions and set $\psi := \lambda_1\,\phi_1 + \lambda_2\,\phi_2$, $\lambda_1, \lambda_2 \in \mathbb{R}$. Then,

$$\dot{\psi}(t) = \lambda_1\,\dot{\phi}_1(t) + \lambda_2\,\dot{\phi}_2(t) = \lambda_1\,A(t)\,\phi_1(t) + \lambda_2\,A(t)\,\phi_2(t) = A(t)\,\psi(t),$$

for all $t \in I$. This implies that $\psi \in \mathcal{S}$. Therefore $\mathcal{S}$ is a linear subspace of $C^1(I, \mathbb{R}^n)$. The solution of the IVP (2.58) corresponding to the initial condition $x(t_0) = x_0$ can be written as (*superposition principle*)

$$\phi(t, t_0, x_0) = \sum_{j=1}^{n} \phi(t, t_0, e_j)(x_0)_j,$$

where $\{e_j\}_{1\leqslant j\leqslant n}$ is the canonical basis of $\mathbb{R}^n$.

3. Using the n solutions $\phi(t,t_0,e_j)$ we define a $\mathfrak{gl}(n,\mathbb{R})$-valued function

$$\Pi(t,t_0) := (\phi(t,t_0,e_1),\ldots,\phi(t,t_0,e_n)),$$

we see that there is a linear map $x_0 \mapsto \phi(t,t_0,x_0) := \Pi(t,t_0)x_0$. It is clear that $\Pi(t,t_0)$ solves the matrix IVP

$$\begin{cases} \dot{\Pi}(t,t_0) = A(t)\,\Pi(t,t_0), \\ \Pi(t_0,t_0) = \mathbb{1}_n. \end{cases} \tag{2.60}$$

Theorem 1.1 implies the existence and uniqueness of the solution, which can be written as (2.59). In fact, an $n \times n$ matrix $X(t)$ satisfies the matrix ODE $\dot{X}(t) = A(t)\,X(t)$ if and only if every column satisfies the ODE $\dot{x} = A(t)\,x$.

4. The first property is obvious. The second follows from the uniqueness of the extension of the solution. Choosing $t = t_0$ in the second formula we see that the third property follows.

The Theorem is proved. ■

Example 2.24 (*A planar non-autonomous IVP*)

Consider the IVP (2.58) with

$$A(t) := \begin{pmatrix} 1 & t \\ 0 & 2 \end{pmatrix}.$$

To construct the principal matrix solution we have to find the solution of the IVP corresponding to the initial conditions $x(t_0) = e_1 := (1,0)$ and $x(t_0) = e_2 := (0,1)$. In the first case the solution reads

$$\phi(t,t_0,e_1) = (\mathrm{e}^{t-t_0},0)\,,$$

while in the second case is

$$\phi(t,t_0,e_2) = \left(\mathrm{e}^{2(t-t_0)}(t-1) - \mathrm{e}^{t-t_0}(t_0-1), \mathrm{e}^{2(t-t_0)}\right).$$

Thus the principal matrix solution is

$$\Pi(t,t_0) = \begin{pmatrix} \mathrm{e}^{t-t_0} & \mathrm{e}^{2(t-t_0)}(t-1) - \mathrm{e}^{t-t_0}(t_0-1) \\ 0 & \mathrm{e}^{2(t-t_0)} \end{pmatrix}.$$

► Remarks:

- If $A(t)$ does not depend explicitly on t the principal matrix solution reduces to

$$\Pi(t,t_0) = \mathrm{e}^{(t-t_0)A},$$

according to Theorem 2.10.

- Let $\phi_1(t), \ldots, \phi_n(t)$ be n solutions of the IVP (2.58). The *Wronskian* of the IVP (2.58) is the function $w : I \to \mathbb{R}$ defined by

$$w(t) := \det\left(\Phi(t)\right), \qquad \Phi(t) := \left(\phi_1(t), \ldots, \phi_n(t)\right).$$

- Let $w : I \to \mathbb{R}$ be the Wronskian of the IVP (2.58). Then either $w(t) \neq 0$ for all $t \in I$ or $w(t) = 0$ for all $t \in I$. In the first case the matrix $\Phi(t)$ is called *fundamental solution* of (2.58). Indeed, $w(t) \neq 0$ for all $t \in I$ is equivalent to say that the n vectors $\phi_1(t), \ldots, \phi_n(t)$ are linearly independent for all $t \in I$.

- If $\Phi(t)$ is a matrix solution of the matrix ODE

$$\dot{X}(t) = A(t)\, X(t),$$

so is $\Phi(t)\, C$, $C \in \mathfrak{gl}(n, \mathbb{R})$. Hence, given two fundamental matrix solutions $\Phi_1(t)$ and $\Phi_2(t)$ we have

$$\Phi_2(t) = \Phi_1(t)\, \Phi_1^{-1}(t_0)\, \Phi_2(t_0),$$

since a matrix solution is uniquely determined by an initial condition. In particular, the principal matrix solution can be obtained from any fundamental matrix solution via

$$\Pi(t, t_0) = \Phi(t)\, (\Phi(t_0))^{-1}. \tag{2.61}$$

▶ The next Theorem shows, in particular, that the condition $w(t_0) \neq 0$ is sufficient to guarantee that $(\phi_1(t), \ldots, \phi_n(t))$ is a fundamental solution of (2.58).

Theorem 2.22 (*Abel-Liouville*)

Let $w : I \to \mathbb{R}$ be the Wronskian of the IVP (2.58). Then there holds:

$$w(t) = w(t_0)\, \exp\left(\int_{t_0}^{t} (\operatorname{Trace} A(s))\, \mathrm{ds}\right). \tag{2.62}$$

Proof. Let $\varepsilon > 0$ be sufficiently small. From (2.59) we have:

$$\Pi(t+\varepsilon, t) = \mathbb{1}_n + \varepsilon\, A(t) + r(\varepsilon), \qquad \lim_{\varepsilon \to 0} \frac{r(\varepsilon)}{\varepsilon} = 0.$$

From (2.61) we have $\Phi(t+\varepsilon) = \Pi(t+\varepsilon, t)\, \Phi(t)$, so that

$$\begin{aligned} \det\left(\Phi(t+\varepsilon)\right) &:= w(t+\varepsilon) = \det\left(\mathbb{1}_n + A(t)\varepsilon + r(\varepsilon)\right) w(t) \\ &= \left(1 + \varepsilon \operatorname{Trace} A(t) + r(\varepsilon)\right) w(t), \end{aligned}$$

implying that the function w obeys the ODE

$$\dot{w}(t) = \operatorname{Trace} A(t)\, w(t).$$

This ODE is separable and the solution is given by (2.62). ■

► We now consider the following *non-autonomous and non-homogeneous linear IVP* in $\mathbb{R}^n$:

$$\begin{cases} \dot{x} = A(t)\,x + g(t), \\ x(t_0) = x_0, \end{cases} \tag{2.63}$$

with $(t_0, x_0) \in I \times \mathbb{R}^n$, $A \in C(I, \mathfrak{gl}(n, \mathbb{R}))$ and $g \in C(I, \mathbb{R}^n)$, where $I \subseteq \mathbb{R}$. In this case the set of solutions of (2.63) is not a vector space.

Theorem 2.23

The solution of the IVP (2.63) reads

$$\phi(t, t_0, x_0) = \Pi(t, t_0)\,x_0 + \int_{t_0}^{t} \Pi(t, s)\,g(s)\,\mathrm{d}s, \qquad t \in I,$$

where $\Pi(t, t_0)$ is the principal matrix solution of the associated homogeneous IVP.

Proof. The difference of two solutions of the non-homogeneous system satisfies the corresponding homogeneous system. Thus it suffices to find one particular solution. This can be done using the following ansatz (*method of variation of constants*):

$$\phi(t, t_0, x_0) = \Pi(t, t_0)\,c(t), \qquad c(t_0) = x_0,$$

where $c(t)$ is an unknown function. Then we have:

$$\begin{aligned} \dot{\phi}(t, t_0, x_0) &= \dot{\Pi}(t, t_0)\,c(t) + \Pi(t, t_0)\,\dot{c}(t) \\ &= A(t)\,\phi(t, t_0, x_0) + \Pi(t, t_0)\,\dot{c}(t), \end{aligned}$$

which compared with (2.63) gives

$$\dot{c}(t) = \Pi^{-1}(t, t_0)\,g(t) = \Pi(t_0, t)\,g(t),$$

namely

$$c(t) = x_0 + \int_{t_0}^{t} \Pi(t_0, s)\,g(s)\,\mathrm{d}s.$$

Using the relation $\Pi(t, t_0)\Pi(t_0, s) = \Pi(t, s)$ the claim follows. ■

2.6.1 Periodic linear IVPs

► We consider the following *periodic homogeneous linear IVP* in $\mathbb{R}^n$:

$$\begin{cases} \dot{x} = A(t)\,x, \\ x(t_0) = x_0, \end{cases} \tag{2.64}$$

with $(t_0, x_0) \in \mathbb{R} \times \mathbb{R}^n$ and $A \in C(\mathbb{R}, \mathfrak{gl}(n, \mathbb{R}))$. Here the $\mathfrak{gl}(n, \mathbb{R})$-valued function A is a periodic function of t with period $T > 0$, i.e.,

$$A(t+T) = A(t) \qquad \forall\, t \in \mathbb{R}.$$

► The periodicity of $A(t)$ implies that $\phi(t+T, t_0, x_0)$ is a solution if $\phi(t, t_0, x_0)$ is, but not that all solutions are T-periodic.

Example 2.25 (*A scalar periodic IVP*)

Consider the following $(2\,\pi)$-periodic scalar IVP:

$$\begin{cases} \dot{x} = (1 + \sin t)\, x, \\ x(0) = x_0, \end{cases}$$

The general solution reads

$$\phi(t, x_0) = x_0\, \mathrm{e}^{1+t-\cos t},$$

which is not $2\,\pi$-periodic.

Theorem 2.24

The principal matrix solution of the IVP (2.64) satisfies

$$\Pi(t+T, t_0+T) = \Pi(t, t_0) \qquad \forall\, t \in \mathbb{R}.$$

Proof. From (2.60) we obtain:

$$\begin{aligned} \dot{\Pi}(t+T, t_0+T) &= A(t+T)\,\Pi(t+T, t_0+T) \\ &= A(t)\,\Pi(t+T, t_0+T). \end{aligned}$$

Furthermore $\Pi(t_0+T, t_0+T) = \mathbb{1}_n$. So $\Pi(t+T, t_0+T) = \Pi(t, t_0)$ by uniqueness of the solution. ■

Definition 2.15

The **monodromy matrix** *of the IVP (2.64) is the $\mathfrak{gl}(n, \mathbb{R})$-valued function defined by*

$$\Lambda(t_0) := \Pi(t_0+T, t_0) = \Lambda(t_0+T).$$

The eigenvalues of $\Lambda(t_0)$, say ρ_j, are called **Floquet multipliers**.

► A naive guess would be that all initial conditions return to their starting values after one period, i.e., $\Lambda(t_0) = \mathbb{1}_n$, and hence all solutions are periodic. However this is too much to hope. This already fails in the one-dimensional case, as shown in Example 2.25. The next Theorem provides a characterization of the principal matrix solution of the IVP (2.64).

Theorem 2.25 (*Floquet*)

Consider the IVP (2.64). The principal matrix solution can be written in the following **Floquet normal form***:*

$$\Pi(t, t_0) = P(t, t_0)\, e^{(t-t_0)\, Q(t_0)}, \tag{2.65}$$

where $P \in C(\mathbb{R}, \mathfrak{gl}(n, \mathbb{R}))$ is periodic with period T, with $P(t_0, t_0) = \mathbb{1}_n$, and $Q(t_0) \in \mathfrak{gl}(n, \mathbb{C})$ is defined in terms of the monodromy matrix by

$$e^{T\, Q(t_0)} := \Lambda(t_0), \qquad Q(t_0 + T) = Q(t_0).$$

The eigenvalues of $Q(t_0)$,

$$\gamma_j := \frac{1}{T} \log \rho_j,$$

are called **Floquet exponents***.*

No Proof.

Example 2.26 (*A scalar periodic IVP*)

Consider the following scalar periodic IVP

$$\begin{cases} \dot{x} = a(t)\, x, \\ x(t_0) = x_0, \end{cases}$$

where $a : \mathbb{R} \to \mathbb{R}$ is a continuous T-periodic function.

- The principal matrix solution is

$$\Pi(t, t_0) = \exp\left(\int_{t_0}^{t} a(s)\, \mathrm{d}s \right),$$

- The monodromy matrix is

$$\Lambda(t_0) = \exp\left(\int_{t_0}^{t_0+T} a(s)\, \mathrm{d}s \right) =: e^{\bar{a}}, \qquad \bar{a} := \int_0^T a(s)\, \mathrm{d}s.$$

- The Floquet normal form (2.65) is obtained by setting

$$P(t, t_0) := \exp\left(\int_{t_0}^{t} (a(s) - \bar{a})\, \mathrm{d}s \right), \qquad Q(t_0) := \bar{a}.$$

- Let $a(t) := 1 + \sin t$ and $t_0 = 0$, as in Example 2.25. We get:

$$\Pi(t, 0) = e^{1+t-\cos t}, \qquad \Lambda(0) = e^{2\pi},$$

and

$$\bar{a} = 2\pi = Q(0), \qquad P(t, 0) = e^{1+t-\cos t} e^{-2\pi t},$$

so that

$$P(t, 0)\, e^{t\, Q(0)} = e^{1+t-\cos t} e^{-2\pi t} e^{2\pi t} = \Pi(t, 0).$$

► Remarks:

- The matrix $Q(t_0)$ will be complex even if $A(t)$ is real unless all real eigenvalues of $\Lambda(t_0)$ are positive.

- The time-dependent change of coordinates $x \mapsto y := P^{-1}(t, t_0)\, x$ transforms (2.64) into $\dot{y} = Q(t_0)y$, that is an autonomous system of ODEs. Therefore, since P is periodic and bounded, the long-time behavior depends only on $Q(t_0)$.

- Computing the Floquet exponents is difficult in general, but the existence of the representation (2.65) is useful to classify the possible behaviors near a periodic orbit.

- Note that the eigenvalues of $A(t)$ provide no information on stability even if they are with negative real part.

Example 2.27 (*A planar periodic IVP*)

Consider the IVP (2.64) with

$$A(t) := \begin{pmatrix} -1 + \frac{3}{2}\cos^2 t & 1 - \frac{3}{2}\sin t \cos t \\ -1 - \frac{3}{2}\sin t \cos t & -1 + \frac{3}{2}\sin^2 t \end{pmatrix}.$$

The eigenvalues of the above matrix, $(-1 \pm \mathrm{i}\sqrt{7})/4$, are time-independent and with negative real part. Nevertheless, there exists a solution $(\phi_1(t), \phi_2(t)) = \mathrm{e}^{t/2}(-\cos t, \sin t)$, which shows that $(0,0)$ is unstable.

► The stability properties of the fixed point $\tilde{x} = 0$ of the IVP (2.64) are described in the following claim.

Theorem 2.26

Consider the IVP (2.64).

1. *$\tilde{x} = 0$ is asymptotically stable if all Floquet multipliers satisfy $|\rho_j| < 1$ for all j (or equivalently, if all Floquet exponents satisfy $\mathfrak{Re}\,\gamma_j < 0$ for all j).*

2. *$\tilde{x} = 0$ is stable if all Floquet multipliers satisfy $|\rho_j| \leqslant 1$ for all j and for all Floquet multipliers with $|\rho_j| = 1$ the algebraic and geometric multiplicities are equal (or equivalently, if all Floquet exponents satisfy $\mathfrak{Re}\,\gamma_j \leqslant 0$ for all j and for all Floquet exponents with $\mathfrak{Re}\,\gamma_j = 0$ the algebraic and geometric multiplicities are equal).*

No Proof.

► Note that if (2.64) admits a T-periodic solution then there must be a Floquet multiplier ρ_j with $|\rho_j| = 1$ (or equivalently, a Floquet exponent γ_j with $\mathfrak{Re}\,\gamma_j = 0$).

2.7 Basic facts on local bifurcation theory

► A *bifurcation* is a qualitative (topological) change of the phase portrait of a given dynamical system under variation of parameters. For example:

- Disappearance of fixed points or of invariant sets.
- Change of stability of fixed points.

► We study *one-parameter local bifurcations*, namely those bifurcations which occur in small neighborhoods of fixed points, for (continuous) dynamical systems defined through IVPs whose vector fields depend on one parameter $\alpha \in \mathbb{R}$.

► The main questions are:

1. What could happen to the phase portrait near a fixed point when a parameter passes a bifurcation value?

   Answer. One has to provide a catalogue of *normal forms*, i.e., canonical systems exhibiting certain typical bifurcations.

   - In many cases one can construct a nice representative of an equivalence class, called *topological normal form*, of parametrized families near a local bifurcation.
   - Any system satisfying the same bifurcation conditions is locally topologically equivalent to the corresponding normal form.

2. How to determine which of the alternatives occurs in a given system?

   Answer. One has to list some *genericity conditions* which guarantee that the given system has a phase portrait which is equivalent to those of a corresponding normal form for corresponding parameter values. Genericity conditions can be given explicitly as inequalities involving partial derivatives of the ODEs w.r.t. variables and parameters.

► The general strategy to attack the above problems can be summarized as follows:

- To study the simplest local bifurcations in one-parameter systems of the lowest possible dimension. Such a dimension is determined by the number of *critical eigenvalues* at the bifurcating fixed point, i.e., those eigenvalues which make the point non-hyperbolic.

- To apply the lowest-dimensional results to higher-dimensional systems. It can be shown that such systems possess smooth (and smoothly dependent on the parameter) invariant center manifolds, with dimension equal to the number of critical eigenvalues. On such center manifolds the bifurcations happen according to the low-dimensional theory.

► Let Φ_t^α and Ψ_t^β, $t \in \mathbb{R}$, $\alpha \in A$, $\beta \in B$, $A, B \subseteq \mathbb{R}^p$, $p \geqslant 1$, be two distinct one-parameter groups of parametric smooth diffeomorphisms on M_1 and M_2 respectively, with $M_1, M_2 \subseteq \mathbb{R}^n$. Here $\alpha := (\alpha_1, \ldots, \alpha_p)$ and $\beta := (\beta_1, \ldots, \beta_p)$ are parameters.

- The infinitesimal generators of Φ_t^α and Ψ_t^β are parametric vector fields on M_1, M_2 respectively, given by

$$\mathbf{v}^\alpha := \sum_{i=1}^n f_i(x,\alpha)\frac{\partial}{\partial x_i}, \qquad \mathbf{w}^\beta := \sum_{i=1}^n g_i(y,\beta)\frac{\partial}{\partial y_i},$$

with

$$f(x,\alpha) := \left.\frac{\mathrm{d}}{\mathrm{d}t}\right|_{t=0} \Phi_t^\alpha(x), \qquad g(y,\beta) := \left.\frac{\mathrm{d}}{\mathrm{d}t}\right|_{t=0} \Psi_t^\beta(y).$$

- Therefore we have two distinct parametric dynamical systems $\{\Phi_t^\alpha, \mathbb{R}, M_1, A\}$ and $\{\Psi_t^\beta, \mathbb{R}, M_2, B\}$.

► Definition 2.13 can be generalized as follows.

Definition 2.16

1. *$\{\Phi_t^\alpha, \mathbb{R}, M_1, A\}$ and $\{\Psi_t^\beta, \mathbb{R}, M_2, B\}$ are* **topologically equivalent** *if there exists a homeomorphism $\Pi : A \longrightarrow B$ and a parametric family of homeomorphisms $\Lambda^\alpha : M_1 \longrightarrow M_2$ which maps the orbits of $\{\Phi_t^\alpha, \mathbb{R}, M_1, A\}$ at $\alpha \in A$ onto the orbits of $\{\Psi_t^\beta, \mathbb{R}, M_2, B\}$ at $\beta = \Pi(\alpha)$ preserving the direction of time. If Λ^α is a parametric family of diffeomorphisms of class C^k then $\mathbf{v}^\alpha$ and $\mathbf{w}^\beta$ (as well as the flows Φ_t^α and Ψ_t^β) are* **C^k-diffeomorphic***.*

2. *$\{\Phi_t^\alpha, \mathbb{R}, M_1, A\}$ and $\{\Psi_t^\beta, \mathbb{R}, M_2, B\}$ are* **topologically conjugate** *if they are topologically equivalent with the same time parametrization.*

► We can now define the notion of bifurcation of a parametric dynamical system.

Definition 2.17

Let $\{\Phi_t^\alpha, \mathbb{R}, M, A\}$, $M \subseteq \mathbb{R}^n$, $A \subseteq \mathbb{R}^p$, $p \geqslant 1$, be a parametric dynamical system.

1. *A point $\widetilde{\alpha} \in A$ is a* **bifurcation point** *of $\{\Phi_t^\alpha, \mathbb{R}, M, A\}$ if in any neighborhood of $\widetilde{\alpha}$ there is a point corresponding to a topologically non-equivalent member of $\{\Phi_t^\alpha, \mathbb{R}, M, A\}$.*

2. *The* **codimension of the bifurcation point** *is the number of independent*

equality conditions determining $\widetilde{\alpha}$.

3. *A* **bifurcation diagram** *of $\{\Phi_t^\alpha, \mathbb{R}, M, A\}$ is a partitioning of the parameter space A induced by topological equivalence, together with the representative phase portrait for each set of the partitioning.*

► One expects that a codimension 1 bifurcation occurs at isolated parameter values for generic dynamical systems $\{\Phi_t^\alpha, \mathbb{R}, M, A\}$ if $p = 1$, on curves in the parameter plane if $p = 2$, on two-dimensional surfaces in the parameter space if $p = 3$, etc. Thus, one would not meet a bifurcation of codimension bigger than p in a generic p-parameter family of dynamical systems $\{\Phi_t^\alpha, \mathbb{R}, M, A\}$, $A \subseteq \mathbb{R}^p$.

Example 2.28 (*A saddle-node bifurcation point*)

Consider the following parametric IVP in $\mathbb{R}^2$:

$$\begin{cases} \dot{x}_1 = x_1\left(1 - x_1^2 - x_2^2\right) - x_2(1 + \alpha + x_1), \\ \dot{x}_2 = x_2\left(1 - x_1^2 - x_2^2\right) + x_1(1 + \alpha + x_1), \\ (x_1(0), x_2(0)) \in \mathbb{R}^2, \end{cases}$$

with $\alpha \in \mathbb{R}$.

- In polar coordinates $(r, \theta) \in (0, \infty) \times [0, 2\pi)$ the above ODEs read

$$\begin{cases} \dot{r} = r\left(1 - r^2\right), \\ \dot{\theta} = 1 + \alpha + r\cos\theta. \end{cases} \tag{2.66}$$

- Depending on the value of α we have three different cases:
  1. $\alpha = 0$. There is a non-hyperbolic stable fixed point at $(r, \theta) = (1, \pi)$.
  2. $\alpha > 0$. There are no fixed points. A stable limit cycle defined by $\{(r, \theta) \,:\, r = 1\}$ appears.
  3. $\alpha < 0$. There are two fixed points on the limit cycle, an unstable saddle and a stable node.
- Thus $\alpha = 0$ corresponds to a codimension 1 bifurcation, called *saddle-node bifurcation*. The bifurcation diagram is:

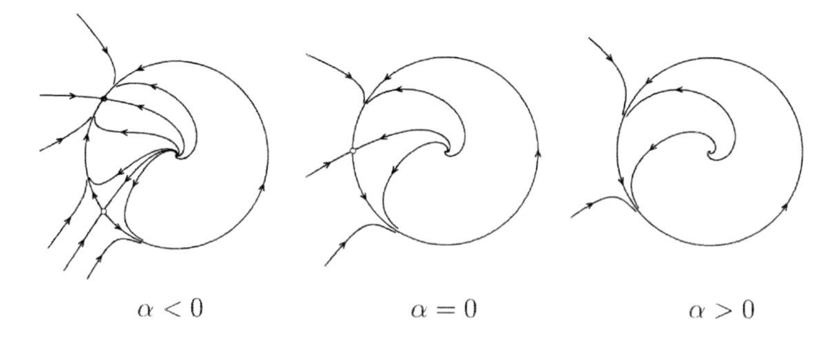

Fig. 2.28. Bifurcation diagram of system (2.66) ([Ku]).

2.7.1 One-parameter local bifurcations

► We consider the following *one-parameter family of IVPs* in $M \subseteq \mathbb{R}^n$:

$$\begin{cases} \dot{x} = f(x, \alpha), \\ x(0) = x_0, \end{cases} \tag{2.67}$$

with $f \in C^k(M \times \mathbb{R}, \mathbb{R}^n)$, $k \geqslant 1$, $\alpha \in \mathbb{R}$.

► We have the following claim.

Theorem 2.27

Local bifurcations of (2.67) can only occur at non-hyperbolic fixed points $\widetilde{x} \in M$ of (2.67).

Proof. We proceed by steps.

- Assume that $\widetilde{x} \in M$ is a fixed point of (2.67) at some parameter value $\widetilde{\alpha} \in \mathbb{R}$, i.e., $f(\widetilde{x}, \widetilde{\alpha}) = 0$.
- Consider the linearization of (2.67) around $\widetilde{x}$:

$$\begin{cases} \dot{x} = A_{\widetilde{\alpha}}\, x, \\ x(0) = x_0, \end{cases} \tag{2.68}$$

  where

$$A_{\widetilde{\alpha}} := \left.\frac{\partial f}{\partial x}\right|_{x=\widetilde{x}}.$$

- If $\widetilde{x}$ is hyperbolic then the eigenvalues of A_α depend continuously on α near $\widetilde{\alpha}$. This is a consequence of the "Implicit function Theorem". Therefore $\widetilde{x}$ remains hyperbolic for all α with sufficiently small $|\alpha - \widetilde{\alpha}|$. Moreover the numbers m and $n - m$ of negative and positive eigenvalues of A_α remain constant near $\widetilde{\alpha}$.
- By Theorem 2.17 the local phase portrait of (2.67) around $\widetilde{x}$ is topologically equivalent to the phase portrait of the linearized system (2.68).
- On the other hand we know that all such linear systems (2.68) are topologically equivalent to the *standard saddle system*

$$\begin{cases} \dot{\xi}_1 = -\xi_1, \\ \dot{\xi}_2 = \xi_2, \end{cases}$$

  where $(\xi_1, \xi_2) \in \mathbb{R}^m \times \mathbb{R}^{n-m}$.
- Therefore the local phase portrait of (2.67) does not bifurcate near $\widetilde{\alpha}$.

The Theorem is proved. ■

► Theorem 2.27 implies that there are only two distinct cases:

1. *Saddle-node bifurcations*: the fixed point $\tilde{x} \in M$ corresponds to a vanishing eingevalue. The lowest possible dimension of M is 1.

2. *Hopf bifurcations*: the fixed point $\tilde{x} \in M$ corresponds to a pair of complex conjugate eigenvalues with vanishing real part. The lowest possible dimension of M is 2.

► We will consider the above two cases in the lowest possible dimension, thus providing the *normal forms* of saddle-node and Hopf bifurcations. The following results are mainly based on examples and we shall give some concrete details only in the simpler case of saddle-node bifurcations.

2.7.2 Saddle-node bifurcations

► We start with the following definition.

Definition 2.18

Consider the IVP (2.67), with $n = 1$. The **normal form of the saddle-node bifurcation** *is defined by*

$$f(x, \alpha) := \alpha \pm x^2, \qquad \alpha \in \mathbb{R}.$$

► Remarks:

- Consider the IVP (2.67), with $n = 1$, and $f(x, \alpha) := \alpha + x^2$.

  (a) If $\alpha = 0$ then the origin is a non-hyperbolic unstable fixed point (with $\lambda = 0$).

  (b) If $\alpha < 0$ there are two fixed points at $x_{1,2}(\alpha) := \mp\sqrt{-\alpha}$, one of which is stable, while the other is unstable.

  (c) If $\alpha > 0$ there are no fixed points.

  While α crosses the zero from negative to positive values the two fixed points $x_{1,2}(\alpha)$ coalesce forming at $\alpha = 0$ a degenerate fixed point at the origin, which then disappears. The bifurcation diagram in the plane (x, α) is the following:

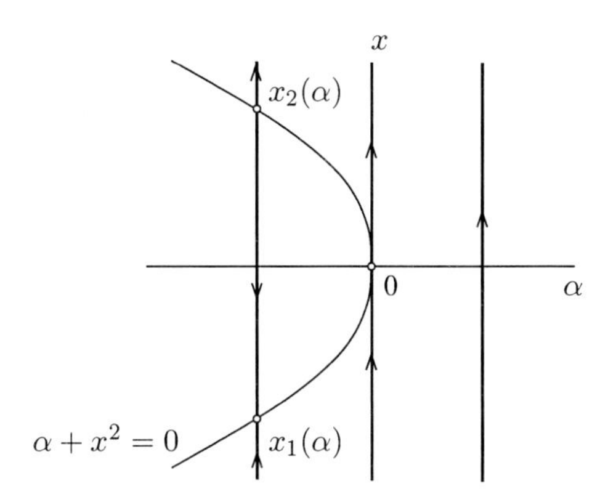

Fig. 2.29. Bifurcation diagram of $\dot{x} = \alpha + x^2$ ([Ku]).

- Similar arguments hold for the IVP (2.67) with $n = 1$ and $f(x, \alpha) := \alpha - x^2$.

Theorem 2.28

The one-parameter family of IVPs in $\mathbb{R}$ given by

$$\begin{cases} \dot{y} = \alpha \pm y^2 + O(y^3), \\ y(0) = x_0, \end{cases} \tag{2.69}$$

is topologically conjugate near the origin to the normal form of the saddle-node bifurcation.

No Proof.

Theorem 2.29

Consider the IVP (2.67), with $n = 1$, and assume that for $\alpha = 0$ there exists a non-hyperbolic fixed point at $\widetilde{x} = 0$, i.e.,

$$\lambda := \left.\frac{\partial f}{\partial x}\right|_{(0,0)} = 0.$$

Moreover assume that the following genericity conditions hold:

$$\left.\frac{\partial \lambda}{\partial x}\right|_{(0,0)} \neq 0, \qquad \left.\frac{\partial f}{\partial \alpha}\right|_{(0,0)} \neq 0.$$

Then the IVP (2.67), with $n = 1$, is topologically conjugate to (2.69).

No Proof.

► As anticipated, we now sketch how to use the low-dimensional theory to investigate higher-dimensional systems. Consider the IVP (2.67), with $n = 2$, and assume that for $\alpha = 0$ there exists a non-hyperbolic fixed point at $\widetilde{x} = 0$ corresponding to one eigenvalue $\lambda_1 = 0$ and one eigenvalue $\lambda_2 < 0$, so that $\dim(E^0) = 1$.

- Under some genericity conditions, it can be shown that there exists a local smooth one-dimensional invariant and attracting center manifold W^0_α for small $|\alpha|$.

- At $\alpha = 0$ the restriction of the original two-dimensional IVP to W^0_0 is topologically conjugate to the scalar ODE

$$\dot{\xi} = \beta \pm \xi^2, \qquad \beta \in \mathbb{R},$$

so that the original two-dimensional IVP is locally topologically conjugate to the decoupled two-dimensional system

$$\begin{cases} \dot{\xi} = \beta \pm \xi^2, \\ \dot{\eta} = -\eta. \end{cases} \tag{2.70}$$

- The resulting phase portraits are presented below for the sign $+$.

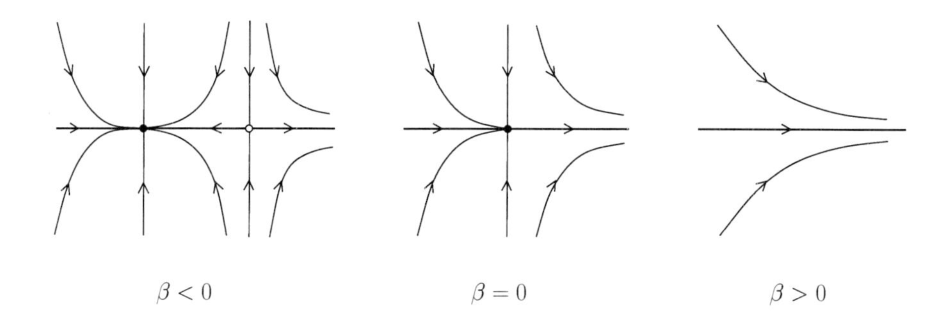

Fig. 2.30. Phase portraits of (2.70) ([Ku]).

For $\beta < 0$ there are two hyperbolic fixed points at $\pm\sqrt{\beta}$ (on the ξ-axis): a stable node and a saddle. They collide at the origin when $\beta = 0$ forming a non-hyperbolic saddle-node point. There are no fixed points for $\beta > 0$.

- The same happens locally for the original two-dimensional IVP on some local center manifold W^0_α.

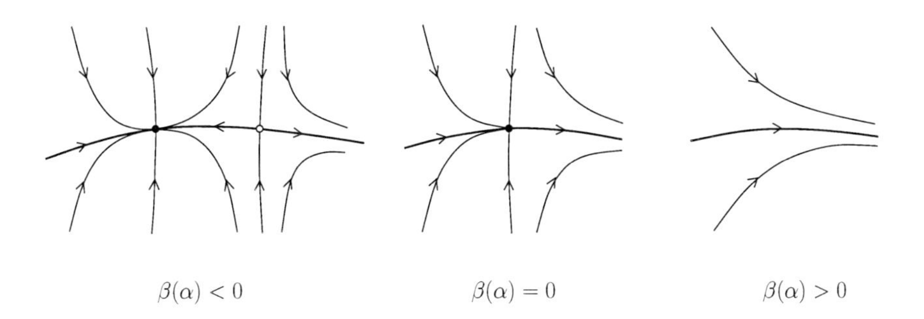

Fig. 2.31. Same phase portraits of Fig. 2.30 if $\alpha \neq 0$ ([Ku]).

2.7.3 Hopf bifurcations

► We start with the following definition.

Definition 2.19

Consider the IVP (2.67), with $n = 2$. The **normal form of the Hopf bifurcation** *is defined by*

$$f(x,\alpha) := \left(\alpha\, x_1 - x_2 \pm x_1 \left(x_1^2 + x_2^2\right), x_1 + \alpha\, x_2 \pm x_2 \left(x_1^2 + x_2^2\right)\right), \qquad \alpha \in \mathbb{R}.$$

► Remarks:

- If $\alpha = 0$ there is a non-hyperbolic fixed point at the origin which corresponds to a pair of complex conjugate eigenvalues with vanishing real part, $\lambda_{1,2} = \pm\mathrm{i}$.
- In polar coordinates $(r,\theta) \in (0,\infty) \times [0,2\,\pi)$ the two ODEs defining the IVP decouple into the following ODEs:

$$\begin{cases} \dot{r} = r(\alpha \pm r^2), \\ \dot{\theta} = 1. \end{cases} \tag{2.71}$$

Then we have two distinct cases:

1. *Supercritical Hopf bifurcation.* If we choose the sign $-$ in (2.71) then the origin is asymptotically stable for $\alpha \leqslant 0$. The fixed point becomes unstable for $\alpha > 0$ and is surrounded by a stable limit cycle of radius $\sqrt{\alpha}$. All non-stationary orbits tend to this cycle when time advances.

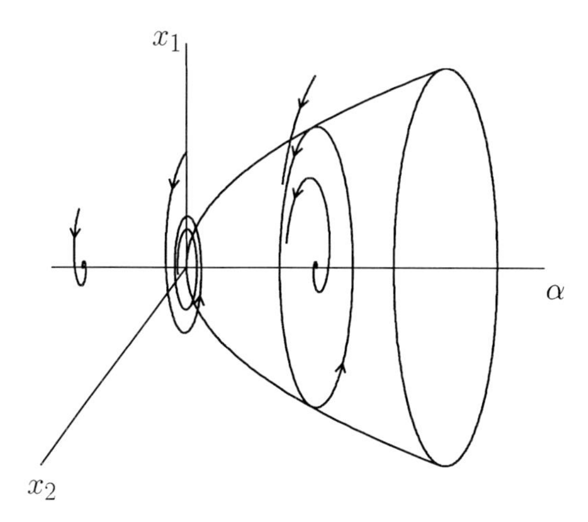

Fig. 2.32. Supercritical Hopf bifurcation ([Ku]).

2. *Subcritical Hopf bifurcation*. If we choose the sign $+$ in (2.71) then the origin is asymptotically stable for $\alpha \leqslant 0$ and it is surrounded by an unstable limit cycle of radius $\sqrt{-\alpha}$. For $\alpha > 0$ the fixed point at the origin becomes unstable. No cycle is present and all non-stationary orbits diverge when time advances.

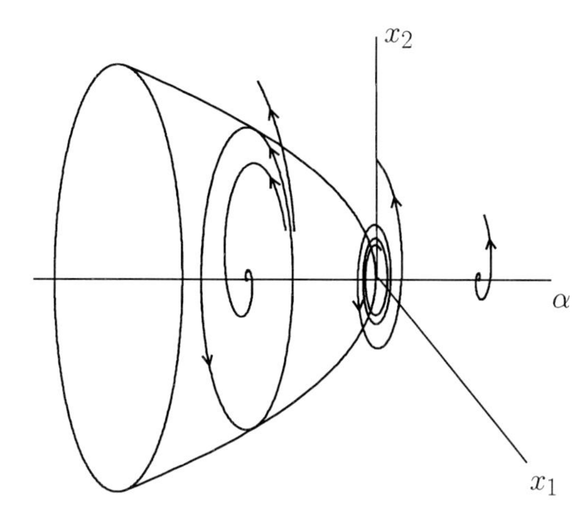

Fig. 2.33. Subcritical Hopf bifurcation ([Ku]).

► Let us increase by one the dimension of M. We consider the IVP (2.67), with $n = 3$, and assume that for $\alpha = 0$ there exists a non-hyperbolic fixed point at $\widetilde{x} = 0$ corresponding to a pair of complex conjugate eigenvalues with vanishing real part, so that $\dim(E^0) = 2$.

- Under some genericity conditions, it can be shown that there exists a local smooth two-dimensional invariant and attracting center manifold W^0_α for small $|\alpha|$.

- At $\alpha = 0$ the restriction of the original three-dimensional IVP to W^0_0 is topologically conjugate to the

$$\begin{cases} \dot{\xi}_1 = \beta\,\xi_1 - \xi_2 \pm \xi_1(\xi_1^2 + \xi_2^2), \\ \dot{\xi}_2 = \xi_1 + \beta\,\xi_2 \pm \xi_2(\xi_1^2 + \xi_2^2), \end{cases}$$

  with $\beta \in \mathbb{R}$, so that the original three-dimensional IVP is locally topologically conjugate to the decoupled system of ODEs

$$\begin{cases} \dot{\xi}_1 = \beta\,\xi_1 - \xi_2 \pm \xi_1(\xi_1^2 + \xi_2^2), \\ \dot{\xi}_2 = \xi_1 + \beta\,\xi_2 \pm \xi_2(\xi_1^2 + \xi_2^2), \\ \dot{\xi}_3 = -\xi_3. \end{cases} \tag{2.72}$$

- The resulting phase portraits are presented below for the sign $-$.

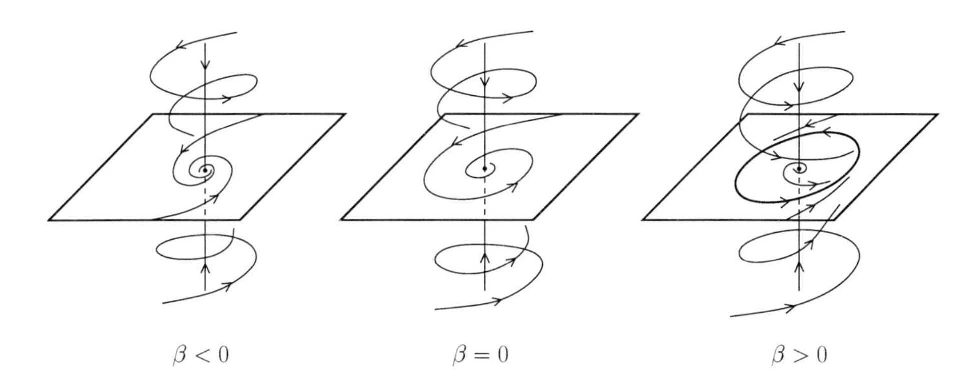

Fig. 2.34. Phase portraits of system (2.72) with sign $-$ ([Ku]).

  A supercritical Hopf bifurcation takes place in the invariant plane $\xi_3 = 0$, which is attracting.

- The same happens locally for the original three-dimensional IVP on some local center manifold W^0_α.

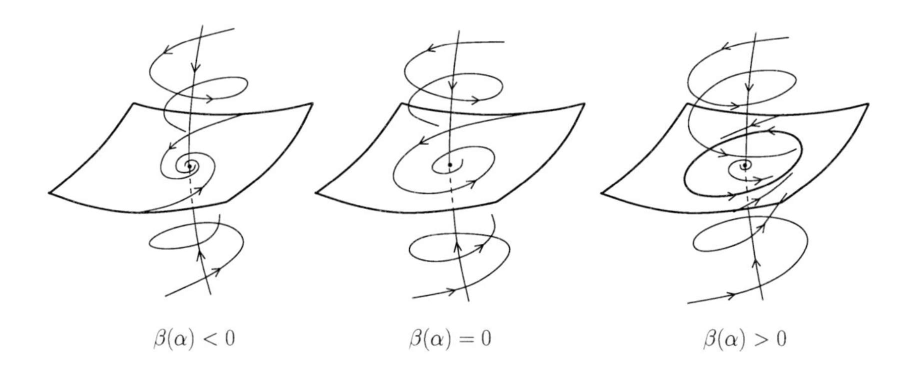

Fig. 2.35. Same phase portraits of Fig. 2.34 if $\alpha \neq 0$ ([Ku]).

2.8 Exercises

Ch2.E1 Consider a one-dimensional system consisting of two masses m and three springs with elasticity constants k, $k/3$ and k, $k > 0$. The springs with elasticity constant k are firmly attached to the wall, while the spring with elasticity constant $k/3$ connects the two masses.

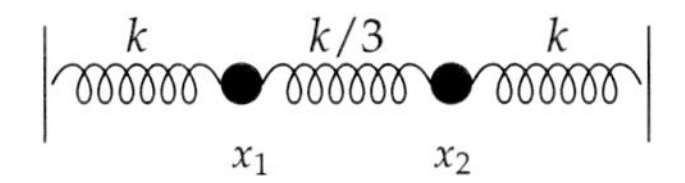

Denote by x_1 and x_2 the positions of the masses when the springs are in their rest points and consider the initial conditions

$$(x_1(0), x_2(0)) = (a, b), \qquad (\dot{x}_1(0), \dot{x}_2(0)) = (c, d),$$

with $b > a > 0$ and $c, d > 0$.

(a) According to Hooke law, express the spring forces acting on each mass (frictional forces are neglected) and write down the coupled system of Newton equations of motion for the two masses.

(Hint: Hooke law states that the spring force is $f(x) := -k\,x$, where $k > 0$ is the elasticity constant and x is the displacement of the end of the spring from its equilibrium position)

(b) Decouple the obtained differential equations by introducing new variables $(\widetilde{x}_1, \widetilde{x}_2) := (x_1 + x_2, x_1 - x_2)$. Solve the IVP in the variables $\widetilde{x}_1$, $\widetilde{x}_2$.

(c) Determine explicitly the flow of the system in terms of the variables x_1, x_2.

Ch2.E2 Let $\{\Phi_t, \mathbb{R}, M\}$, $M \subset \mathbb{R}^n$, be a continuous dynamical system. Consider a point $x_0 \in M$ and consider the orbit $\mathcal{O}(x_0) := \{x \in M \,:\, x = \Phi_t(x_0)\ \forall\, t \in \mathbb{R}\}$. Prove that if $\mathcal{O}(x_0)$ is a periodic orbit then all points $x \in \mathcal{O}(x_0)$ have the same period.

Ch2.E3 Consider the map $\Phi_t : \mathbb{R} \times \mathbb{R}^3 \to \mathbb{R}^3$ defined by

$$\Phi_t : (t, (x_1, x_2, x_3)) \mapsto \left(x_1 + 2\,t\,x_2, x_2, \mathrm{e}^{-t\,x_1 - t^2 x_2}\,x_3\right).$$

(a) Prove that Φ_t defines a one-parameter global Lie group of smooth diffeomorphisms.

(b) Compute the infinitesimal generator of Φ_t.

(c) Write down the IVP corresponding to Φ_t.

Ch2.E4 Let $M := \{(x_1, x_2) \in \mathbb{R}^2 \,:\, x_1 > 0\}$. Consider the map $\Phi_t : [0, 2\,\pi) \times M \to M$ defined by

$$\Phi_t : (t, (x_1, x_2)) \mapsto \frac{\left(2\,x_1, 2\,x_2 \cos t + \left(1 - x_1^2 - x_2^2\right) \sin t\right)}{1 + x_1^2 + x_2^2 + \left(1 - x_1^2 - x_2^2\right) \cos t - 2\,x_2 \sin t}.$$

(a) Prove that Φ_t defines a one-parameter global Lie group of smooth diffeomorphisms.

(b) Compute the infinitesimal generator of Φ_t.

(c) Write down the IVP corresponding to Φ_t.

Ch2.E5 Consider the following vector fields acting on $\mathbb{R}^2$:

$$\mathbf{v}_1 := x_1 x_2 \frac{\partial}{\partial x_1} + x_2^2 \frac{\partial}{\partial x_2}, \qquad \mathbf{v}_2 := x_1 \frac{\partial}{\partial x_1}, \qquad \mathbf{v}_3 := x_2 \frac{\partial}{\partial x_1}.$$

Denote by Φ_t, Ψ_s, Θ_w, $t, s, w \in \mathbb{R}$, the flows generated by $\mathbf{v}_1$, $\mathbf{v}_2$ and $\mathbf{v}_3$.

(a) Are Φ_t and Ψ_s commuting flows?

(b) Are Ψ_s and Θ_w commuting flows?

(c) Are Θ_w and Φ_t commuting flows?

(d) Is it true that $(\Phi_t \circ \Psi_s \circ \Theta_w)(x) = (\Theta_w \circ \Psi_s \circ \Phi_t)(x)$ for all $x \in \mathbb{R}^2$?

Ch2.E6 Consider the following Lie group of smooth diffeomorphisms:

$$\Phi_t : (t, (x_1, x_2)) \mapsto \left(x_1 + t, \frac{x_1 x_2}{x_1 + t} \right).$$

(a) Compute the infinitesimal generator of Φ_t.

(b) Write down the IVP corresponding to Φ_t.

(c) Determine Φ_t in terms of its Lie series.

Ch2.E7 Let $z_1 \in \mathbb{C}$ and $\alpha \in \mathbb{R} \setminus \{0\}$. Define

$$z_2 := e^{2\pi i \alpha} z_1, \; z_3 := e^{4\pi i \alpha} z_2, \ldots, \; z_{n+1} := e^{2\pi i \alpha n} z_n, \ldots$$

This recurrence defines a discrete dynamical system. Prove that the system has periodic orbits if and only if $\alpha \in \mathbb{Q}$. In this case, compute the (minimal) period.

Ch2.E8 Consider the discrete dynamical system defined in terms of iterations of the map $\Phi : \mathbb{R}^2 \setminus \{1 + x_1 + x_2 = 0\} \to \mathbb{R}^2$ defined by

$$\Phi : (x_1, x_2) \mapsto \left(\frac{x_2}{1 + x_1 + x_2}, -\frac{x_1}{1 + x_1 + x_2} \right).$$

Show that each orbit is periodic. Compute the (minimal) period.

Ch2.E9 Consider the discrete dynamical system defined in terms of iterations of the map $\Phi : \mathbb{R}^2 \setminus \{x_1 x_2 = 0\} \to \mathbb{R}^2$ defined by

$$\Phi : (x_1, x_2) \mapsto \left(x_2, \frac{\alpha}{x_1 x_2} \right), \qquad \alpha \in \mathbb{R} \setminus \{0\}.$$

(a) Show that each orbit is periodic. Compute the (minimal) period.

(b) Show that the two functions

$$F_1(x_1,x_2) := x_1 + x_2 + \frac{\alpha}{x_1 x_2}, \qquad F_2(x_1,x_2) := x_1^2 + x_2^2 + \frac{\alpha^2}{x_1^2 x_2^2},$$

are invariant functions for Φ. Are they functionally independent?

Ch2.E10 Let us denote a sequence $x_k : \mathbb{Z} \to \mathbb{R}$ by x. The iterate of $x \equiv x_k$ is $\widetilde{x} \equiv x_{k+1}$. Consider the dynamical system defined by the planar system of difference equations

$$\begin{cases} \widetilde{x} - x = \alpha(\widetilde{x}\, y + x\, \widetilde{y}), \\ \widetilde{y} - y = -2\, x\, \widetilde{x}, \end{cases} \tag{2.73}$$

with $\alpha > 0$. The above system defines an explicit map $(x,y) \mapsto (\widetilde{x},\widetilde{y})$. Without finding the explicit form of the map, use its implicit form, given by (2.73), to prove that the relation

$$\frac{x^2 + \alpha\, y^2}{1 + \alpha\, x^2} = \frac{\widetilde{x}^2 + \alpha\, \widetilde{y}^2}{1 + \alpha\, \widetilde{x}^2}$$

holds identically on solutions of (2.73). In other words, the function

$$F(x,y) := \frac{x^2 + \alpha\, y^2}{1 + \alpha\, x^2}$$

is an integral of motion of the discrete dynamical system.

Ch2.E11 Consider the following IVPs in $\mathbb{R}$:

$$\begin{cases} \dot{x} = (x+\alpha)^2\,(x^2 - \alpha)\,, \\ x(0) \in \mathbb{R}, \end{cases}$$

and

$$\begin{cases} \dot{x} = x^2 - 3\,\alpha\, x + 2\,\alpha^2, \\ x(0) \in \mathbb{R}, \end{cases}$$

where $\alpha \in \mathbb{R}$. The solutions to these IVPs have qualitatively different behaviors depending on α. For both IVPs:

(a) Find the fixed points.

(b) Study the stability nature of the fixed points.

Ch2.E12 Consider the following IVP in $\mathbb{R}$:

$$\begin{cases} \dot{x} = \alpha\, x(1-x), \\ x(0) \in \mathbb{R}, \end{cases}$$

where $\alpha > 0$.

(a) Find the flow and its domain of definition.

(b) Let $\alpha = 1$. Draw some trajectory curves in the plane (t, x) for $t \geqslant 0$ and $x \geqslant 0$.

Ch2.E13 Consider the following second-order ODEs in $\mathbb{R}$:

$$\ddot{x} + \alpha^2 x = 0, \qquad \ddot{x} - \alpha^2 x = 0, \qquad \alpha > 0.$$

For both ODEs:

(a) Write down the corresponding IVP with initial condition $(x(0), \dot{x}(0)) \in \mathbb{R}^2$.

(b) Find an integral of motion.

(c) Find the fixed points.

(d) Draw the phase portrait. Is there any periodic orbit?

(e) What can you say about the stability of fixed points (just by looking at the phase portrait)?

Ch2.E14 Consider the following IVP in $\mathbb{R}^2$:

$$\begin{cases} \dot{x}_1 = x_1 - \alpha \sin x_2, \\ \dot{x}_2 = \alpha + x_1^4, \\ (x_1(0), x_2(0)) \in \mathbb{R}^2, \end{cases}$$

with $\alpha > 0$. Prove that the system cannot possess a periodic orbit.

Ch2.E15 (a) Consider the following IVP in $M \subset \mathbb{R}^2$:

$$\begin{cases} \dot{x}_1 = f_1(x_1, x_2), \\ \dot{x}_2 = f_2(x_1, x_2), \\ (x_1(0), x_2(0)) \in M, \end{cases}$$

where $f_1, f_2 \in \mathscr{F}(M, \mathbb{R})$. Define the quantity

$$F(x_1, x_2) := \frac{\partial f_1}{\partial x_1} + \frac{\partial f_2}{\partial x_2}$$

and assume that F has constant sign for all $(x_1, x_2) \in M$. Use the two-dimensional "Green Theorem" to prove that the IVP cannot possess a closed orbit in M.

(b) Determine a non-empty region $M \subset \mathbb{R}^2$ where the IVP

$$\begin{cases} \dot{x}_1 = x_1(x_1^2 + x_2^2 - 2\,x_1 - 3) - x_2, \\ \dot{x}_2 = x_2(x_1^2 + x_2^2 - 2\,x_1 - 3) + x_1, \\ (x_1(0), x_2(0)) \in M, \end{cases}$$

does not have closed orbits.

Ch2.E16 Consider the discrete dynamical system defined in terms of iterations of the map $\Phi : \mathbb{R} \setminus \{1/2\} \to \mathbb{R}$ defined by

$$\Phi : x \mapsto \frac{3\,x - 2}{2\,x - 1}.$$

(a) Construct the n-th iteration of Φ.

(b) Show that $\lim_{n\to\infty} \Phi^n(x) = 1$.

(c) Is the point $x = 1$ asymptotically stable?

Ch2.E17 Consider the discrete dynamical system defined in terms of iterations of the map $\Phi : (0,\infty) \times (0,\infty) \to \mathbb{R}^2$ defined by

$$\Phi : (x_1, x_2) \mapsto \left(\frac{2\,x_1 x_2}{x_1 + x_2}, \frac{x_1 + x_2}{2} \right).$$

(a) Find an invariant function of Φ.

(b) Sketch the phase portrait.

(c) Consider the initial condition $(1,5)$. Use the map Φ to find a good approximation (say up to the third decimal digit) of $\sqrt{5} = 2.236...$

Ch2.E18 Consider the following IVP in $\mathbb{R}/(2\,\pi\,\mathbb{Z})$:

$$\begin{cases} \dot{x} = \cos x - 1, \\ x(0) \in \mathbb{R}/(2\,\pi\,\mathbb{Z}). \end{cases}$$

Prove that $x = 0$ is an attracting point and find the basin of attraction.

Ch2.E19 Consider the following IVP in $\mathbb{R}$:

$$\begin{cases} \dot{x} = -x\,|x|, \\ x(0) \in \mathbb{R}. \end{cases}$$

(a) Does it admit a unique solution?

(b) Find the solution(s) and the maximal interval(s) of existence.

(c) Prove that $x = 0$ is asymptotically stable and determine its basin of attraction.

Ch2.E20 Consider the following IVP in $\mathbb{R}^3$:

$$\begin{cases} \dot{x}_1 = -x_1 x_3, \\ \dot{x}_2 = x_2 x_3, \\ \dot{x}_3 = x_1^2 - x_2^2, \\ (x_1(0), x_2(0), x_3(0)) \in \mathbb{R}^3. \end{cases}$$

Determine two functionally independent (polynomial) integrals of motion of the system.

Ch2.E21 Fix $n \geqslant 3$. Let $D \subset \mathbb{R}^n$ be a compact set. On D consider the dynamical system defined through the flow of an n-dimensional IVP whose ODEs are

$$\dot{x}_i = \prod_{j \neq i}^{n} x_j, \qquad i = 1, \dots, n. \tag{2.74}$$

(a) Prove that the flow is volume preserving.

(b) Prove that the functions

$$F_{ij}(x_i, x_j) := x_i^2 - x_j^2, \qquad i, j = 1, \dots, n.$$

are integrals of motion. How many functionally independent integrals of motion does the system admit?

(c) Define n new coordinates by the transformation $\Phi : x \mapsto y$ defined by

$$y_i := \frac{1}{x_i} \prod_{j \neq i}^{n} x_j, \qquad i = 1, \dots, n.$$

Prove that the system of ODEs (2.74) is transformed into the quadratic system

$$\dot{y}_i = y_i \left(-2 y_i + \sum_{j=1}^{n} y_j \right), \qquad i = 1, \dots, n. \tag{2.75}$$

(d) Find the functions $K_{ij}(y) := F_{ij}(\Phi^{-1}(y))$. Are the functions K_{ij}, $i, j = 1, \dots, n$, integrals of motion of the system of ODEs (2.75)?

Ch2.E22 (a) Consider the following IVP in $\mathbb{R}^2$:

$$\begin{cases} \dot{x}_1 = f_1(x_1, x_2), \\ \dot{x}_2 = f_2(x_1, x_2), \\ (x_1(0), x_2(0)) \in \mathbb{R}^2, \end{cases}$$

where $f_1, f_2 \in \mathscr{F}(\mathbb{R}^2, \mathbb{R})$. Note that the ODE

$$\frac{\mathrm{d}x_2}{\mathrm{d}x_1} = \frac{f_2(x_1, x_2)}{f_1(x_1, x_2)} \tag{2.76}$$

can be formally obtained from the original system by dividing the second ODE by the first and then canceling the time differential $\mathrm{d}t$. Let $F(x_1, x_2) = c$, $c \in \mathbb{R}$, be the solution of (2.76) in implicit form. Prove that F is an integral of motion of the IVP.

(b) Use the procedure described in (a) to construct an integral of motion in the case

$$f_1(x_1, x_2) := 2\, x_1 - x_1 x_2, \qquad f_2(x_1, x_2) := -9\, x_2 + 3\, x_1 x_2.$$

Sketch the phase portrait in the first quadrant.

Ch2.E23 Let $M := \{(x_1, x_2) \in \mathbb{R}^2 \,:\, x_2 > 0\}$. On M consider the following IVP:

$$\begin{cases} \dot{x}_1 = \frac{1}{2}\left(x_1^2 - x_2^2\right), \\ \dot{x}_2 = x_1 x_2, \\ (x_1(0), x_2(0)) \in M. \end{cases}$$

(a) Find an integral of motion.

(b) Determine the flow of the system.

(c) Sketch the phase portrait.

Ch2.E24 Consider the following IVP in $\mathbb{R}^3$:

$$\begin{cases} \dot{x}_1 = x_1(-x_1 + x_2 + x_3), \\ \dot{x}_2 = x_2(-x_2 + x_3 + x_1), \\ \dot{x}_3 = x_3(-x_3 + x_1 + x_2), \\ (x_1(0), x_2(0), x_3(0)) \in \mathbb{R}^3. \end{cases}$$

Denote by $\mathbf{v}$ the infinitesimal generator of the flow and define the functions

$$\begin{aligned} F_1(x_1, x_2, x_3) &:= x_1(x_2 - x_3), \\ F_2(x_1, x_2, x_3) &:= x_2(x_3 - x_1), \\ F_3(x_1, x_2, x_3) &:= x_3(x_1 - x_2). \end{aligned}$$

(a) Prove that

$$(\mathfrak{L}_{\mathbf{v}} F_i)(x_1, x_2, x_3) = 0, \qquad i = 1, 2, 3, \quad \forall\, (x_1, x_2, x_3) \in \mathbb{R}^3,$$

namely that F_1, F_2, F_3 are integrals of motion.

(b) Are F_1, F_2, F_3 functionally independent?

Ch2.E25 Consider a gradient system in $M \subset \mathbb{R}^n$:

$$\begin{cases} \dot{x} = -\operatorname{grad}_x G(x), \\ x(0) \in M, \end{cases}$$

where $G \in C^2(M, \mathbb{R})$.

(a) Prove that if $\tilde{x} \in M$ is a regular point of the system, i.e., $\operatorname{grad}_x G(\tilde{x}) \neq 0$, then the vector field generating the flow Φ_t is orthogonal to each level set $S_h := \{x \in M \,:\, G(x) = h\}$, with $h \in \mathbb{R}$, at the point $\tilde{x}$.

(b) Fix $n = 2$ and

$$G(x_1, x_2) := x_1^2(x_1 - 1)^2 + x_2^2.$$

Find the fixed points of the corresponding gradient system. Prove that there exists at least one asymptotically stable fixed point.

Ch2.E26 Consider the following IVP in $\mathbb{R}^2$:

$$\begin{cases} \dot{x}_1 = x_2, \\ \dot{x}_2 = -x_2 - \sin x_1, \\ (x_1(0), x_2(0)) \in \mathbb{R}^2. \end{cases}$$

Consider the fixed point $(0,0)$.

(a) Is the function

$$F(x_1, x_2) := (1 - \cos x_1) + \frac{1}{2} x_2^2$$

a Lyapunov function. What can you say about the stability of $(0,0)$?

(b) Is the function

$$F(x_1, x_2) := 2(1 - \cos x_1) + \frac{1}{2} x_2^2 + \frac{1}{2}(x_1 + x_2)^2$$

a Lyapunov function. What can you say about the stability of $(0,0)$?

Ch2.E27 Consider the following IVP in $\mathbb{R}^2$:

$$\begin{cases} \dot{x}_1 = -x_1^3 - 2\,x_1 x_2^2, \\ \dot{x}_2 = x_1^2 x_2 - x_2^3, \\ (x_1(0), x_2(0)) \in \mathbb{R}^2. \end{cases}$$

Consider the fixed point $(0,0)$. Is the function

$$F(x_1, x_2) := x_1^2 + x_1^2\, x_2^2 + x_2^4$$

a Lyapunov function. What can you say about the stability of $(0,0)$?

Ch2.E28 Consider the following IVP in $M \subset \mathbb{R}^n$:

$$\begin{cases} \dot{x} = f(x), \\ x(0) \in M, \end{cases}$$

where $f \in \mathscr{F}(M, \mathbb{R}^n)$. Assume that $x = 0$ is a fixed point and suppose that the Jacobian matrix of f at $x = 0$ is $\operatorname{diag}(\nu_1, \ldots, \nu_n)$, with $\nu_i < 0$, $1 \leqslant i \leqslant n$. Prove that $x = 0$ is asymptotically stable.

Ch2.E29 Consider the following IVP in $\mathbb{R}^3$:

$$\begin{cases} \dot{x}_1 = 3\,x_2(x_3 - 1), \\ \dot{x}_2 = -x_1(x_3 - 1), \\ \dot{x}_3 = -x_3^3\left(x_1^2 + 1\right), \\ (x_1(0), x_2(0), x_3(0)) \in \mathbb{R}^2. \end{cases}$$

Determine the stability of the the fixed point $(0,0,0)$ by finding a proper (quadratic) Lyapunov function.

Ch2.E30 Consider the following second-order ODE in $\mathbb{R}$:

$$\ddot{x} - \varepsilon\left(1 - x^2\right)\dot{x} + x = 0, \qquad \varepsilon \in \mathbb{R}.$$

Fix an initial condition $(x(0), \dot{x}(0)) \in \mathbb{R}^2$.

(a) Prove that $(0, 0)$ is a fixed point.

(b) Fix $\varepsilon < 0$. Discuss the stability of $(0, 0)$ by finding a proper Lyapunov function.

(c) Prove that the transformation $\varepsilon \mapsto -\varepsilon$ is equivalent to the time reversal $t \mapsto -t$. Using this result, what can you say about the stability of $(0, 0)$ if $\varepsilon > 0$?

Ch2.E31 Let $A \in \mathfrak{gl}(n, \mathbb{R})$ and define the exponential of a A as

$$e^A := \sum_{k=0}^{\infty} \frac{A^k}{k!}.$$

(a) Prove that, if $[\,A,\,B\,] := A\,B - B\,A = 0$, then

$$e^{A+B} = e^A e^B, \qquad A, B \in \mathfrak{gl}(n, \mathbb{R}).$$

(b) Prove that

$$\det\left(e^A\right) = e^{\operatorname{Trace} A}.$$

Ch2.E32 Consider the following ODE in $\mathbb{R}$:

$$\ddot{x} - 4\,x + 2\,\alpha\,x = 0, \qquad \alpha \in \mathbb{R},$$

together with two conditions $x(0) = 0$ and $x(\pi) = 0$. Find α such that there exists a solution which is not the trivial solution.

Ch2.E33 Consider the following linear IVP in $\mathbb{R}^3$:

$$\begin{cases} \dot{x}_1 = x_1 + x_2, \\ \dot{x}_2 = x_2 - x_3, \\ \dot{x}_3 = x_2 + x_3, \\ (x_1(0), x_2(0), x_3(0)) \in \mathbb{R}^3. \end{cases}$$

(a) Find the solution.

(b) Discuss the stability of the fixed point $(0, 0, 0)$.

Ch2.E34 Consider the following linear IVPs in $\mathbb{R}^2$:

$$\begin{cases} \dot{x}_1 = x_1 \pm 2\,x_2, \\ \dot{x}_2 = 4\,x_1 + x_2, \\ (x_1(0), x_2(0)) = (-1, \sqrt{2}). \end{cases}$$

(a) Find the solution.

(b) Discuss the stability of the fixed point $(0,0)$.

(c) Sketch the phase portrait.

Ch2.E35 (a) Let $A \in \mathbf{GL}(n,\mathbb{R})$ be a matrix such that $A^{\ell} = \alpha\,A$ for some $\alpha \in \mathbb{R}$ and $\ell \in \mathbb{N}$, $\ell \geqslant 2$. Prove that e^{A} is a polynomial of degree $\ell - 1$ in A and compute explicitly its coefficients.

(b) Use the result (a) to find the solution of the IVP

$$\begin{cases} \dot{x} = A\,x, \\ x(0) = (2,\ldots,2), \end{cases}$$

where $A \in \mathbf{GL}(n,\mathbb{R})$ is matrix whose all entries are given by 2.

Ch2.E36 Consider the following linear IVPs in $\mathbb{R}^2$:

$$\begin{cases} \dot{x}_1 = x_1, \\ \dot{x}_2 = -x_1 + x_2, \end{cases} \quad \begin{cases} \dot{x}_1 = -x_1 + x_2, \\ \dot{x}_2 = -x_2, \end{cases} \quad \begin{cases} \dot{x}_1 = -x_2, \\ \dot{x}_2 = x_1, \end{cases} \quad \begin{cases} \dot{x}_1 = 2\,x_1 + x_2, \\ \dot{y} = 6\,x_1 + 3\,x_2, \end{cases}$$

with $(x_1(0), x_2(0)) \in \mathbb{R}^2$. For each of them:

(a) Find the solution.

(b) Study the stability of the fixed points.

(c) Sketch the phase portrait.

Ch2.E37 Consider the following linear IVPs in $\mathbb{R}^3$:

$$\begin{cases} \dot{x}_1 = -2\,x_1 - x_2, \\ \dot{x}_2 = x_1 - 2\,x_2, \\ \dot{x}_3 = 3\,x_3, \end{cases} \qquad \begin{cases} \dot{x}_1 = -x_2, \\ \dot{x}_2 = x_1, \\ \dot{x}_3 = x_3, \end{cases}$$

with $(x_1(0), x_2(0), x_3(0)) \in \mathbb{R}^3$. For both of them:

(a) Find the stable, unstable and center subspaces. What can you say about the stability of $(0,0,0)$?

(b) Sketch the phase portrait.

Ch2.E38 Consider the following IVP in $\mathbb{R}^2$:

$$\begin{cases} \dot{x} = A\,x + \varepsilon\,x\,\|x\|^2, \\ x(0) \in \mathbb{R}^2, \end{cases} \qquad A := \begin{pmatrix} 0 & 1 \\ -1 & 0 \end{pmatrix},$$

where $\varepsilon \in \mathbb{R}$.

(a) Assume $\varepsilon = 0$. Study the stability of the fixed point $(0,0)$.

(b) Assume $\varepsilon > 0$. Study the stability of the fixed point $(0,0)$.

(c) Assume $\varepsilon < 0$. Study the stability of the fixed point $(0,0)$.

~~~~~~~~~~

Ch2.E39 Consider the following IVP in $\mathbb{R}^2$:

$$\begin{cases} \dot{x}_1 = x_1\left(1 - x_1^2 - x_2^2\right) - x_2\left(1 + x_1^2 + x_2^2\right), \\ \dot{x}_2 = x_1\left(1 + x_1^2 + x_2^2\right) + x_2\left(1 - x_1^2 - x_2^2\right), \\ (x_1(0), x_2(0)) \in \mathbb{R}^2. \end{cases}$$

(a) Rewrite the system of ODEs using polar coordinates, $(x_1, x_2) := r(\cos\theta, \sin\theta)$, $r > 0$, $\theta \in [0, 2\pi)$.

(b) Find a periodic solution.

~~~~~~~~~~

Ch2.E40 Consider the following planar linear systems of ODEs:

$$\begin{cases} \dot{x}_1 = -2\,x_1, \\ \dot{x}_2 = -3\,x_2, \end{cases} \qquad \begin{cases} \dot{x}_1 = -2\,x_1, \\ \dot{x}_2 = -2\,x_2. \end{cases}$$

They have respectively the following phase portraits:

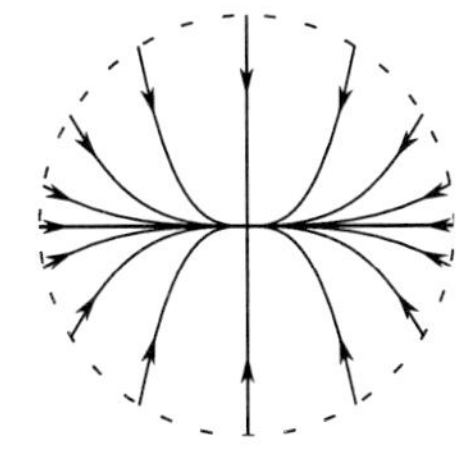

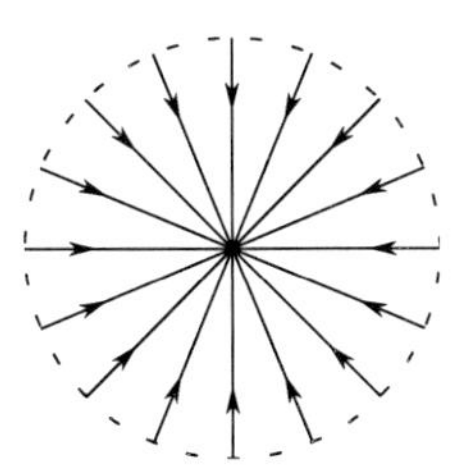

Prove that the systems are topologically conjugate by constructing explicitly the map between the orbits.

~~~~~~~~~~
~~~~~~~~~~

Ch2.E41 (a) Let $T > 0$. Consider the following non-autonomous IVP in $\mathbb{R}$:

$$\begin{cases} \dot{x} = a(t)\, x, \\ x(0) \in \mathbb{R}, \end{cases}$$

where $a : \mathbb{R} \to \mathbb{R}$ is a continuous T-periodic function. For which values of

$$\alpha := \int_0^T a(t)\, \mathrm{d}t$$

does the IVP admit non-trivial T-periodic solutions?

(b) Let $T > 0$. Consider the following non-autonomous IVP in $\mathbb{R}$:

$$\begin{cases} \dot{x} = a(t)\, x + b(t), \\ x(0) \in \mathbb{R}, \end{cases}$$

where $a, b : \mathbb{R} \to \mathbb{R}$ are continuous T-periodic functions. Find the set of maximal solutions that are T-periodic.

Ch2.E42 Consider the following linear homogeneous and periodic IVP in $\mathbb{R}^n$:

$$\begin{cases} \dot{x} = A(t)\, x, \\ x(t_0) = x_0, \end{cases}$$

with $(t_0, x_0) \in \mathbb{R} \times \mathbb{R}^n$ and $A \in C(\mathbb{R}, \mathfrak{gl}(n, \mathbb{R}))$. Here the $\mathfrak{gl}(n, \mathbb{R})$-valued function A is a periodic function of t with period $T > 0$, i.e., $A(t + T) = A(t)$ for all $t \in \mathbb{R}$. "Floquet Theorem" 2.25 claims that the principal matrix solution can be written in the form

$$\Pi(t, t_0) = P(t, t_0)\, \mathrm{e}^{(t-t_0)\, Q(t_0)},$$

where $P \in C(\mathbb{R}, \mathfrak{gl}(n, \mathbb{R}))$ is periodic with period T, with $P(t_0, t_0) = \mathbb{1}$, and $Q(t_0) \in \mathfrak{gl}(n, \mathbb{C})$ is defined in terms of the monodromy matrix $\Lambda(t_0)$ by

$$\mathrm{e}^{T\, Q(t_0)} := \Lambda(t_0), \qquad Q(t_0 + T) = Q(t_0).$$

Prove that the time-dependent change of coordinates $x \mapsto y := P^{-1}(t, t_0)\, x$ maps $\dot{x} = A(t)\, x$ into the autonomous system of ODEs $\dot{y} = Q(t_0)\, y$.

Ch2.E43 Consider the following non-autonomous IVP in $\mathbb{R}^2$:

$$\begin{cases} \dot{x}_1 = x_1 + x_2, \\ \dot{x}_2 = a(t)\, x_2, \\ (x_1(0), x_2(0)) \in \mathbb{R}^2, \end{cases}$$

where

$$a(t) := \frac{\cos t + \sin t}{2 + \sin t - \cos t}.$$

(a) Find the principal matrix solution.

(b) Compute the Floquet multipliers. What can you say about the stability of $(0, 0)$?

Ch2.E44 Consider the following IVP in $\mathbb{R}^3$:

$$\begin{cases} \dot{x}_1 = -x_2, \\ \dot{x}_2 = -x_1 - x_3^3, \\ \dot{x}_3 = -\sin(x_3^3) + x_2, \\ (x_1(0), x_2(0), x_3(0)) = (0,0,0). \end{cases}$$

(a) Explain why the "Poincaré-Lyapunov Theorem" gives no information about the stability of the fixed point $(0,0,0)$.

(b) Find suitable $\alpha, \beta, \gamma \in \mathbb{R}$ such that

$$F(x_1, x_2, x_3) := \alpha\, x_1^2 + \beta\, x_2^2 + \gamma\, x_3^4$$

is a Lyapunov function. Is F a strict Lyapunov function?

(c) What is the largest domain of stability of $(0,0,0)$?

Ch2.E45 Consider the following IVP in $\mathbb{R}^2$:

$$\begin{cases} \dot{x}_1 = 2\, x_2 \mathrm{e}^{-x_1^2 - x_2^2}, \\ \dot{x}_2 = (3\, x_1^2 - 6)\, \mathrm{e}^{-x_1^2 - x_2^2}, \\ (x_1(0), x_2(0)) = (0,0). \end{cases}$$

(a) Find the fixed points.

(b) Linearize the system around the fixed points. Only on the basis of the "Poincaré-Lyapunov Theorem", what can you say about the stability of the fixed points for the nonlinear system?

Ch2.E46 Consider the following gradient system in $\mathbb{R}^2$:

$$\begin{cases} \dot{x} = -\operatorname{grad}_x G(x), \\ x(0) \in \mathbb{R}^2, \end{cases}$$

where

$$G(x_1, x_2) := x_1^4 - 2\, x_1^2 + 2\left(x_2^2 + 1\right)\left(x_1^2 - 1\right) + x_2^4 - 2\, x_2^2.$$

(a) Find the fixed points and study their stability.

(b) Sketch the phase portrait.

Ch2.E47 Consider the following IVP in $\mathbb{R}^2$:

$$\begin{cases} \dot{x}_1 = x_1 - x_2 - x_1\left(x_1^2 + x_2^2\right), \\ \dot{x}_2 = x_1 + x_2 - x_2\left(3\, x_1^2 + x_2^2\right), \\ (x_1(0), x_2(0)) = (0,0). \end{cases}$$

(a) Discuss the stability of the fixed point $(0,0)$.

(b) Rewrite the system of ODEs using polar coordinates, $(x_1, x_2) := r(\cos\theta, \sin\theta)$, $r > 0$, $\theta \in [0, 2\pi)$.

(c) Find the inequality which the radius r_1 of a circle centered at $(0,0)$ must obey if all orbits have an outward radial component on it.

(d) Find the inequality which the radius r_2 of a circle centered at $(0,0)$ must obey if all orbits have an inward radial component on it.

Ch2.E48 Consider the following IVP in $\mathbb{R}^2$:

$$\begin{cases} \dot{x}_1 = -x_1, \\ \dot{x}_2 = \sin(x_1 + x_2), \\ (x_1(0), x_2(0)) = (0,0). \end{cases}$$

(a) Find all fixed points.

(b) Discuss the stability of fixed points.

Ch2.E49 Consider the following IVP in $\mathbb{R}^2$:

$$\begin{cases} \dot{x}_1 = 4\,x_2, \\ \dot{x}_2 = -x_1\left(x_1^3 + 1\right), \\ (x_1(0), x_2(0)) = (0,0). \end{cases}$$

(a) Write the above system in the form

$$\begin{cases} \dot{x}_1 = \dfrac{\partial F}{\partial x_2}, \\ \dot{x}_2 = -\dfrac{\partial F}{\partial x_1}, \end{cases}$$

for some function $F = F(x_1, x_2)$ to be determined. Prove that F is an integral of motion.

(b) Prove that the fixed point $(0,0)$ is stable by finding a Lyapunov function.

(c) Consider the following perturbation of the original system

$$\begin{cases} \dot{x}_1 = 4\,x_2 - \varepsilon\,x_1, \\ \dot{x}_2 = -x_1(x_1^3 + 1) - \varepsilon x_2, \\ (x_1(0), x_2(0)) = (0,0), \end{cases}$$

with $\varepsilon > 0$. Use the "Poincaré-Lyapunov Theorem" to prove that the point $(0,0)$ is asymptotically stable.

Ch2.E50 Consider the following IVP in $\mathbb{R}^2$:

$$\begin{cases} \dot{x}_1 = x_1^4 + x_1 x_2, \\ \dot{x}_2 = -2\,x_2 - x_1^2 + x_1 x_2^2, \\ (x_1(0), x_2(0)) = (0,0). \end{cases} \tag{2.77}$$

(a) Linearize (2.77) around the fixed point $(0,0)$. What can you say about the stability of system of $(0,0)$ on the basis of the "Poincaré-Lyapunov Theorem"?

(b) Find the center (linear) space $E^0(0,0)$. Construct an approximation of the center manifold $W^0(0,0)$.

(Hint: The center manifold is parametrized, in a neighborhood of $(0,0)$, by $x_2 = h(x_1)$ for some function h. An approximation of $W^0(0,0)$ is given by the series expansion - say up to $O(x_1^4)$ - of the function h around $x_1 = 0$)

(c) Find the first four terms of the series expansion of the system obtained by reducing (2.77) on $W^0(0,0)$. What can you say now about the stability of the fixed point $(0,0)$ for system (55)?

Ch2.E51 Consider the following IVP in $\mathbb{R}^3$:

$$\begin{cases} \dot{x}_1 = -x_2 - x_1 x_3, \\ \dot{x}_2 = x_1 - x_2^3, \\ \dot{x}_3 = -x_3 - 2\,x_1 x_2 - 2\,x_1^4 + x_1^2, \\ (x_1(0), x_2(0), x_3(0)) = (0,0,0). \end{cases} \tag{2.78}$$

(a) Linearize (2.78) around the fixed point $(0,0,0)$ and find the center (linear) space $E^0(0,0,0)$.

(b) Find the center manifold $W^0(0,0,0)$.

(c) Reduce (2.78) on $W^0(0,0,0)$.

(d) What can you say about the stability of the fixed point $(0,0,0)$ for system (2.78)?

Ch2.E52 Consider the following IVP in $\mathbb{R}^2$:

$$\begin{cases} \dot{x}_1 = x_1 x_2, \\ \dot{x}_2 = -x_2 - x_1^2, \\ (x_1(0), x_2(0)) = (0,0). \end{cases}$$

(a) Construct the center manifold $W^0(0,0)$ (up to the sixth order).

(b) Find the reduced system on $W^0(0,0)$.

Ch2.E53 Consider the following IVP in $\mathbb{R}^3$:

$$\begin{cases} \dot{x}_1 = \frac{1}{2}(x_3 - x_1) - x_2 + 2\,x_2^3, \\ \dot{x}_2 = \frac{1}{2}(x_3 - x_1), \\ \dot{x}_3 = x_3 - x_2^2(x_1 + x_3), \\ (x_1(0), x_2(0), x_3(0)) = (0,0,0). \end{cases} \tag{2.79}$$

(a) Find all fixed points.

(b) Linearize system (2.79) around the fixed point $(0,0,0)$ and discuss the stability of $(0,0,0)$. Write the general solution of the linearized system.

(c) Prove that

$$\dot{x}_1 - 2\,\dot{x}_2 + \dot{x}_3 = \frac{1}{2}(x_1 - 2\,x_2 + x_3)\left(1 - 2\,x_2^2\right),$$

and deduce that $M := \{(x_1, x_2, x_3) \in \mathbb{R}^3 \,:\, x_1 - 2\,x_2 + x_3 = 0\}$ is an invariant manifold for (2.79).

(d) Reduce system (2.79) on M by eliminating the variable x_1, thus getting a first-order IVP for x_2 and x_3. Then eliminate the variable x_3, thus obtaining a second-order ODE for x_2. Find an integral of motion of the resulting ODE.

Ch2.E54 Consider the following IVP in $\mathbb{R}^2$:

$$\begin{cases} \dot{x}_1 = -x_1, \\ \dot{x}_2 = x_2 + x_1^2, \\ (x_1(0), x_2(0)) \in \mathbb{R}^2. \end{cases}$$

(a) Verify that its flow is

$$\Phi_t(x_1(0), x_2(0)) = \left(x_1(0)\,\mathrm{e}^{-t}, x_2(0)\,\mathrm{e}^{t} + \frac{1}{3}x_1^2(0)\left(\mathrm{e}^{t} - \mathrm{e}^{-2t}\right)\right).$$

(b) Linearize the system around the fixed point $(0,0)$ and find the stable and unstable linear subspaces.

(c) Construct the stable and unstable manifolds of the nonlinear system.

(d) Plot the phase portraits of both systems.

Ch2.E55 Consider the following two scalar ODEs:

$$\dot{x} = x - \alpha\,x(1 - x), \qquad \dot{x} = x + \frac{\alpha\,x}{1 + x^2}, \qquad \alpha \in \mathbb{R}.$$

For both of them:

(a) Find and classify the bifurcations that occur as α is varied.

(b) Draw the bifurcation diagram.

Ch2.E56 Consider the scalar ODE

$$\dot{x} = \alpha\, x - \sin x, \qquad \alpha \geqslant 0.$$

(a) Fix $\alpha = 0$. Find the fixed points, study their stability and sketch the phase portrait.

(b) Fix $\alpha > 1$. Show that there is only one fixed point. Is it stable?

(c) Find and classify the bifurcations that occur as α is varied and $-\pi < x < \pi$.

Ch2.E57 Consider the planar system of ODEs

$$\begin{cases} \dot{x}_1 = x_2 - 2\,x_1, \\ \dot{x}_2 = \alpha + x_1^2 - x_2, \end{cases}$$

with $\alpha \geqslant 0$. Find and classify the bifurcations that occur as α is varied.

3

Lagrangian and Hamiltonian Mechanics on Euclidean Spaces

3.1 Introduction

▶ Mechanics has two main points of view, the *Lagrangian formulation* and the *Hamiltonian formulation*.

- In one sense, Lagrangian mechanics is more fundamental, since it is based on variational principles and it is what generalizes most directly to the general relativistic context. The Lagrangian formulation of mechanics is based on the observation that there are variational principles behind the fundamental laws of force balance as given by Newton law.

- In another sense, Hamiltonian mechanics is more fundamental, since it is based directly on the energy concept and it is what is more closely tied to quantum mechanics.

Fortunately, in many cases these approaches are equivalent.

▶ We start our study of classical mechanics with the Lagrangian formulation. To facilitate the understanding of a number of points we begin with mechanics on Euclidean spaces, say $M = \mathbb{R}^n$. Such a simplification allows to be acquainted with the machinery we are going to present, but it has two evident limitations which will be removed with the formulation of mechanics on manifolds (see Chapter 4):

1. The *configuration space*, where spatial coordinates live, is a linear space where one can define a *global* system of coordinates. As a matter of fact, it is not suitable for a global description of configuration spaces of many natural systems, as for instance a particle constrained to move on a two-dimensional closed surface embedded in the ambient space $\mathbb{R}^3$.

2. The *phase spaces* (both Lagrangian and Hamiltonian) we are going to consider are by construction even-dimensional. Roughly speaking they consist of the cartesian product of two copies of $\mathbb{R}^n$: a first copy for the spatial coordinates, a second one for the corresponding velocities or momenta. In particular, it turns out that a canonical Hamiltonian system is governed by an IVP consisting of an even number of ODEs. However there are mechanical systems, not considered in this Chapter, which are described by IVPs consisting of an odd number of ODEs. For the latter case the Hamiltonian theory we are going to present does not apply directly.

► It is important to emphasize that, despite the above two limitations, the Lagrangian and Hamiltonian theory we are going to present is indeed the correct *local* theory also when the configuration space is a *symplectic manifold* (instead of a linear space). The corresponding phase space will be another smooth manifold, whose dimension is twice the dimension of the configuration manifold. We will consider such aspects and generalizations in Chapter 4.

3.2 Definition of a mechanical system and Newton equations

► The set of (time-dependent, $t \in \mathbb{R}$) coordinates on $M = \mathbb{R}^n$ is denoted by $q := (q_1, \ldots, q_n)$.

- Coordinates $q := (q_1, \ldots, q_n)$ are called *Lagrangian coordinates*, while their time-derivatives $\dot{q} := (\dot{q}_1, \ldots, \dot{q}_n)$ are called *Lagrangian velocities*. Lagrangian coordinates and velocities $(q, \dot{q}) \in \mathbb{R}^{2n}$ are coordinates on the *Lagrangian phase space*.
- We use the following notation:

$$\mathsf{grad}_q := \left(\frac{\partial}{\partial q_1}, \ldots, \frac{\partial}{\partial q_n}\right), \qquad \mathsf{grad}_{\dot{q}} := \left(\frac{\partial}{\partial \dot{q}_1}, \ldots, \frac{\partial}{\partial \dot{q}_n}\right).$$

► We start with the following definition.

Definition 3.1

A **mechanical system** *in $\mathbb{R}^n$ is a dynamical system defined in terms of the system of n second-order ODEs (***Newton equations***):*

$$A\,\ddot{q} = f(q, \dot{q}, t), \tag{3.1}$$

*where $t \in \mathbb{R}$, $A \in \mathbf{GL}(n, \mathbb{R})$ is positive definite and symmetric (***mass matrix***) and $f \in C^1(\mathbb{R}^{2n} \times \mathbb{R}, \mathbb{R}^n)$ defines a vector field on $\mathbb{R}^{2n} \times \mathbb{R}$ called* **force field**.

► System (3.1) is equivalent to a non-autonomous system of $2\,n$ first-order ODEs:

$$\begin{cases} \dot{q} = A^{-1}p, \\ \dot{p} = f\left(q, A^{-1}p, t\right), \end{cases} \tag{3.2}$$

where $p := A\,\dot{q}$. To properly define a dynamical system from (3.2) we always assume that some initial conditions $(q(t_0), p(t_0)) \in \mathbb{R}^{2n}$, $t_0 \in \mathbb{R}$, are prescribed.

Example 3.1 (*The N-body gravitational problem*)

The N-body gravitational problem is the problem of predicting the motion of a group of celestial objects that interact with each other gravitationally. Solving this problem has been motivated by the need to understand the motion of the sun, planets and the visible stars. Its first complete mathematical formulation appeared in Isaac Newton's "Principia".

- The 2-body problem (a special case of which is the *Kepler problem*) is completely solvable. Indeed, it can be reduced to the motion of a single body in a central potential field. If $N \geqslant 3$, the resulting motion is much more complicated.
- The problem can be stated as follows: Consider N point masses $m_i > 0$, $i = 1, \ldots, N$, in $\mathbb{R}^3$ with time-dependent coordinates
$$q^{(j)} := \left(q_1^{(j)}, q_2^{(j)}, q_3^{(j)}\right) \in \mathbb{R}^3.$$
Define
$$Q := \left(q^{(1)}, \ldots, q^{(N)}\right).$$
The total configuration space is $M = \mathbb{R}^n$, with $n = 3N$. Suppose that the force of attraction experienced between each pair of particles is Newtonian. Then, if the initial positions in space and initial velocities are specified for every particle at $t = 0$, the problem is to determine the position of each particle at $t > 0$.
- Specifically, Newton equations form a system of $3N$ ODEs of the form
$$m_j \ddot{q}^{(j)} = f_j(Q) := G \sum_{k \neq j} m_j m_k \frac{q^{(k)} - q^{(j)}}{\left\| q^{(k)} - q^{(j)} \right\|^3}, \tag{3.3}$$
where $j = 1, \ldots, N$ and G is the gravitational constant. The corresponding autonomous IVP, which has the form (3.2), is a dynamical system for $6N$ variables $\left(q^{(j)}, \dot{q}^{(j)}\right) \in \mathbb{R}^6, j = 1, \ldots, N$.
- System (3.3) admits ten functionally independent integrals of motion:
  1. Six integrals of motion because the *center of mass*
  $$Q_{CM}(Q) := \left(\sum_{j=1}^{N} m_j\right)^{-1} \sum_{j=1}^{N} m_j q^{(j)},$$
  moves uniformly in a straight line, that is
  $$\ddot{Q}_{CM}(Q) = 0.$$
  In particular, this implies the conservation of *total linear momentum*,
  $$P(\dot{Q}) := \sum_{j=1}^{N} m_j \dot{q}^{(j)},$$
  that is
  $$\dot{P}(\dot{Q}) = 0.$$
  2. Three for the *total angular momentum*,
  $$\ell(Q, \dot{Q}) := \sum_{j=1}^{N} m_i q^{(j)} \times \dot{q}^{(j)},$$
  that is
  $$\dot{\ell}(Q, \dot{Q}) = 0.$$
  3. One for the *total energy*,
  $$E(Q, \dot{Q}) := T(\dot{Q}) + U(Q),$$
  where T is the *kinetic energy*:
  $$T(\dot{Q}) := \frac{1}{2} \sum_{i=1}^{N} m_i \left\| \dot{q}^{(i)} \right\|^2,$$

and U is the *gravitational potential energy*:

$$U(Q) := -G \sum_{1 \leqslant k < j \leqslant N} \frac{m_j\, m_k}{\| q^{(k)} - q^{(j)} \|}.$$

One has

$$\dot{E}(Q, \dot{Q}) = 0.$$

The existence of ten integrals of motions allows one the reduction of variables to $6\,N - 10$.

► A fundamental class of mechanical systems is given by conservative systems.

Definition 3.2

Consider a mechanical system (3.1).

1. *The system (3.1) is* **conservative** *if the force field can be expressed as the gradient w.r.t. q of a function $U \in C^2(\mathbb{R}^n, \mathbb{R})$* **(potential energy)***:*

$$f(q) := -\operatorname{grad}_q U(q) \qquad \forall\, q \in \mathbb{R}^n,$$

   so that Newton equations (3.1) read

$$A\,\ddot{q} = -\operatorname{grad}_q U(q), \tag{3.4}$$

   and correspondingly, (3.2) is the autonomous system of $2\,n$ ODEs

$$\begin{cases} \dot{q} = A^{-1} p, \\ \dot{p} = -\operatorname{grad}_q U(q). \end{cases} \tag{3.5}$$

2. *Let $(q(0), p(0)) \in \mathbb{R}^{2n}$ be the initial conditions of (3.5) at time $t = 0$. We denote by $\Phi_t : \mathbb{R} \times \mathbb{R}^{2n} \to \mathbb{R}^{2n}$ the* **flow** *of (3.5).*

3. *The (mechanical)* **total energy** *of (3.4) is a function $E \in C^2(\mathbb{R}^{2n}, \mathbb{R})$ defined by*

$$E(q, \dot{q}) := T(\dot{q}) + U(q), \tag{3.6}$$

   where

$$T(\dot{q}) := \frac{1}{2} \langle \dot{q}, A\,\dot{q} \rangle,$$

   is called **kinetic energy***.*

► Conservative vector fields admit several important characterizations. In the next example we consider conservative vector fields in $\mathbb{R}^3$.

Example 3.2 (*Vector fields in* $\mathbb{R}^3$)

Fix $n = 3$. Let $f \in C^1(M, \mathbb{R}^3)$, $M \subseteq \mathbb{R}^3$, define a vector field:

$$f(q) := (f_1(q), f_2(q), f_3(q)), \qquad q := (q_1, q_2, q_3).$$

- The curl of f is another vector field defined by

$$\mathsf{curl}_q f(q) := \left(\frac{\partial f_3}{\partial q_2} - \frac{\partial f_2}{\partial q_3}, \frac{\partial f_1}{\partial q_3} - \frac{\partial f_3}{\partial q_1}, \frac{\partial f_2}{\partial q_1} - \frac{\partial f_1}{\partial q_2} \right).$$

  If $\mathsf{curl}_q f(q) \equiv 0$ for all $q \in M$ then f is called *irrotational*.
- The divergence of f is a scalar function defined by

$$\mathsf{div}_q f(q) := \frac{\partial f_1}{\partial q_1} + \frac{\partial f_2}{\partial q_2} + \frac{\partial f_3}{\partial q_3}.$$

  If $\mathsf{div}_q f(q) \equiv 0$ for all $q \in M$ then f is called *divergence-free*.
- There hold the following formulas:

$$\mathsf{curl}_q(\,\mathsf{grad}_q U(q)) = 0, \qquad \mathsf{div}_q(\mathsf{curl}_q f(q)) = 0, \tag{3.7}$$

  for all $U \in C^2(M, \mathbb{R})$ and $f \in C^1(M, \mathbb{R}^3)$. The second formula implies that a divergence-free vector field can be expressed as the curl of some vector field.
- The following claim holds: If M is simply-connected, then $f \in C^1(M, \mathbb{R}^3)$ is irrotational if and only if f is conservative, i.e., if and only if f can be expressed as the gradient w.r.t. q of a function $U \in C^2(M, \mathbb{R})$:

$$f(q) := -\,\mathsf{grad}_q U(q) \qquad \forall q \in M.$$

  Note that every conservative vector field f is irrotational, because of the first formula in (3.7). The converse is true only if M is a simply-connected.
- The following claim holds: If $f \in C^1(M, \mathbb{R}^3)$ is a conservative vector field then its integral along a rectifiable path depends only on the endpoints of the path.

► From a mechanical point of view, the term "conservative" refers to the fact that the total energy (3.6) is an integral of motion.

Theorem 3.1

The total energy (3.6) of a conservative mechanical system is an integral of motion.

Proof. It is enough to compute the time derivative of E and use the equations of motion (3.5):

$$\begin{aligned}
\dot{E}(q, \dot{q}) &= \langle \dot{q}, A\,\ddot{q} \rangle + \left\langle \,\mathsf{grad}_q U(q), \dot{q} \right\rangle \\
&= \left\langle A\,\ddot{q} + \mathsf{grad}_q U(q), \dot{q} \right\rangle \\
&= \langle A\,\ddot{q} - f(q), \dot{q} \rangle = 0.
\end{aligned}$$

The claim is proved. ■

Example 3.3 (*The N-body gravitational problem*)

The N-body gravitational problem (see Example 3.2) described by Newton equations

$$m_j \ddot{q}^{(j)} = f_j(Q) := G \sum_{k \neq j} m_j m_k \frac{q^{(k)} - q^{(j)}}{\|q^{(k)} - q^{(j)}\|^3}, \tag{3.8}$$

with $j = 1, \ldots, N$, is a conservative system. Indeed,

$$f_j(Q) = -\operatorname{grad}_{q^{(j)}} U(Q),$$

with

$$U(Q) := -G \sum_{1 \leqslant k < j \leqslant N} m_j m_k \frac{1}{\|q^{(k)} - q^{(j)}\|},$$

which is the gravitational potential energy.

Example 3.4 (*Central force fields*)

The fact the the gravitational force is conservative reflects a more general fact: if $f \in C^1(\mathbb{R}^n, \mathbb{R})$ is a *central force field* (or *spherically symmetric*),

$$f(q) := g(\|q\|) \frac{q}{\|q\|}, \qquad g \in C^1(\mathbb{R}, \mathbb{R}),$$

then f defines a conservative vector field. Define $r := \|q\|$. Then

$$g(r) = -\frac{\mathrm{d}U}{\mathrm{d}r}, \qquad U(r) := -\int g(r)\, \mathrm{d}r.$$

Recall that the potential energy is defined up to an additive constant.

► Remarks:

- A definition more general than 3.2 is the following. A mechanical system, i.e., a system of Newton equations, is *conservative* if it admits a total energy of the form

$$E(q, \dot{q}) := T(q, \dot{q}) + U(q) := \frac{1}{2} \langle \dot{q}, A(q)\, \dot{q} \rangle + U(q), \tag{3.9}$$

which is an integral of motion. Here $A \in C^2(\mathbb{R}^n, \mathbf{GL}(n, \mathbb{R}))$ is symmetric and positive definite and $U \in C^2(\mathbb{R}^n, \mathbb{R})$.

(a) It can be verified that the system of n Newton equations which has the total energy (3.9) as integral of motion is given by

$$\sum_{i=1}^{n} A_{ik}(q)\, \ddot{q}_i = \sum_{i,j=1}^{n} \left(\frac{1}{2} \frac{\partial A_{ij}}{\partial q_k} - \frac{\partial A_{ik}}{\partial q_j} \right) \dot{q}_i \dot{q}_j - \frac{\partial U}{\partial q_k}, \tag{3.10}$$

with $k = 1, \ldots, n$, or, in a compact form,

$$A(q)\, \ddot{q} = f(q, \dot{q}),$$

where

$$f_k(q,\dot{q}) := \sum_{i,j=1}^{n} \left(\frac{1}{2} \frac{\partial A_{ij}}{\partial q_k} - \frac{\partial A_{ik}}{\partial q_j} \right) \dot{q}_i \, \dot{q}_j - \frac{\partial U}{\partial q_k}, \qquad k = 1, \ldots, n. \tag{3.11}$$

(b) The corresponding autonomous system of $2\,n$ ODEs is

$$\begin{cases} \dot{q} = p, \\ \dot{p} = A^{-1}(q) f(q,\dot{q}). \end{cases} \tag{3.12}$$

- A *total energy* can be defined also in the case of non-conservative systems (3.1) as a function
$$E(q,\dot{q},t) := T(q,\dot{q},t) + U(q,t),$$
where T is the *kinetic energy* and U is the *potential energy*. In general, for such systems, E is not an integral of motion. Usually, the forms of T and U are prescribed by the physical problem under investigation.

3.3 Lagrangian mechanics

▶ From "Derivation of the laws of motion and equilibrium from a metaphysical principle" (1746) by Pierre Louis Moreau de Maupertuis: *"After so many great men have worked on this subject, I almost do not dare to say that I have discovered the universal principle upon which all these laws are based, a principle that covers both elastic and inelastic collisions and describes the motion and equilibrium of all material bodies. This is the* principle of least action, *a principle so wise and so worthy of the supreme Being, and intrinsic to all natural phenomena; one observes it at work not only in every change, but also in every constancy that Nature exhibits. In the collision of bodies, motion is distributed such that the quantity of action is as small as possible, given that the collision occurs. At equilibrium, the bodies are arranged such that, if they were to undergo a small movement, the quantity of action would be smallest. The laws of motion and equilibrium derived from this principle are exactly those observed in Nature. We may admire the applications of this principle in all phenomena: the movement of animals, the growth of plants, the revolutions of the planets, all are consequences of this principle. The spectacle of the universe seems all the more grand and beautiful and worthy of its Author, when one considers that it is all derived from a small number of laws laid down most wisely. Only thus can we gain a fitting idea of the power and wisdom of the supreme Being, not from some small part of creation for which we know neither the construction, usage nor its relationship to other parts. What satisfaction for the human spirit in contemplating these laws of motion and equilibrium for all bodies in the universe, and in finding within them proof of the existence of Him who governs the universe!"*

▶ Our empirical observation of physical laws should convince us about Nature's propensity to optimize. Minimization principles form one of the most powerful tools to formulate mathematical models governing the equilibrium configurations of physical systems.

- We will present the basic mathematical analysis of nonlinear minimization principles on infinite-dimensional function spaces, a subject known as *calculus of variations*.
- Classical solutions to minimization problems in the calculus of variations are prescribed by boundary value problems involving certain types of differential equations, known as *Euler-Lagrange equations*.

3.3.1 *Euler-Lagrange equations*

► We want to provide a variational formulation of Newton equations governing the motion of a mechanical system. We start with the following definition.

Definition 3.3

Consider a mechanical system (3.1). Let $T \in C^2(\mathbb{R}^{2n} \times \mathbb{R}, \mathbb{R})$ be its kinetic energy and $U \in C^2(\mathbb{R}^n \times \mathbb{R}, \mathbb{R})$ its potential energy. The **Lagrangian** *of (3.1) is the function $\mathscr{L} \in C^2(\mathbb{R}^{2n} \times \mathbb{R}, \mathbb{R})$ defined by*

$$\mathscr{L}(q, \dot{q}, t) := T(q, \dot{q}, t) - U(q, t).$$

► Remarks:

- More generally, one can say that a *Lagrangian system* is a mechanical system defined in terms of its Lagrangian $\mathscr{L} \in C^2(\mathbb{R}^{2n} \times \mathbb{R}, \mathbb{R})$. Theorems 3.2 and 3.3 will clarify us how the system is effectively defined in terms of $\mathscr{L}$.
- As follows from the proof of Theorem 3.2 the regularity conditions on $\mathscr{L}$ can be relaxed by requiring: $U \in C^1(\mathbb{R}^n \times \mathbb{R}, \mathbb{R})$ and T of class C^2 in $\dot{q}$ and of class C^1 in q and t. Nevertheless, the condition $U \in C^2(\mathbb{R}^n \times \mathbb{R}, \mathbb{R})$ is necessary to have uniqueness of solution curves (see Theorem 1.1). Moreover in many problems of physical interest the Lagrangian is a smooth function of its arguments.

Example 3.5 (***Planar pendulum***)

Consider the planar pendulum described in Examples 2.1 and 2.15.

- Newton equation of motion is the scalar second-order ODE

$$\ddot{q} = -\sin q, \qquad q \in [0, 2\pi).$$

- The kinetic and potential energies are given respectively by

$$T(\dot{q}) := \frac{\dot{q}^2}{2}, \qquad U(q) := -\cos q,$$

where U is defined up to an additive constant. The total energy is:

$$E(q,\dot{q}) := \frac{\dot{q}^2}{2} - \cos q,$$

which is an integral of motion (see Example 2.15).

- The Lagrangian of the planar pendulum is:

$$\mathscr{L}(q,\dot{q}) := \frac{\dot{q}^2}{2} + \cos q.$$

We will see in Chapter 4 that the planar pendulum is indeed an example of a mechanical system defined on a manifold. Such a manifold is the configuration space of the system, which, in the present case, is diffeomorphic to the circle S^1. Anyway, in this Chapter, we will neglect these geometrical aspects.

► To provide a variational formulation of a mechanical system (3.1) we need the following definitions.

- Let K be the affine space of all trajectory curves $t \mapsto q(t)$ of class C^1 joining two fixed points $q(t_1), q(t_2) \in \mathbb{R}^n$ in a fixed time $t_2 - t_1 > 0$. An element $\gamma \in K$ is

$$\gamma := \left\{(t,q) \,:\, q = q(t),\, q \in C^1(\mathbb{R},\mathbb{R}^n),\, t_1 \leqslant t \leqslant t_2\right\}.$$

- Let K_0 be the vector space of all curves $t \mapsto h(t)$ of class C^1, with $h(t_1) = h(t_2) = 0$. Curves in K_0 are *deformations* of curves in K. More precisely, given two curves in K, $t \mapsto q(t)$, say γ, and $t \mapsto \widetilde{q}(t)$, say $\widetilde{\gamma}$, there exists $h(t) := \widetilde{q}(t) - q(t)$ in K_0, with $h(t_1) = h(t_2) = 0$. We write $\widetilde{\gamma} \equiv \gamma + h$.

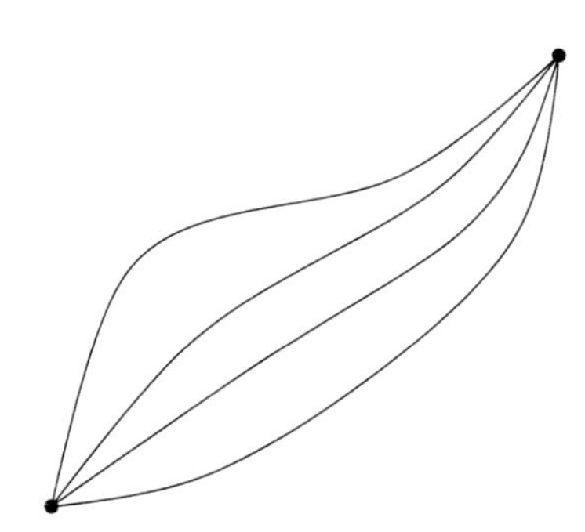

Fig. 3.1. Curves in K ([Ge]).

- A *functional* $\psi : K \to \mathbb{R}$ is *differentiable* if

$$\delta\psi := \psi(\gamma + h) - \psi(\gamma) = D(h) + R(\gamma, h),$$

where $D(h)$ depends linearly on $h \in K_0$ and $R(\gamma,h)$ is such that for $\|h\| < \varepsilon$ and $\|\dot{h}\| < \varepsilon$, one has $|R(\gamma,h)| < \alpha\,\varepsilon^2$ for some $\alpha > 0$. The quantity $D(h)$ is called *differential of* ψ (or *variation of* ψ), which is uniquely defined if ψ is differentiable.

- An *extremal* (or *stationary point*) of a differentiable functional $\psi : K \to \mathbb{R}$ is a curve $\gamma \in K$ such that $D(h) = 0$ for all $h \in K_0$. To find extremals of a differentiable functional $\psi : K \to \mathbb{R}$ means to solve a *variational problem.*

Definition 3.4

Consider a mechanical system (3.1) with Lagrangian $\mathscr{L} \in C^2\left(\mathbb{R}^{2n} \times \mathbb{R}, \mathbb{R}\right)$. The **action** *of (3.1) is the functional $\psi : K \to \mathbb{R}$ defined by*

$$\psi(\gamma) := \int_{t_1}^{t_2} \mathscr{L}(q(t), \dot{q}(t), t)\,\mathrm{d}t, \tag{3.13}$$

where $\gamma \in K$ is a trajectory curve $t \mapsto q(t)$ of class C^1 joining two fixed points $q(t_1), q(t_2) \in \mathbb{R}^n$ in a fixed time $t_2 - t_1 > 0$.

► To find extremals of the action (3.13) means to solve a first-order variational problem.

► We give the following Lemma.

Lemma 3.1

The action (3.13) is differentiable. Its variation is given by

$$D(h) = \int_{t_1}^{t_2} \left(\left\langle \operatorname{grad}_q \mathscr{L}(q, \dot{q}, t), h \right\rangle + \left\langle \operatorname{grad}_{\dot{q}} \mathscr{L}(q, \dot{q}, t), \dot{h} \right\rangle \right) \mathrm{d}t, \tag{3.14}$$

with $h \in K_0$.

Proof. For $\gamma \in K$ and $h \in K_0$ we have

$$\begin{aligned} \delta\psi &:= \psi(\gamma + h) - \psi(\gamma) \\ &= \int_{t_1}^{t_2} \left(\mathscr{L}(q(t) + h(t), \dot{q}(t) + \dot{h}(t), t) - \mathscr{L}(q(t), \dot{q}(t), t)\right) \mathrm{d}t \\ &= \int_{t_1}^{t_2} \sum_{k=1}^{n} \left(\frac{\partial \mathscr{L}}{\partial q_k} h_k + \frac{\partial \mathscr{L}}{\partial \dot{q}_k} \dot{h}_k \right) \mathrm{d}t + O(\|h\|^2) \\ &= D(h) + O(\|h\|^2), \end{aligned}$$

from which the claim follows. ■

► The next technical Lemma is useful as well.

Lemma 3.2

Let $f \in C([t_1, t_2], \mathbb{R})$. If for any function $g \in C^k([t_1, t_2], \mathbb{R})$, $k \geqslant 0$, with $g(t_1) =$

$g(t_2) = 0$, there holds

$$\int_{t_1}^{t_2} f(t)g(t)\,\mathrm{d}t = 0,$$

then $f(t) \equiv 0$ in $[t_1, t_2]$.

No Proof.

► We are now ready to prove the following fundamental statement.

Theorem 3.2

Consider a mechanical system (3.1) with Lagrangian $\mathscr{L} \in C^2(\mathbb{R}^{2n} \times \mathbb{R}, \mathbb{R})$ and action (3.13). Then a trajectory curve $\gamma \in K$, $t \mapsto q(t)$, $q \in C^2([t_1, t_2], \mathbb{R}^n)$, is an extremal of the action (3.13) if and only if the **Euler-Lagrange equations**

$$\frac{\mathrm{d}}{\mathrm{d}t}(\mathrm{grad}_{\dot{q}} \mathscr{L}(q, \dot{q}, t)) = \mathrm{grad}_q \mathscr{L}(q, \dot{q}, t), \tag{3.15}$$

are satisfied along $\gamma \in K$.

Proof. A trajectory γ defined by $t \mapsto q(t)$ is a stationary point of (3.13) if and only if (see Lemma 3.1)

$$D(h) = \int_{t_1}^{t_2} \sum_{k=1}^{n} \left(\frac{\partial \mathscr{L}}{\partial q_k} h_k + \frac{\partial \mathscr{L}}{\partial \dot{q}_k} \dot{h}_k \right) \mathrm{d}t = 0 \qquad \forall\, h \in K_0. \tag{3.16}$$

Note that $\partial \mathscr{L} / \partial \dot{q}_k$ is of class C^1 in $\dot{q}$, where $\dot{q}$ is of class C^1 in t because q is of class C^2. Thus $\partial \mathscr{L} / \partial \dot{q}_k$ is differentiable in t. Integrating by parts (3.16) and using the fact that $h(t_1) = h(t_2) = 0$ we have:

$$\begin{aligned} D(h) &= \int_{t_1}^{t_2} \sum_{k=1}^{n} \left(\frac{\partial \mathscr{L}}{\partial q_k} - \frac{\mathrm{d}}{\mathrm{d}t} \frac{\partial \mathscr{L}}{\partial \dot{q}_k} \right) h_k \,\mathrm{d}t + \sum_{k=1}^{n} \frac{\partial \mathscr{L}}{\partial \dot{q}_k} h_k \Bigg|_{t_1}^{t_2} \\ &= \int_{t_1}^{t_2} \sum_{k=1}^{n} \left(\frac{\partial \mathscr{L}}{\partial q_k} - \frac{\mathrm{d}}{\mathrm{d}t} \frac{\partial \mathscr{L}}{\partial \dot{q}_k} \right) h_k \,\mathrm{d}t = 0. \end{aligned}$$

If we fix $i \in \{1, \ldots, n\}$ and we choose $h \in K_0$ such that $h_j(t) = 0$ for all $j \neq i$ and for all $t \in [t_1, t_2]$, we can use Lemma 3.2 to conclude that

$$\frac{\mathrm{d}}{\mathrm{d}t} \frac{\partial \mathscr{L}}{\partial \dot{q}_i} = \frac{\partial \mathscr{L}}{\partial q_i} \qquad \forall\, t \in [t_1, t_2],$$

because $h_i(t_1) = h_i(t_2) = 0$. The same holds true for all $i = 1, \ldots, n$. ■

► Remarks:

- The trajectories $\gamma \in K$ obtained from Theorem 3.2 are solutions of a system of n second-order differential equations with prescribed boundary conditions (*boundary value problem*). This is *not* an IVP where some initial conditions for q and $\dot{q}$ at $t = 0$ are prescribed.

- The fact that an extremal $\gamma \in K$ satisfies the Euler-Lagrange equations (3.15) and the prescribed boundary conditions merely gives a candidate for minimizing the variational problem. But an extremal is also a candidate for maximizing the variational problem. The nature of extremals can only be distinguished by a second derivative test.

- Since the Euler-Lagrange equations determine the extremals of a variational problem, their solution set should remain unchanged by a change of variables. Indeed, it can be proved that the condition for a solution curve $\gamma \in K$ to be an extremal of ψ does not depend on the choice of the coordinate system.

- Occasionally, for a given variational problem, the Euler-Lagrange equations vanish identically, so that every function is a possible extremal of the problem. In this context, this happens if $\mathscr{L}$ can be expressed as the total time-derivative of another function. In such a case $\mathscr{L}$ is called *null Lagrangian*. From this observation there follows that two Lagrangians $\mathscr{L}, \widetilde{\mathscr{L}} \in C^2\left(\mathbb{R}^{2n} \times \mathbb{R}, \mathbb{R}\right)$ have the same Euler-Lagrange equations if and only if they differ by a total time-derivative. In other words, any Lagrangian is defined up to a total time-derivative.

► Let us consider a simple example of a null Lagrangian.

Example 3.6 (*A null Lagrangian*)

Define the Lagrangian

$$\mathscr{L}(q,\dot{q}) := q\,\dot{q}, \qquad q \in \mathbb{R},$$

and the corresponding action

$$\psi(\gamma) := \int_{t_1}^{t_2} \mathscr{L}(q,\dot{q})\mathrm{d}t.$$

- The Euler-Lagrange equation is

$$0 = \frac{\mathrm{d}}{\mathrm{d}t}\frac{\partial \mathscr{L}}{\partial \dot{q}} - \frac{\partial \mathscr{L}}{\partial q} = \dot{q} - \dot{q} \equiv 0.$$

- The above result is not surprising because

$$\mathscr{L}(q,\dot{q}) = \frac{\mathrm{d}}{\mathrm{d}t}\left(\frac{1}{2}q^2\right),$$

so that

$$\psi(\gamma) = \frac{1}{2}(t_2^2 - t_1^2).$$

Any function $t \mapsto q(t)$ will give the same value of ψ.

▶ The following Theorem is a formulation of the famous "principle of least action". The *Legendre condition* (3.17) on $\mathscr{L}$ will guarantee that if $\gamma \in K$ is an extremal of ψ then γ will be a local minimum.

Theorem 3.3 (*Principle of least action*)

Consider a mechanical system (3.1) with Lagrangian $\mathscr{L} \in C^2\left(\mathbb{R}^{2n} \times \mathbb{R}, \mathbb{R}\right)$ and action (3.13). If

$$\det\left(\frac{\partial^2 \mathscr{L}}{\partial \dot{q}_i \partial \dot{q}_j}\right)_{1 \leqslant i,j \leqslant n} > 0 \qquad \forall\,(q, \dot{q}, t) \in \mathbb{R}^{2n} \times \mathbb{R}, \tag{3.17}$$

(**Legendre condition**) *then there exists $t_2 > t_1$ such that if the trajectory curve $\gamma \in K$, defined by $t \mapsto q(t) \in C^2([t_1, t_2], \mathbb{R}^n)$, is an extremal of ψ for $t \in [t_1, t_2]$, then γ is a local minimum of ψ.*

No Proof.

▶ At this point we can give the following definition.

Definition 3.5

A mechanical system (3.1) is a **Lagrangian system** *if there exists a Lagrangian $\mathscr{L} \in C^2\left(\mathbb{R}^{2n} \times \mathbb{R}, \mathbb{R}\right)$, satisfying (3.17), such that Newton equations (3.1) are equivalent to Euler-Lagrange equations (3.15).*

Example 3.7 (*Planar pendulum*)

The planar pendulum, which is a conservative system, is a Lagrangian system. Its Lagrangian

$$\mathscr{L}(q, \dot{q}) := \frac{\dot{q}^2}{2} + \cos q,$$

gives the Euler-Lagrange equation:

$$0 = \frac{\mathrm{d}}{\mathrm{d}t}\frac{\partial \mathscr{L}}{\partial \dot{q}} - \frac{\partial \mathscr{L}}{\partial q} = \ddot{q} + \sin q,$$

which is exactly its Newton equation.

3.3.2 *Lagrangians for conservative systems*

▶ Example 3.7 is just a simple instance of a more general fact valid for conservative systems.

Theorem 3.4

Consider the conservative mechanical system (3.10), (or, equivalently, the dynam-

ical system (3.12)), whose total energy is

$$E(q,\dot{q}) := T(q,\dot{q}) + U(q) := \frac{1}{2}\langle \dot{q}, A(q)\,\dot{q}\rangle + U(q). \tag{3.18}$$

Here $A \in C^2(\mathbb{R}^n, \mathbf{GL}(n,\mathbb{R}))$ is symmetric and positive definite and $U \in C^2(\mathbb{R}^n, \mathbb{R})$ is the potential energy.

1. *The Lagrangian of the system is:*

$$\mathscr{L}(q,\dot{q}) := \frac{1}{2}\langle \dot{q}, A(q)\,\dot{q}\rangle - U(q). \tag{3.19}$$

2. *The Lagrangian (3.19) satisfies the Legendre condition. Therefore a solution curve of (3.10) is a local minimum of the action ψ.*
3. *The total energy (3.18) is given in terms of $\mathscr{L}$ by*

$$E(q,\dot{q}) := \left\langle \dot{q}, \operatorname{grad}_{\dot{q}}\mathscr{L}(q,\dot{q}) \right\rangle - \mathscr{L}(q,\dot{q}). \tag{3.20}$$

Proof. We prove all claims.

1. We have:

$$\mathscr{L}(q,\dot{q}) := \frac{1}{2}\sum_{i,j=1}^{n} A_{ij}(q)\,\dot{q}_i\,\dot{q}_j - U(q).$$

The Euler-Lagrange equations are in components ($k = 1,\dots,n$):

$$\begin{aligned} 0 &= \frac{\mathrm{d}}{\mathrm{d}t}\frac{\partial \mathscr{L}}{\partial \dot{q}_k} - \frac{\partial \mathscr{L}}{\partial q_k} \\ &= \frac{\mathrm{d}}{\mathrm{d}t}\sum_{i=1}^{n} A_{ik}(q)\,\dot{q}_i - \frac{1}{2}\sum_{i,j=1}^{n}\frac{\partial A_{ij}}{\partial q_k}\,\dot{q}_i\,\dot{q}_j + \frac{\partial U}{\partial q_k} \\ &= \sum_{i,j=1}^{n}\frac{\partial A_{ik}}{\partial q_j}\,\dot{q}_i\,\dot{q}_j + \sum_{i=1}^{n} A_{ik}(q)\,\ddot{q}_i - \frac{1}{2}\sum_{i,j=1}^{n}\frac{\partial A_{ij}}{\partial q_k}\,\dot{q}_i\,\dot{q}_j + \frac{\partial U}{\partial q_k}, \end{aligned}$$

which coincides with (3.10).

2. The first claim follows from the fact that A is positive definite. The second from Theorem 3.3.
3. Substituting the Lagrangian (3.19) in (3.20) we get:

$$\begin{aligned} E(q,\dot{q}) &= \langle \dot{q}, A(q)\,\dot{q}\rangle - \frac{1}{2}\langle \dot{q}, A(q)\,\dot{q}\rangle + U(q) \\ &= \frac{1}{2}\langle \dot{q}, A(q)\,\dot{q}\rangle + U(q). \end{aligned}$$

which is the function (3.18).

The Theorem is proved. ■

► The following Theorem describes the stability properties of conservative mechanical systems.

Theorem 3.5 (*Dirichlet*)

Consider the conservative mechanical system (3.10), (or, equivalently, the dynamical system (3.12)). Let $\Phi_t : \mathbb{R} \times \mathbb{R}^{2n} \to \mathbb{R}^{2n}$ be the flow of (3.12).

1. *The point $(\widetilde{q}, \widetilde{\dot{q}}) \in \mathbb{R}^{2n}$ is a fixed point of Φ_t if and only if $\widetilde{q}$ is a critical point of U and $\widetilde{\dot{q}} = 0$.*
2. *If $\widetilde{q} \in \mathbb{R}^n$ is a local minimum of U then $(\widetilde{q}, 0) \in \mathbb{R}^{2n}$ is a stable fixed point of the dynamical system defined by (3.10).*

Proof. We prove both claims.

1. From (3.12) it is clear that a fixed point of Φ_t corresponds to $\widetilde{\dot{q}} = 0$. If $\widetilde{\dot{q}} = 0$ we see from (3.11) that $f(q, 0) = 0$ if and only if $\widetilde{q}$ is a critical point of U.
2. If $\widetilde{q} \in \mathbb{R}^n$ is a local minimum of U we can define the Lyapunov function
$$F(q, \dot{q}) := E(q, \dot{q}) - U(\widetilde{q}),$$
where E is the total energy (3.18). From Theorem 2.8 the claim follows.

The Theorem is proved. ■

► We now consider the *linearization* of the Euler-Lagrange equations of a conservative system around the fixed points $(\widetilde{q}, 0) \in \mathbb{R}^{2n}$, where $\widetilde{q}$ is a critical point of U. It is not restrictive to consider $\widetilde{q} = 0$.

Theorem 3.6

Consider the conservative mechanical system (3.10), (or, equivalently, the dynamical system (3.12)). Let $(0, 0) \in \mathbb{R}^{2n}$ be a fixed point of Φ_t.

1. *The linearized Euler-Lagrange equations around $(0, 0)$ are given by*
$$A(0)\, \ddot{q} = -B\, q, \tag{3.21}$$
where
$$B := \left(\left. \frac{\partial^2 U}{\partial q_i \partial q_j} \right|_{q=0} \right)_{1 \leqslant i,j \leqslant n}.$$

2. *The linearized Lagrangian around* $(0,0)$*, corresponding to Newton equations (3.21), is*

$$\widetilde{\mathscr{L}}(q,\dot{q}) = \frac{1}{2}\langle \dot{q}, A(0)\,\dot{q}\rangle - \frac{1}{2}\langle q, B\,q\rangle\,. \tag{3.22}$$

Proof. We prove both claims.

1. It is evident that the linearization of Newton equations (3.10) around $(0,0)$ produces Newton equations (3.21).

2. This follows easily from Euler-Lagrange equation corresponding to the linearized Lagrangian (3.22).

The Theorem is proved. ■

Example 3.8 (*Planar pendulum and harmonic oscillator*)

We already know from Example 2.15 that the linearization of the planar pendulum gives the harmonic oscillator system. Let us verify this from a Lagrangian point of view.

- The planar pendulum admits the Lagrangian

$$\mathscr{L}(q,\dot{q}) := \frac{\dot{q}^2}{2} + \cos q - 1,$$

where the potential energy $U(q) := 1 - \cos q$ has been shifted in such a way that $(0,0)$ is a stable fixed point (i.e., $\widetilde{q} = 0$ is a local minimum of U). This Lagrangian gives the Euler-Lagrange equation:

$$0 = \frac{\mathrm{d}}{\mathrm{d}t}\frac{\partial \mathscr{L}}{\partial \dot{q}} - \frac{\partial \mathscr{L}}{\partial q} = \ddot{q} + \sin q.$$

- From Theorem 3.6 we conclude that the linearized Lagrangian of the planar pendulum around the (stable) fixed point $(0,0)$ is:

$$\widetilde{\mathscr{L}}(q,\dot{q}) = \frac{1}{2}(\dot{q}^2 - q^2).$$

This Lagrangian gives the Euler-Lagrange equation:

$$0 = \frac{\mathrm{d}}{\mathrm{d}t}\frac{\partial \widetilde{\mathscr{L}}}{\partial \dot{q}} - \frac{\partial \widetilde{\mathscr{L}}}{\partial q} = \ddot{q} + q,$$

which is the Newton equation of the harmonic oscillator.

3.3.3 *Symmetries of Lagrangians and Noether Theorem*

► We start with the following claim. It illustrates how the existence of some evident symmetries of $\mathscr{L}$ leads to the existence of some integrals of motion. As a byproduct one gets a reduction of the dimension of the configuration space, i.e., a reduction of the number of *degrees of freedom* of the problem.

Theorem 3.7

Consider a mechanical system with Lagrangian $\mathscr{L} \in C^2\left(\mathbb{R}^{2n} \times \mathbb{R}, \mathbb{R}\right)$.

1. *If $\mathscr{L}$ does not depend explicitly on time, then the function*

$$E(q,\dot{q}) := \left\langle \dot{q}, \operatorname{grad}_{\dot{q}} \mathscr{L}(q,\dot{q}) \right\rangle - \mathscr{L}(q,\dot{q})$$

is an integral of motion.

2. *If $\mathscr{L}$ does not depend explicitly on the coordinate q_i, then the function*

$$p_i := \frac{\partial \mathscr{L}}{\partial \dot{q}_i} \tag{3.23}$$

is an integral of motion. If $\partial^2 \mathscr{L} / \partial \dot{q}_i^2 \neq 0$, then the dynamics of the remaining $n-1$ coordinates is determined by the Euler-Lagrange equations associated with the reduced Lagrangian

$$\begin{aligned}\widetilde{\mathscr{L}}&(q_1,\ldots,\not{q_i},\ldots,q_n,\dot{q}_1,\ldots,p_i,\ldots,\dot{q}_n,t)\\ &= \left(\mathscr{L}(q,\dot{q},t) - p_i\,\dot{q}_i\right)\Big|_{\dot{q}_i = F(q_1,\ldots,\not{q_i},\ldots,q_n,\dot{q}_1,\ldots,p_i,\ldots,\dot{q}_n,t)},\end{aligned}$$

where the function F is obtained by inverting (3.23).

Proof. We prove both claims, but only the first statement of the second claim.

1. It is enough to compute the time derivative of E and use the Euler-Lagrange equations (3.15):

$$\begin{aligned}\dot{E}(q,\dot{q}) &= \left\langle \ddot{q}, \operatorname{grad}_{\dot{q}} \mathscr{L}(q,\dot{q}) \right\rangle + \left\langle \dot{q}, \frac{\mathrm{d}}{\mathrm{d}t}\left(\operatorname{grad}_{\dot{q}} \mathscr{L}(q,\dot{q})\right) \right\rangle - \dot{\mathscr{L}}(q,\dot{q})\\ &= \left\langle \ddot{q}, \operatorname{grad}_{\dot{q}} \mathscr{L}(q,\dot{q}) \right\rangle + \left\langle \dot{q}, \operatorname{grad}_{q} \mathscr{L}(q,\dot{q}) \right\rangle - \dot{\mathscr{L}}(q,\dot{q}),\end{aligned}$$

where

$$\dot{\mathscr{L}}(q,\dot{q}) = \left\langle \operatorname{grad}_q \mathscr{L}(q,\dot{q}), \dot{q} \right\rangle + \left\langle \operatorname{grad}_{\dot{q}} \mathscr{L}(q,\dot{q}), \ddot{q} \right\rangle.$$

Therefore $\dot{E}(q,\dot{q}) = 0$.

2. It is enough to compute the time derivative of p_i and use the Euler-Lagrange equations (3.15):

$$\dot{p}_i = \frac{\mathrm{d}}{\mathrm{d}t}\frac{\partial \mathscr{L}}{\partial \dot{q}_i} = \frac{\partial \mathscr{L}}{\partial q_i} = 0.$$

The Theorem is proved ■

Example 3.9 (*A planar Lagrangian system*)

Consider a point of mass $m > 0$ moving on $\mathbb{R}^2$, with coordinates (q_1, q_2), under the influence of a (smooth) potential energy $U = U(q_1, q_2)$. Its Lagrangian is:

$$\mathscr{L}(q_1, q_2, \dot{q}_1, \dot{q}_2) := \frac{m}{2}\left(\dot{q}_1^2 + \dot{q}_2^2\right) - U(q_1, q_2).$$

- The Euler-Lagrange equations are:

$$m\,\ddot{q}_1 = -\frac{\partial U}{\partial q_1}, \qquad m\,\ddot{q}_2 = -\frac{\partial U}{\partial q_2},$$

  which are Newton equations for a *two-dimensional* mechanical system.
- Since $\mathscr{L}$ does not depend on t we have that the total energy

$$E(q_1, q_2, \dot{q}_1, \dot{q}_2) = \frac{m}{2}\left(\dot{q}_1^2 + \dot{q}_2^2\right) + U(q_1, q_2)$$

  is an integral of motion.
- Assume that U is the potential energy of a central force field. In polar coordinates $(q_1, q_2) := r(\cos\theta, \sin\theta)$, $(r, \theta) \in (0, \infty) \times [0, 2\pi)$, we have $U = U(r)$. Moreover,

$$\dot{q}_1 = \dot{r}\cos\theta - r\sin\theta\,\dot{\theta}, \qquad \dot{q}_2 = \dot{r}\sin\theta + r\cos\theta\,\dot{\theta},$$

  so that

$$\dot{q}_1^2 + \dot{q}_2^2 = \dot{r}^2 + r^2\,\dot{\theta}^2.$$

  The Lagrangian takes the form

$$\mathscr{L}(r, \dot{r}, \dot{\theta}) = \frac{1}{2}m\left(\dot{r}^2 + r^2\,\dot{\theta}^2\right) - U(r).$$

- The Lagrangian $\mathscr{L}(r, \dot{r}, \dot{\theta})$ does not depend explicitly on θ. By Theorem 3.7 the quantity

$$p_\theta := \frac{\partial \mathscr{L}}{\partial \dot{\theta}} = m\,r^2\,\dot{\theta}$$

  is an integral of motion. It is called *angular momentum*.
- The reduced Lagrangian is

$$\begin{aligned}
\widetilde{\mathscr{L}}(r, \dot{r}) &= \mathscr{L}(r, \dot{r}, \dot{\theta}) - p_\theta\dot{\theta} \\
&= \frac{1}{2}m\left(\dot{r}^2 + \frac{p_\theta^2}{m^2\,r^2}\right) - U(r) - \frac{p_\theta^2}{m\,r^2} \\
&= \frac{1}{2}m\,\dot{r}^2 - U_{\text{eff}}(r),
\end{aligned}$$

  where

$$U_{\text{eff}}(r) := U(r) + \frac{p_\theta^2}{2\,m\,r^2}$$

  is the *effective potential energy*.
- The reduced Euler-Lagrange equation

$$m\,\ddot{r} = -\frac{\mathrm{d}U_{\text{eff}}}{\mathrm{d}r}$$

  describes a *one-dimensional* mechanical problem.

► Theorem 3.7 is a simple instance of a general and beautiful mechanism, discovered by E. Noether, which shows that symmetries (more formally, *symmetry Lie groups*) and integrals of motion are two faces of the same medal.

- For example, invariance of a variational principle under a group of time translations implies the conservation of the total energy for the solutions of the associated Euler-Lagrange equations, invariance under a group of spatial translations implies the conservation of linear momentum, invariance under a group of spatial rotations implies the conservation of angular momentum.

- This basic principle constitutes the first fundamental result in the study of classical and quantum-mechanical systems with prescribed groups of symmetries. Noether's method is the principal systematic procedure for constructing integrals of motion for complicated systems of ODEs.

▶ Let us give a formal definition of a (one-parameter) *symmetry Lie group* of an autonomous mechanical system with Lagrangian $\mathscr{L} \in C^2\left(\mathbb{R}^{2n}, \mathbb{R}\right)$.

- Let

$$\Psi_s : \mathbb{R} \times \mathbb{R}^n \to \mathbb{R}^n \; : \; (s, q) \mapsto \widetilde{q} := \Psi_s(q) \tag{3.24}$$

  be a (global) one-parameter Lie group of diffeomorphisms on $\mathbb{R}^n$. Let

$$\mathbf{v} := \sum_{i=1}^{n} f_i(q) \frac{\partial}{\partial q_i}, \qquad f(q) := \left.\frac{\mathrm{d}}{\mathrm{d}s}\right|_{s=0} \Psi_s(q),$$

  be the infinitesimal generator of Ψ_s. We know from Theorem 2.2 that the *infinitesimal transformation*

$$\widetilde{q} := \Psi_s(q) = q + s\, f(q) + O(s^2), \tag{3.25}$$

  contains the essential information to characterize Ψ_s.

- From (3.25) we see that the induced infinitesimal transformation for the Lagrangian velocities is

$$\dot{\widetilde{q}} := \frac{\mathrm{d}}{\mathrm{d}t} \Psi_s(q) = \dot{q} + s\, \frac{\partial f}{\partial q} \dot{q} + O(s^2).$$

Definition 3.6

The one-parameter Lie group of diffeomorphisms (3.24) is a **symmetry Lie group** *of $\mathscr{L}$ if*

$$\mathscr{L}(q, \dot{q}) = \mathscr{L}\left(\widetilde{q}, \dot{\widetilde{q}}\right) \qquad \forall\, q \in \mathbb{R}^n,\; s \in \mathbb{R}. \tag{3.26}$$

▶ Note that the invariance property (3.26) implies that if $t \mapsto q(t)$ is a solution of Euler-Lagrange equations (3.15) then $\widetilde{q} := \Psi_s(q)$ is also a solution of (3.15) for all $s \in \mathbb{R}$.

▶ We now give a formulation of "Noether Theorem".

Theorem 3.8 (*Noether*)

Consider an autonomous mechanical system with Lagrangian $\mathscr{L} \in C^2(\mathbb{R}^{2n}, \mathbb{R})$. Assume that (3.24) is a symmetry group of $\mathscr{L}$. Then the quantity

$$F(q, \dot{q}) := \langle\, p, f(q)\,\rangle, \tag{3.27}$$

where

$$p := \operatorname{grad}_{\dot{q}} \mathscr{L}(q, \dot{q}), \tag{3.28}$$

is an integral of motion. The quantity p is called **momentum conjugated to** q.

Proof. We proceed by steps.

- The invariance condition (3.26) implies

$$\mathscr{L}\left(q + s\, f(q), \dot{q} + s\frac{\partial f}{\partial q}\dot{q}\right) = \mathscr{L}(q, \dot{q}) + O(s^2),$$

- The l.h.s. can be expanded in Taylor series around $s = 0$ as

$$\mathscr{L}\left(q + s\, f(q)\dot{q} + s\frac{\partial f}{\partial q}\dot{q}\right) = \mathscr{L}(q, \dot{q})$$
$$+ s\left(\left\langle\, \operatorname{grad}_q \mathscr{L}(q, \dot{q}), f(q)\,\right\rangle + \left\langle\, \operatorname{grad}_{\dot{q}} \mathscr{L}(q, \dot{q}), \frac{\partial f}{\partial q}\dot{q}\,\right\rangle\right) + O(s^2).$$

- Therefore there must hold

$$\left\langle\, \operatorname{grad}_q \mathscr{L}(q, \dot{q}), f(q)\,\right\rangle + \left\langle\, \operatorname{grad}_{\dot{q}} \mathscr{L}(q, \dot{q}), \frac{\partial f}{\partial q}\dot{q}\,\right\rangle = 0. \tag{3.29}$$

- Introduce the function F by (3.27) and the variable p by (3.28) . Note that

$$\left\langle\, p, \frac{\partial f}{\partial q}\dot{q}\,\right\rangle = \dot{F}(q, \dot{q}) - \langle\, \dot{p}, f(q)\,\rangle.$$

- Therefore (3.29) takes the form

$$\left\langle\, \operatorname{grad}_q \mathscr{L}(q, \dot{q}), f(q)\,\right\rangle - \langle\, \dot{p}, f(q)\,\rangle = -\dot{F}(q, \dot{q}),$$

that is

$$\left\langle \operatorname{grad}_q \mathscr{L}(q,\dot q), f(q) \right\rangle - \left\langle \frac{\mathrm{d}}{\mathrm{d}t} \operatorname{grad}_{\dot q} \mathscr{L}(q,\dot q), f(q) \right\rangle = -\dot F(q,\dot q),$$

namely

$$\left\langle \frac{\mathrm{d}}{\mathrm{d}t} \operatorname{grad}_{\dot q} \mathscr{L}(q,\dot q) - \operatorname{grad}_q \mathscr{L}(q,\dot q), f(q) \right\rangle = \dot F(q,\dot q),$$

which vanishes thanks to Euler-Lagrange equations (3.15).

The Theorem is proved. ■

Example 3.10 (*A planar Lagrangian system*)

Consider the planar system of Example 3.9. Its Lagrangian is:

$$\mathscr{L}(q_1,q_2,\dot q_1,\dot q_2) := \frac{m}{2}\left(\dot q_1^2 + \dot q_2^2\right) - U(q_1,q_2).$$

- The Lie group of spatial translations acting on $\mathbb{R}^2$ is given by

$$\Psi_s : \mathbb{R}\times\mathbb{R}^2 \to \mathbb{R}^2 \; : \; (s,(q_1,q_2)) \mapsto (\tilde q_1, \tilde q_2) := (q_1 + s\,a_1, q_2 + s\,a_2),$$

where $(a_1,a_2)\in\mathbb{R}^2$.

- The infinitesimal generator is

$$\mathbf{v} := a_1 \frac{\partial}{\partial q_1} + a_2 \frac{\partial}{\partial q_2},$$

so that $f := (a_1,a_2)$.

- It is evident that $\mathscr{L}$ admits the invariance condition (3.26) under the action of Ψ_s if and only if the potential energy U is invariant under the action of Ψ_s, namely

$$U(q_1,q_2) := U(a_2\,q_1 - a_1\,q_2).$$

Indeed

$$\begin{aligned} (\mathfrak{L}_{\mathbf{v}} U)(q_1,q_2) &= a_1 \frac{\partial U}{\partial q_1} + a_2 \frac{\partial U}{\partial q_2} \\ &= a_1\,a_2\,U'(a_2\,q_1 - a_1\,q_2) - a_2\,a_1\,U'(a_2\,q_1 - a_1\,q_2) \\ &= 0. \end{aligned}$$

If this is the case, then Theorem 3.8 implies that the function

$$F(q_1,q_2) := m\,(a_1\,\dot q_1 + a_2\,\dot q_2)$$

is an integral of motion.

Example 3.11 (*A rotational-invariant Lagrangian*)

Consider a Lagrangian $\mathscr{L}\in C^2(\mathbb{R}^6,\mathbb{R})$ which is invariant under the action of the one-parameter Lie group of rotations around the axis q_1:

$$\Psi_s(q_1,q_2,q_3) := (q_1, q_2\cos s + q_3 \sin s, -q_2\sin s + q_3\cos s), \qquad s\in[0,2\pi).$$

- The infinitesimal generator is

$$\mathbf{v} := q_3 \frac{\partial}{\partial q_2} - q_2 \frac{\partial}{\partial q_3},$$

so that $f(q_2, q_3) := (0, q_3, -q_2)$.

- Theorem 3.8 implies that the function

$$F(q_2, q_3, \dot{q}_2, \dot{q}_3) := \langle\, p, f(q_2, q_3)\,\rangle = p_2\, q_3 - p_3\, q_2,$$

with

$$p_2 := \frac{\partial \mathscr{L}}{\partial \dot{q}_2}, \qquad p_3 := \frac{\partial \mathscr{L}}{\partial \dot{q}_3},$$

is an integral of motion.

3.4 Canonical Hamiltonian mechanics

► Let us recall the following facts from Analysis.

- Let $F \in C^2(\mathbb{R}^n, \mathbb{R})$ be a convex function, i.e.,

$$\det\left(\frac{\partial^2 F}{\partial q_i \partial q_j}\right)_{1\leqslant i,j\leqslant n} > 0 \qquad \forall\, q \in \mathbb{R}^n. \tag{3.30}$$

The function

$$F^*(p) := \sup_{q\in\mathbb{R}^n} (\langle\, q, p\,\rangle - F(q)), \tag{3.31}$$

is called *Legendre transformation* of the function F.

- If in (3.31) the supremum is also a maximum then such maximum corresponds to a point $q = q(p)$ such that $\operatorname{grad}_q F(q) = p$. Hence we have:

$$F^*(p) := \langle\, p, q(p)\,\rangle - F(q(p)),$$

with

$$p := \operatorname{grad}_q F(q).$$

Note that condition (3.30) assures that it is possible to invert $\operatorname{grad}_q F(q) = p$ to get $q = q(p)$.

- The main properties of Legendre transformations are:

  1. F^* is a convex function.
  2. The Legendre transformation is involutive, i.e., $(F^*(p))^* = F(q)$.

3.4.1 *Hamilton equations and Hamiltonian flows*

► We can now define the Hamiltonian of a mechanical system.

Definition 3.7

Consider a Lagrangian system with Lagrangian $\mathscr{L} \in C^2(\mathbb{R}^{2n} \times \mathbb{R}, \mathbb{R})$, where $\mathscr{L}$ is a convex function in the Lagrangian velocities $\dot{q}$, i.e., the Legendre condition (3.17) is satisfied.

1. *The* **canonical Hamiltonian** *of the mechanical system is a function $\mathscr{H} \in C^2(\mathbb{R}^{2n} \times \mathbb{R}, \mathbb{R})$ defined as the Legendre transformation (w.r.t. $\dot{q}$) of the Lagrangian:*
$$\mathscr{H}(q,p,t) := \langle p, \dot{q} \rangle - \mathscr{L}(q,\dot{q},t). \tag{3.32}$$
where
$$p := \mathsf{grad}_{\dot{q}} \mathscr{L}(q,\dot{q},t) \tag{3.33}$$
is called **momentum conjugated to** q.

2. *The space $\mathbb{R}^{2n}$ parametrized by the* **canonical coordinates** (q,p) *is called* **canonical Hamiltonian phase space.**

► Remarks:

- The Legendre condition (3.17) implies that (3.33) is invertible, so that we can express $\dot{q}$ as a function of p and the Hamiltonian (3.32) is indeed a function of (q,p,t).
- While the Lagrangian is defined up to a total time-derivative, the Hamiltonian is defined up to an additive constant.
- The Legendre transformation is involutive, so that the Lagrangian is the Legendre transformation of the Hamiltonian:
$$\mathscr{L}(q,\dot{q},t) = \langle p, \dot{q} \rangle - \mathscr{H}(q,p,t), \tag{3.34}$$
where
$$\dot{q} = \mathsf{grad}_p \mathscr{H}(q,p,t). \tag{3.35}$$

► It is natural to expect that the Euler-Lagrange equations (3.15) will induce on the canonical Hamiltonian phase space $\mathbb{R}^{2n}$ a time evolution of the canonical coordinates (q,p). Such a time evolution, which will be again equivalent to Newton equation, is described by an IVP on $\mathbb{R}^{2n}$ as described by the next Theorem.

Theorem 3.9

Let $\mathscr{H} \in C^2(\mathbb{R}^{2n} \times \mathbb{R}, \mathbb{R})$ be the Hamiltonian of a mechanical system as described in Definition 3.7. The canonical coordinates $(q,p) \in \mathbb{R}^{2n}$ obey the follow-

ing system of $2\,n$ *ODEs:*

$$\begin{cases} \dot{q} = \operatorname{grad}_p \mathscr{H}(q,p,t), \\ \dot{p} = -\operatorname{grad}_q \mathscr{H}(q,p,t), \end{cases} \tag{3.36}$$

called **canonical Hamilton equations**. *We here assume that some initial conditions* $(q(t_0), p(t_0)) \in \mathbb{R}^{2n}$, $t_0 \in \mathbb{R}$, *are prescribed.*

Proof. As expected the result is just a consequence of the Euler-Lagrange equations (3.15). From (3.35) we already know the system of n ODEs for the time evolution of the variables q. Then, from (3.33), (3.15) and (3.34) we have:

$$\begin{aligned} \dot{p}_k &= \frac{\mathrm{d}}{\mathrm{d}t}\frac{\partial \mathscr{L}}{\partial \dot{q}_k} = \frac{\partial \mathscr{L}}{\partial q_k} = \frac{\partial}{\partial q_k}\left(\langle p, \dot{q}\rangle - \mathscr{H}(q,p,t)\right) \\ &= \left\langle \frac{\partial p}{\partial q_k}, \dot{q} \right\rangle - \frac{\partial \mathscr{H}}{\partial q_k} - \left\langle \operatorname{grad}_p \mathscr{H}(q,p,t), \frac{\partial p}{\partial q_k} \right\rangle \\ &= \left\langle \frac{\partial p}{\partial q_k}, \dot{q} \right\rangle - \frac{\partial \mathscr{H}}{\partial q_k} - \left\langle \dot{q}, \frac{\partial p}{\partial q_k} \right\rangle = -\frac{\partial \mathscr{H}}{\partial q_k}, \end{aligned}$$

with $k = 1, \ldots, n$. ■

Definition 3.8

1. *Let* $\mathscr{H} \in C^2\left(\mathbb{R}^{2n} \times \mathbb{R}, \mathbb{R}\right)$ *be the Hamiltonian of a mechanical system as described in Definition 3.7 so that the time evolution of the canonical coordinates* $(q,p) \in \mathbb{R}^{2n}$ *is described by the IVP (3.36). We say that such a mechanical system is a* **canonical Hamiltonian system**.
2. *Consider a canonical Hamiltonian system with time-independent Hamiltonian* $\mathscr{H} \in C^2\left(\mathbb{R}^{2n}, \mathbb{R}\right)$. *The flow* $\Phi_t : \mathbb{R} \times \mathbb{R}^{2n} \to \mathbb{R}^{2n}$ *of the IVP*

$$\begin{cases} \dot{q} = \operatorname{grad}_p \mathscr{H}(q,p), \\ \dot{p} = -\operatorname{grad}_q \mathscr{H}(q,p), \\ (q(0), p(0)) \in \mathbb{R}^{2n}, \end{cases} \tag{3.37}$$

is called **canonical Hamiltonian flow**. *Its infinitesimal generator*

$$\mathbf{v}_{\mathscr{H}} := \sum_{k=1}^{n} \left(\frac{\partial \mathscr{H}}{\partial p_k}\frac{\partial}{\partial q_k} - \frac{\partial \mathscr{H}}{\partial q_k}\frac{\partial}{\partial p_k} \right), \tag{3.38}$$

defines a vector field on the canonical Hamiltonian phase space $\mathbb{R}^{2n}$ *called* **canonical Hamiltonian vector field**:

$$f(q,p) := \left(\operatorname{grad}_p \mathscr{H}(q,p), -\operatorname{grad}_q \mathscr{H}(q,p)\right) = \left.\frac{\mathrm{d}}{\mathrm{d}t}\right|_{t=0} \Phi_t(q,p).$$

► We now present two immediate fundamental consequences of the Hamiltonian formulation of an autonomous dynamical system.

Theorem 3.10

Let Φ_t be a canonical Hamiltonian flow defining a canonical Hamiltonian vector field f.

1. *f is divergence-free, i.e.,*

$$\mathrm{div}_{(q,p)} f(q,p) = 0 \qquad \forall\, (q,p) \in \mathbb{R}^{2n}.$$

   As a consequence, if D is a compact region of the canonical phase space $\mathbb{R}^{2n}$, then Φ_t preserves the volume of D.

2. *The Hamiltonian $\mathscr{H}$ is an integral of motion.*

Proof. We prove both claims.

1. We have:

$$\begin{aligned}\mathrm{div}_{(q,p)} f(q,p) &= \sum_{k=1}^{n} \left(\frac{\partial}{\partial q_k} \frac{\partial \mathscr{H}}{\partial p_k} + \frac{\partial}{\partial p_k} \left(-\frac{\partial \mathscr{H}}{\partial q_k} \right) \right) \\ &= \sum_{k=1}^{n} \left(\frac{\partial^2 \mathscr{H}}{\partial q_k \partial p_k} - \frac{\partial^2 \mathscr{H}}{\partial p_k \partial q_k} \right) = 0,\end{aligned}$$

   for all $(q,p) \in \mathbb{R}^{2n}$. The second claim follows from Theorem 2.6 .

2. We have:

$$\left(\mathfrak{L}_{\mathbf{v}_{\mathscr{H}}} \mathscr{H}\right)(q,p) = \sum_{k=1}^{n} \left(\frac{\partial \mathscr{H}}{\partial p_k} \frac{\partial \mathscr{H}}{\partial q_k} - \frac{\partial \mathscr{H}}{\partial q_k} \frac{\partial \mathscr{H}}{\partial p_k} \right) = 0,$$

   for all $(q,p) \in \mathbb{R}^{2n}$.

■

► Remarks:

- As a consequence of Theorems 3.10 and 2.7 we conclude that Hamiltonian flows defined in compact regions of the canonical Hamiltonian phase space cannot have asymptotically stable limit cycles and asymptotically stable fixed points.
- Canonical Hamilton equations are less generic than volume-preserving systems of ODEs. Indeed, all projections $(\partial\mathscr{H}/\partial p_k, -\partial\mathscr{H}/\partial q_k)$ of f on each of the planes (q_k, p_k) have zero divergence in the geometry of the respective planes.

► We here present the Hamiltonian formulation of some mechanical systems previously considered in the Lagrangian framework.

Example 3.12 (*Canonical Hamiltonian systems*)

1. Consider the conservative mechanical system (3.10), whose Lagrangian is

$$\mathscr{L}(q,\dot{q}) := \frac{1}{2}\langle \dot{q}, A(q)\,\dot{q}\rangle - U(q).$$

where $A \in C^2(\mathbb{R}^n, \mathbf{GL}(n,\mathbb{R}))$ is symmetric and positive definite and $U \in C^2(\mathbb{R}^n,\mathbb{R})$ is the potential energy. The conjugated momenta are

$$p := \operatorname{grad}_{\dot{q}}\mathscr{L}(q,\dot{q}) = A(q)\,\dot{q},$$

so that

$$\dot{q} = A^{-1}(q)\,p.$$

From (3.32) we get

$$\begin{aligned}\mathscr{H}(q,p) &:= \left\langle p, A^{-1}(q)\,p\right\rangle - \frac{1}{2}\left\langle A^{-1}(q)\,p, A(q)\,A^{-1}(q)\,p\right\rangle + U(q)\\ &= \frac{1}{2}\left\langle p, A^{-1}(q)p\right\rangle + U(q).\end{aligned}$$

Assume for simplicity that A does not depend on q. Then,

$$\mathscr{H}(q,p) = \frac{1}{2}\left\langle p, A^{-1}p\right\rangle + U(q),$$

and the canonical Hamiltonian equations are

$$\begin{cases} \dot{q} = \operatorname{grad}_p\mathscr{H}(q,p) = A^{-1}\,p,\\ \dot{p} = -\operatorname{grad}_q\mathscr{H}(q,p) = -\operatorname{grad}_q U(q),\end{cases}$$

which coincide with (3.5) as expected.

2. As a special case consider the planar pendulum, whose Lagrangian is

$$\mathscr{L}(q,\dot{q}) := \frac{\dot{q}^2}{2} + \cos q.$$

The corresponding Hamiltonian is

$$\mathscr{H}(q,p) := \frac{p^2}{2} - \cos q,$$

which gives the canonical Hamilton equations

$$\begin{cases} \dot{q} = \dfrac{\partial\mathscr{H}}{\partial p} = p,\\[2ex] \dot{p} = -\dfrac{\partial\mathscr{H}}{\partial q} = -\sin q.\end{cases}$$

3.4.2 *Symplectic structure of the canonical Hamiltonian phase space*

► The Hamiltonian phase space has a rich geometric structure. In This Subsection we start investigating some of the most important geometric aspects of the Hamiltonian formulation of mechanics. Such aspects will become even more fundamental in the study of Hamiltonian mechanics on differentiable manifolds (see Chapter 4).

▶ Let us recall the following facts (see Chapter 4, Subsection 4.2.7, for further details on Lie algebras and Lie groups):

- In Chapter 2 we introduced two fundamental spaces: the *general linear Lie algebra on* $\mathbb{R}^n$, $\mathfrak{gl}(n,\mathbb{R})$, and the *general linear Lie group on* $\mathbb{R}^n$, $\mathbf{GL}(n,\mathbb{R})$. In the language of Differential Geometry one says that the linear vector space $\mathfrak{gl}(n,\mathbb{R})$ is the (matrix) Lie algebra of the (matrix) Lie group $\mathbf{GL}(n,\mathbb{R})$, which is a differentiable manifold equipped with a group structure. The link between $\mathfrak{gl}(n,\mathbb{R})$ and $\mathbf{GL}(n,\mathbb{R})$ is given by the *exponential map*. For the moment we just need a very elementary understanding of these two spaces (a more geometric interpretation will be given in Chapter 4).

- $\mathfrak{gl}(n,\mathbb{R})$ is the n^2-dimensional vector space of all linear maps from $\mathbb{R}^n$ to $\mathbb{R}^n$. In other words, $A \in \mathfrak{gl}(n,\mathbb{R})$ can be represented as a $n \times n$ matrix, not necessarily invertible, with real entries. The space $\mathfrak{gl}(n,\mathbb{R})$ is equipped with the commutator bracket

$$\begin{aligned} [\,\cdot,\cdot\,] : \mathfrak{gl}(n,\mathbb{R}) \times \mathfrak{gl}(n,\mathbb{R}) &\to \mathfrak{gl}(n,\mathbb{R}) \\ (A_1, A_2) &\mapsto [\,A_1, A_2\,] := A_1\,A_2 - A_2\,A_1. \end{aligned}$$

  One can define subspaces of $\mathfrak{gl}(n,\mathbb{R})$ which are closed under the commutator bracket. Such subspaces are called *subalgebras of* $\mathfrak{gl}(n,\mathbb{R})$.

- $\mathbf{GL}(n,\mathbb{R})$ is a n^2-dimensional group of matrices defined by

$$\mathbf{GL}(n,\mathbb{R}) := \{A \in \mathfrak{gl}(n,\mathbb{R}) \;:\; \det A \neq 0\}.$$

  Matrices of $\mathbf{GL}(n,\mathbb{R})$ form a group under the (associative) matrix multiplication, the identity element being the $n \times n$ identity matrix, the inverse element of $A \in \mathbf{GL}(n,\mathbb{R})$ being the inverse matrix A^{-1}. If $A \in \mathfrak{gl}(n,\mathbb{R})$ then $e^{t\,A} \in \mathbf{GL}(n,\mathbb{R})$ for all $t \in \mathbb{R}$. As for subalgebras one can define *subgroups of* $\mathbf{GL}(n,\mathbb{R})$.

▶ The matrix Lie group which arises naturally in the study of Hamiltonian mechanics is the *symplectic Lie group on* $\mathbb{R}^{2n}$, $\mathbf{SP}(n,\mathbb{R})$, which is a (closed) subgroup of $\mathbf{GL}(2\,n,\mathbb{R})$. Using again an elementary approach, thus avoiding the language of Differential Geometry, we present here the main definitions and characterizations of $\mathbf{SP}(n,\mathbb{R})$ and its Lie algebra.

- The *canonical symplectic matrix* is the defined by

$$\mathbb{J} := \begin{pmatrix} \mathbb{O}_n & \mathbb{1}_n \\ -\mathbb{1}_n & \mathbb{O}_n \end{pmatrix}, \tag{3.39}$$

where $\mathbb{1}_n$ is the $n \times n$ identity matrix and $\mathbb{O}_n$ is the $n \times n$ zero matrix. One has

$$\mathbb{J}^\top = \mathbb{J}^{-1} = -\mathbb{J}, \qquad \mathbb{J}^2 = -\mathbb{1}_{2n}.$$

- The *symplectic Lie group* $\mathbf{SP}(n, \mathbb{R})$ is defined as the following group of matrices:

$$\mathbf{SP}(n, \mathbb{R}) := \left\{ A \in \mathbf{GL}(2\,n, \mathbb{R}) \, : \, A^\top \mathbb{J}\, A = \mathbb{J} \right\}.$$

Equivalently, if we define the skew-symmetric bilinear non-degenerate form $\omega : \mathbb{R}^{2n} \times \mathbb{R}^{2n} \to \mathbb{R}$,

$$\langle\, x, y \,\rangle_{\mathbb{J}} := \langle\, x, \mathbb{J}\, y \,\rangle = \sum_{k=1}^{n} (x_k\, y_{n+k} - x_{n+k} y_k), \qquad x, y \in \mathbb{R}^{2n}, \tag{3.40}$$

which is called *canonical symplectic product*, we define $\mathbf{SP}(n, \mathbb{R})$ as the group of matrices which preserves ω:

$$\mathbf{SP}(n, \mathbb{R}) := \left\{ A \in \mathbf{GL}(2\,n, \mathbb{R}) \, : \, \langle\, A\, x, A\, y \,\rangle_{\mathbb{J}} = \langle\, x, y \,\rangle_{\mathbb{J}} \; \forall\, x, y \in \mathbb{R}^{2n} \right\}.$$

One has the following properties:

(a) $\dim(\mathbf{SP}(n, \mathbb{R})) = n(2\,n + 1)$.

(b) If $A \in \mathbf{SP}(n, \mathbb{R})$ then $\det A = 1$.

(c) Define a $2\,n \times 2\,n$ block matrix

$$A := \begin{pmatrix} a & b \\ c & d \end{pmatrix} \in \mathbf{GL}(2\,n, \mathbb{R}).$$

Then $A \in \mathbf{SP}(n, \mathbb{R})$ if and only if $a^\top d - c^\top b = \mathbb{1}_n$ and $a^\top c$, $b^\top d$ are symmetric.

(d) If $A \in \mathbf{GL}(2, \mathbb{R})$ then

$$A^\top \mathbb{J}\, A = \begin{pmatrix} 0 & \det A \\ -\det A & 0 \end{pmatrix},$$

so that $A \in \mathbf{SP}(1, \mathbb{R})$ if and only if $\det A = 1$.

- The vector space $\mathbb{R}^{2n}$ equipped with the canonical symplectic product (3.40) is called *canonical symplectic vector space*.

(a) The canonical symplectic product has an interesting geometric characterization. Given two vectors $x, y \in \mathbb{R}^{2n}$, the canonical symplectic product (3.40) corresponds to the sum of the (oriented) areas of the projection of the parallelogram with sides x, y on the n planes $(x_1, x_{n+1}), \ldots, (x_n, x_{2n})$.

(b) The non-degeneracy of the canonical symplectic product means by definition that $\langle x, y \rangle_{\mathbb{J}} = 0$ for all $y \in \mathbb{R}^{2n}$ necessarily implies $x = 0$.

(c) The notion of symplectic product can be extended to any vector space. But only vector spaces of even dimension admit a symplectic structure. Indeed, all bilinear skew-symmetric forms are necessarily degenerate in a space of odd dimension.

- The *symplectic Lie algebra* $\mathfrak{sp}(n, \mathbb{R})$ (which is the Lie algebra of $\mathbf{SP}(n, \mathbb{R})$) is defined as the following vector space:

$$\mathfrak{sp}(n, \mathbb{R}) := \left\{ A \in \mathfrak{gl}(2n, \mathbb{R}) \,:\, \mathbb{J}\, A^{\top} \mathbb{J} = A \right\}.$$

One has the following properties:

(a) $\dim(\mathfrak{sp}(n, \mathbb{R})) = n(2n+1)$.

(b) If $S \in \mathfrak{gl}(2n, \mathbb{R})$ is symmetric then $\mathbb{J}\, S \in \mathfrak{sp}(n, \mathbb{R})$. That is equivalent to say that a symmetric matrix can be written as $\mathbb{J}\, A$, $A \in \mathfrak{sp}(n, \mathbb{R})$.

(c) Define a $2n \times 2n$ block matrix

$$A := \begin{pmatrix} a & b \\ c & d \end{pmatrix} \in \mathfrak{gl}(2n, \mathbb{R}).$$

Then $A \in \mathfrak{sp}(n, \mathbb{R})$ if and only if $a^{\top} + d = \mathbb{O}_n$ and b, c are symmetric matrices.

(d) If $A \in \mathfrak{gl}(2, \mathbb{R})$ then $A \in \mathfrak{sp}(1, \mathbb{R})$ if and only if $\mathrm{Trace}\, A = 0$.

(e) If $A \in \mathfrak{sp}(n, \mathbb{R})$ then $\mathrm{e}^{t\,A} \in \mathbf{SP}(n, \mathbb{R})$ for all $t \in \mathbb{R}$.

► Consider a canonical Hamiltonian system with time-independent Hamiltonian $\mathscr{H} \in C^2\left(\mathbb{R}^{2n}, \mathbb{R}\right)$ (what follows holds true also in the case of Hamiltonians which depend on time, but we here consider the autonomous case only for a more transparent construction). Let $x := (q, p) \in \mathbb{R}^{2n}$ be the set of time-dependent coordinates on the canonical Hamiltonian phase space. Then, following Definition 3.8 we have:

- The flow

$$\Phi_t : \mathbb{R} \times \mathbb{R}^{2n} \to \mathbb{R}^{2n} \,:\, (t, x(0)) \mapsto x(t, x(0)),$$

of the IVP

$$\begin{cases} \dot{x} = \mathbb{J}\, \mathrm{grad}_x \mathscr{H}(x), \\ x(0) \in \mathbb{R}^{2n}, \end{cases} \tag{3.41}$$

is the canonical Hamiltonian flow. Here $\mathbb{J}$ is the canonical symplectic matrix (3.39) and

$$\mathrm{grad}_x \mathscr{H}(x) := \left(\mathrm{grad}_q \mathscr{H}(q, p), \mathrm{grad}_p \mathscr{H}(q, p) \right).$$

- The infinitesimal generator of Φ_t is

$$\mathbf{v}_{\mathscr{H}} := \sum_{i=1}^{2n} f_i(x) \frac{\partial}{\partial x_i}, \tag{3.42}$$

where $f_i(x)$ is the i-th component of the canonical Hamiltonian vector field:

$$f(x) := \mathbb{J}\, \mathsf{grad}_x \mathscr{H}(x) = \left.\frac{\mathsf{d}}{\mathsf{d}t}\right|_{t=0} \Phi_t(x).$$

From the above observations we argue that the canonical Hamiltonian phase space has a natural structure of a symplectic vector space.

► The above characterization of a canonical vector field is not merely a more compact way to write down formulas. Indeed, one has the following claim.

Theorem 3.11

Let $f \in C^1(\mathbb{R}^{2n} \times \mathbb{R}, \mathbb{R}^{2n})$ define a time-dependent vector field on $\mathbb{R}^{2n} \times \mathbb{R}$. Then f is a canonical Hamiltonian vector field if and only if

$$\frac{\partial f}{\partial x} \in \mathfrak{sp}(n, \mathbb{R}) \qquad \forall\, (x,t) \in \mathbb{R}^{2n} \times \mathbb{R}.$$

Proof. Suppose that f is a Hamiltonian vector field. Then we can write

$$f(x,t) = \mathbb{J}\ \mathsf{grad}_x \mathscr{H}(x,t).$$

Note that

$$\frac{\partial f_i}{\partial x_j} = \sum_{k=1}^{2n} \mathbb{J}_{ik} \frac{\partial^2 \mathscr{H}}{\partial x_k \partial x_j}.$$

We see that the Jacobian $\partial f / \partial x$ of f w.r.t. x can be written as $\mathbb{J}$ times the Hessian matrix of $\mathscr{H}$, which is evidently symmetric. Thus $\partial f / \partial x \in \mathfrak{sp}(n, \mathbb{R})$ for all $(x,t) \in \mathbb{R}^{2n} \times \mathbb{R}$.

Conversely, suppose that the matrix $\partial f / \partial x \in \mathfrak{sp}(n, \mathbb{R})$ for all for all $(x,t) \in \mathbb{R}^{2n} \times \mathbb{R}$. Define a new vector field $g(x,t) := \mathbb{J}\, f(x,t)$. Then

$$\frac{\partial g_i}{\partial x_j} = \frac{\partial g_j}{\partial x_i},$$

due to the fact that $\mathbb{J}\,(\partial f / \partial x)$ is a symmetric matrix. Therefore there exists a function $\mathscr{H} = \mathscr{H}(x,t)$ such that

$$g(x,t) = -\,\mathsf{grad}_x \mathscr{H}(x,t).$$

Hence we have:

$$f(x,t) = -\mathbb{J}^{-1}\, \mathsf{grad}_x \mathscr{H}(x,t) = \mathbb{J}\ \mathsf{grad}_x \mathscr{H}(x,t).$$

The claim is proved. ■

Example 3.13 (*A linear canonical Hamiltonian system*)

A linear canonical Hamiltonian system with constant coefficients is described by a system of $2n$ ODEs of the form (3.41) where

$$\mathscr{H}(x) := \frac{1}{2}\langle x, S\,x\rangle, \qquad x := (q,p) \in \mathbb{R}^{2n},$$

where $S \in \mathbf{GL}(2n, \mathbb{R})$ is symmetric.

- The canonical Hamilton equations read
$$\dot{x} = f(x) = \mathbb{J}\,S\,x.$$
Note that $\mathbb{J}\,S \in \mathfrak{sp}(n, \mathbb{R})$.
- The canonical Hamiltonian flow is
$$\Phi_t(x(0)) = e^{t\,\mathbb{J}\,S}x(0).$$

3.4.3 Canonical Poisson brackets

► We now introduce a new fundamental object which plays a crucial role in the Hamiltonian formulation of mechanics.

Definition 3.9

Let $F, G \in C^1(\mathbb{R}^{2n} \times \mathbb{R}, \mathbb{R})$ be two functions defined on the canonical Hamiltonian phase space with coordinates $x := (q,p)$.

1. *The* **canonical Poisson bracket** *of F and G is a function that we denote by $\{F(x,t), G(x,t)\} \in C^1(\mathbb{R}^{2n} \times \mathbb{R}, \mathbb{R})$ defined by the canonical symplectic product of the gradients w.r.t. x of F and G:*
$$\begin{aligned}\{F(x,t), G(x,t)\} &:= \langle\, \mathrm{grad}_x F(x,t), \mathrm{grad}_x G(x,t)\,\rangle_{\mathbb{J}} \\ &= \langle\, \mathrm{grad}_x F(x,t), \mathbb{J}\ \mathrm{grad}_x G(x,t)\,\rangle\,. \end{aligned} \qquad (3.43)$$
2. *If $\{F(x,t), G(x,t)\} = 0$ for all $(x,t) \in \mathbb{R}^{2n} \times \mathbb{R}$ then we say that F and G are in* **(Poisson) involution**.

► Before listing the (mechanical) consequences of the introduction of the canonical Poisson bracket on the canonical Hamiltonian phase space let us consider some characteristic features of this bracket:

- It is not difficult to verify directly that the Poisson bracket (3.43) satisfies the following properties (we omit the arguments of functions):
  1. *(Bi)linearity*: $\{\lambda_1 F + \lambda_2 G, K\} = \lambda_1\{F,K\} + \lambda_2\{G,K\}$,
  2. *Skew-symmetry*: $\{F,G\} = -\{G,F\}$,

3. *Leibniz rule*: $\{ F\,G, K \} = F\{ G, K \} + G\{ F, K \}$,
4. *Jacobi identity*: $\{ F, \{ G, K \} \} + \circlearrowleft (F, G, H) = 0$,

for all $F, G, K \in C^1\left(\mathbb{R}^{2n} \times \mathbb{R}, \mathbb{R}\right)$ and $\lambda_1, \lambda_2 \in \mathbb{R}$. One says that the canonical Poisson bracket on $\mathbb{R}^{2n} \times \mathbb{R}$ defines a Lie algebra structure on $C^1\left(\mathbb{R}^{2n} \times \mathbb{R}, \mathbb{R}\right)$.

- The non-degeneracy of the canonical symplectic product implies that the canonical Poisson bracket is non-degenerate, i.e., $\{ F(x,t), G(x,t) \} = 0$ for all $F \in C^1\left(\mathbb{R}^{2n} \times \mathbb{R}, \mathbb{R}\right)$ necessarily implies that G is a constant function.
- The canonical Poisson bracket depends on the coordinate system. Indeed, an important branch of Hamiltonian Theory is to construct those maps on $\mathbb{R}^{2n}$ which preserve the Poisson structure (see Subsection 3.4.4).

► We now present a list of immediate and fundamental consequences induced by the notion of canonical Poisson bracket.

- In canonical coordinates $(q, p) \in \mathbb{R}^{2n}$ we have:

$$\begin{aligned} \{ F(q,p,t), G(q,p,t) \} &:= \left\langle \operatorname{grad}_q F(q,p,t), \operatorname{grad}_p G(q,p,t) \right\rangle \\ &- \left\langle \operatorname{grad}_p F(q,p,t), \operatorname{grad}_q G(q,p,t) \right\rangle \\ &= \sum_{k=1}^{n} \left(\frac{\partial F}{\partial q_k}\frac{\partial G}{\partial p_k} - \frac{\partial F}{\partial p_k}\frac{\partial G}{\partial q_k} \right). \end{aligned} \tag{3.44}$$

- Equivalently, in coordinates x, we have:

$$\{ F(x,t), G(x,t) \} := \sum_{i,j=1}^{2n} \mathbb{J}_{ij} \frac{\partial F}{\partial x_i}\frac{\partial G}{\partial x_j}. \tag{3.45}$$

- Using (3.44) with $(F, G) = (q_i, \mathscr{H}(q,p,t))$ and $(F, G) = (p_i, \mathscr{H}(q,p,t))$ we can write the *canonical Hamilton equations* (3.36) in the following symmetric way:

$$\begin{cases} \dot{q}_k = \dfrac{\partial \mathscr{H}}{\partial p_k} = \{ q_k, \mathscr{H}(q,p,t) \}, \\ \dot{p}_k = -\dfrac{\partial \mathscr{H}}{\partial q_k} = \{ p_k, \mathscr{H}(q,p,t) \}, \end{cases}$$

with $k = 1, \ldots, n$.

- Using (3.45) with $(F, G) = (x_k, \mathscr{H}(x))$ we obtain an equivalent formulation of (3.41):

$$\dot{x}_k = \sum_{j=1}^{2n} \mathbb{J}_{kj} \frac{\partial \mathscr{H}}{\partial x_j} = \{ x_k, \mathscr{H}(x) \}, \tag{3.46}$$

with $k = 1, \ldots, n$.

- From (3.38) and (3.42) we see that the infinitesimal generator of the *canonical Hamiltonian flow* can be formally written as

$$\mathbf{v}_{\mathscr{H}} := \{ \cdot, \mathscr{H} \}.$$

- Using (3.44) with $(F,G) = (q_i, q_j)$, $(F,G) = (p_i, p_j)$ and $(F,G) = (q_i, p_j)$ we obtain

$$\{ p_i, p_j \} = \{ q_i, q_j \} = 0, \qquad \{ q_i, p_j \} = \delta_{ij}, \qquad i,j = 1, \dots, n.$$

  They define the so called *canonical Poisson structure* of the canonical Hamiltonian phase space.

- Equivalently, using (3.45) with $(F,G) = (x_i, x_j)$ we have:

$$\{ x_i, x_j \} = \mathbb{J}_{ij}, \qquad i,j = 1, \dots, n.$$

- Let $F \in C^1(\mathbb{R}^{2n} \times \mathbb{R}, \mathbb{R})$. Then the total-time derivative of F is

$$\begin{aligned} \dot{F}(x,t) &= \langle \operatorname{grad}_x F(x,t), \dot{x} \rangle + \frac{\partial F}{\partial t} \\ &= \langle \operatorname{grad}_x F(x,t), \mathbb{J} \operatorname{grad}_x \mathscr{H}(x,t) \rangle + \frac{\partial F}{\partial t} \\ &= \{ F(x,t), \mathscr{H}(x,t) \} + \frac{\partial F}{\partial t}. \end{aligned} \tag{3.47}$$

  Therefore, if F does not depend explicitly on time, F is an integral of motion if and only if F and $\mathscr{H}$ are in Poisson involution. An equivalent way to get the same conclusion is the following. Let $F \in C^1(\mathbb{R}^{2n}, \mathbb{R})$ and $\mathbf{v}_{\mathscr{H}}$ be the infinitesimal generator of a Hamiltonian flow. Then the Lie derivative of F along $\mathbf{v}_{\mathscr{H}}$ is:

$$(\mathfrak{L}_{\mathbf{v}_{\mathscr{H}}} F)(x) = \mathbf{v}_{\mathscr{H}}[F(x)] = \sum_{i,j=1}^{2n} \mathbb{J}_{ij} \frac{\partial \mathscr{H}}{\partial x_j} \frac{\partial F}{\partial x_i} = \{ F(x,t), \mathscr{H}(x,t) \}, \tag{3.48}$$

  so that $(\mathfrak{L}_{\mathbf{v}_{\mathscr{H}}} F)(x) = 0$ if and only if F and $\mathscr{H}$ are in Poisson involution.

- Let $F, G \in C^1(\mathbb{R}^{2n} \times \mathbb{R}, \mathbb{R})$. From (3.47) we have

$$\begin{aligned} \dot{F}(x,t) &= \{ F(x,t), \mathscr{H}(x,t) \} + \frac{\partial F}{\partial t}, \\ \dot{G}(x,t) &= \{ G(x,t), \mathscr{H}(x,t) \} + \frac{\partial G}{\partial t}, \end{aligned}$$

  so that

$$\{ \dot{F}(x,t), G(x,t) \} = \{ \{ F(x,t), \mathscr{H}(x,t) \}, G(x,t) \} + \left\{ \frac{\partial F}{\partial t}, G(x,t) \right\}$$

and

$$\{ F(x,t), \dot{G}(x,t) \} = \{ F(x,t), \{ G(x,t), \mathscr{H}(x,t) \} \} + \left\{ F(x,t), \frac{\partial G}{\partial t} \right\}.$$

On the other hand

$$\begin{aligned}\frac{\mathrm{d}}{\mathrm{d}t}\{ F(x,t), G(x,t) \} &= \{ \{ F(x,t), G(x,t) \}, \mathscr{H}(x,t) \} \\ &+ \left\{ \frac{\partial F}{\partial t}, G(x,t) \right\} + \left\{ F(x,t), \frac{\partial G}{\partial t} \right\}.\end{aligned}$$

Now, the Jacobi identity implies that

$$\frac{\mathrm{d}}{\mathrm{d}t}\{ F(x,t), G(x,t) \} = \{ \dot{F}(x,t), G(x,t) \} + \{ F(x,t), \dot{G}(x,t) \}. \tag{3.49}$$

- Let $F, G \in C^1(\mathbb{R}^{2n} \times \mathbb{R}, \mathbb{R})$ be two integrals of motion of a canonical Hamiltonian flow. Then from (3.49) we immediately conclude that also $\{ F, G \}$ is an integral of motion. This result is known as "Poisson Theorem".

► The next claim will be used to prove an important result about commutation of canonical Hamiltonian flows.

Theorem 3.12

Let $F, G \in C^1(\mathbb{R}^{2n}, \mathbb{R})$ and $\mathbf{v}_F := \{ \cdot, F \}$, $\mathbf{v}_G := \{ \cdot, G \}$, be the infinitesimal generators of two canonical Hamiltonian flows, or, equivalently, two canonical Hamiltonian vector fields. Then the Lie bracket between $\mathbf{v}_F$ and $\mathbf{v}_G$ is given by

$$[\mathbf{v}_F, \mathbf{v}_G] = -\mathbf{v}_{\{F,G\}}.$$

Proof. It is a consequence of the Jacobi identity. Let $K \in C^1(\mathbb{R}^{2n}, \mathbb{R})$. Then we have:

$$\begin{aligned}[\mathbf{v}_F, \mathbf{v}_G][K(x)] &= \mathbf{v}_F[\mathbf{v}_G[K(x)]] - \mathbf{v}_G[\mathbf{v}_F[K(x)]] \\ &= \mathbf{v}_F[\{ K(x), G(x) \}] - \mathbf{v}_G[\{ K(x), F(x) \}] \\ &= \{\{ K(x), G(x) \}, F(x) \} - \{\{ K(x), F(x) \}, G(x) \} \\ &= -\{K(x), \{ F(x), G(x) \}\} \\ &= -\mathbf{v}_{\{ F,G \}}[K(x)].\end{aligned}$$

The claim is proved. ■

► Combining Theorems 3.12 and 2.5 we obtain the following result.

Theorem 3.13

Let $\Phi_t, \Psi_s : \mathbb{R} \times \mathbb{R}^{2n} \to \mathbb{R}^{2n}$ be two canonical Hamiltonian flows whose infinitesimal generators are the vector fields $\mathbf{v}_F := \{\cdot, F\}$, $\mathbf{v}_G := \{\cdot, G\}$, $F, G \in C^1(\mathbb{R}^{2n}, \mathbb{R})$. Then

$$(\Phi_t \circ \Psi_s)(x) = (\Psi_s \circ \Phi_t)(x) \qquad \forall\, t, s \in \mathbb{R},\ x \in \mathbb{R}^{2n}, \tag{3.50}$$

if and only if

$$\{F(x), G(x)\} = \text{const.} \qquad \forall\, x \in \mathbb{R}^{2n}.$$

Proof. From Theorem 2.5 we know that (3.50) holds true if and only if $[\mathbf{v}_F, \mathbf{v}_G] = 0$ for all $x \in \mathbb{R}^{2n}$. Then Theorem 3.12 finishes the proof. ■

► In the language of symmetries one says that a flow $\Psi_s : \mathbb{R} \times \mathbb{R}^{2n} \to \mathbb{R}^{2n}$, with infinitesimal generator $\mathbf{v}_F$, $F \in C^1(\mathbb{R}^{2n}, \mathbb{R})$, determines a (one-parameter) *symmetry Lie group* of the Hamiltonian system whose flow is $\Phi_t : \mathbb{R} \times \mathbb{R}^{2n} \to \mathbb{R}^{2n}$, with infinitesimal generator $\mathbf{v}_{\mathscr{H}}$, $\mathscr{H} \in C^1(\mathbb{R}^{2n}, \mathbb{R})$, if $[\mathbf{v}_F, \mathbf{v}_{\mathscr{H}}] = 0$ for all $x \in \mathbb{R}^{2n}$. This is indeed the Hamiltonian analog of what we stated in Definition 3.6.

► Theorem 3.13 implies the next statement, which is the Hamiltonian formulation of "Noether Theorem" 3.8.

Theorem 3.14

Let $F \in C^1(\mathbb{R}^{2n}, \mathbb{R})$ be an integral of motion of a canonical Hamiltonian flow $\Phi_t : \mathbb{R} \times \mathbb{R}^{2n} \to \mathbb{R}^{2n}$ whose infinitesimal generator is the vector field $\mathbf{v}_{\mathscr{H}} := \{\cdot, \mathscr{H}\}$, $\mathscr{H} \in C^1(\mathbb{R}^{2n}, \mathbb{R})$. Then the flow generated by F is a one-parameter symmetry group of diffeomorphisms of the canonical Hamiltonian system.

Proof. If F is an integral of motion then its flow, say Ψ_t, is generated by a Hamiltonian vector field $\mathbf{v}_F := \{\cdot, F\}$. Furthermore we have that F and $\mathscr{H}$ are in Poisson involution, so that, by Theorem 3.13 we have $[\mathbf{v}_F, \mathbf{v}_{\mathscr{H}}] = 0$ for all $x \in \mathbb{R}^{2n}$. ■

Example 3.14 (*A planar canonical Hamiltonian system*)

Consider the planar system of Examples 3.9 and 3.10. Its Hamiltonian is:

$$\mathscr{H}(q_1, q_2, p_1, p_2) := \frac{m}{2}\left(p_1^2 + p_2^2\right) + U(q_1, q_2).$$

- We know that if

$$U(q_1, q_2) := U(a_2 q_1 - a_1 q_2), \tag{3.51}$$

then the system is invariant under the action of the Lie group of spatial translations $\Psi_s : \mathbb{R} \times \mathbb{R}^4 \to \mathbb{R}^4$:

$$(s, (q_1, q_2, p_1, p_2)) \mapsto (\tilde{q}_1, \tilde{q}_2, \tilde{p}_1, \tilde{p}_2) := (q_1 + s\, a_1, q_2 + s\, a_2, p_1, p_2),$$

where $(a_1, a_2) \in \mathbb{R}^2$. The infinitesimal generator of Ψ_s is

$$\mathbf{v}_F := a_1 \frac{\partial}{\partial q_1} + a_2 \frac{\partial}{\partial q_2} + 0 \frac{\partial}{\partial p_1} + 0 \frac{\partial}{\partial p_2},$$

and the corresponding integral of motion is

$$F(p_1, p_2) := m\,(a_1\,p_1 + a_2\,p_2).$$

- Let us verify that F and $\mathscr{H}$ are in Poisson involution:

$$\begin{aligned}\{F, \mathscr{H}\} &= m\,a_1\{p_1, U\} + m\,a_2\{p_2, U\} \\ &= -m\,a_1\,a_2\,U'(a_2\,q_1 - a_1\,q_2) + m\,a_2\,a_1\,U'(a_2\,q_1 - a_1\,q_2) = 0.\end{aligned}$$

- Under assumption (3.51) the canonical Hamiltonian equations are:

$$\begin{cases} \dot{q}_1 = \dfrac{\partial \mathscr{H}}{\partial p_1} = \{q_1, \mathscr{H}\} = p_1, \\ \dot{q}_2 = \dfrac{\partial \mathscr{H}}{\partial p_2} = \{q_2, \mathscr{H}\} = p_2, \\ \dot{p}_1 = -\dfrac{\partial \mathscr{H}}{\partial q_1} = \{p_1, \mathscr{H}\} = -a_2\,U'(a_2\,q_1 - a_1\,q_2), \\ \dot{p}_2 = -\dfrac{\partial \mathscr{H}}{\partial q_2} = \{p_2, \mathscr{H}\} = a_1\,U'(a_2\,q_1 - a_1\,q_2). \end{cases}$$

Therefore the infinitesimal generator of the canonical Hamiltonian flow is

$$\mathbf{v}_{\mathscr{H}} := p_1\frac{\partial}{\partial q_1} + p_2\frac{\partial}{\partial q_2} - a_2\,U'(a_2\,q_1 - a_1\,q_2)\frac{\partial}{\partial p_1} + a_1\,U'(a_2\,q_1 - a_1\,q_2)\frac{\partial}{\partial p_2}.$$

- A tedious but straightforward computation shows that the Lie bracket of $\mathbf{v}_F$ and $\mathbf{v}_{\mathscr{H}}$ vanishes for all $(q_1, q_2, p_1, p_2) \in \mathbb{R}^4$.

3.4.4 *Canonical and symplectic transformations*

► We established a well-defined way to write down canonical Hamiltonian equations for a given mechanical system. We ended up with an IVP in $\mathbb{R}^{2n}$ which admits several characterizing features. Nevertheless, no direct general integration method of the canonical Hamilton equations is known. There exist indirect methods which allow to highly simplify the integration problem.

- One of them is the method of *coordinate transformations*, whose goal is to find new coordinates in $\mathbb{R}^{2n}$ for which the resulting canonical Hamilton equations are somehow "simpler" to integrate.

- For a generic coordinate transformation, the canonical Hamilton equations are not form-invariant, i.e., the form of the ODEs in the IVP is not preserved by the coordinate transformation.

- Therefore, the interesting preliminary problem is to characterize the set of invertible differentiable transformations on $\mathbb{R}^{2n}$ (i.e., diffeomorphisms on $\mathbb{R}^{2n}$) which preserve the canonical form of Hamilton equations.

- It turns out that the family of diffeomorphisms which preserves the form of canonical Hamilton equations plays a crucial role in Hamiltonian Theory. Such transformations, called *canonical transformations*, admit several equivalent formulations, such as:

1. Preservation of the form of canonical Hamilton equations.
2. Preservation of the canonical Poisson structure.
3. Preservation of the canonical symplectic 2-form.
4. Existence of generating functions.

► Let $\mathscr{H} \in C^2\left(\mathbb{R}^{2n} \times \mathbb{R}, \mathbb{R}\right)$ be the Hamiltonian of a mechanical system in the canonical Hamiltonian phase space $\mathbb{R}^{2n}$. Its canonical Hamilton equations are

$$\dot{x} = \mathbb{J}\,\mathsf{grad}_x \mathscr{H}(x,t), \qquad x := (q,p) \in \mathbb{R}^{2n}. \tag{3.52}$$

- Let
$$\Psi_t : \mathbb{R} \times \mathbb{R}^{2n} \to \mathbb{R}^{2n} \; : \; (t,x) \mapsto \widetilde{x} := \Psi_t(x)$$
be a one-parameter family of C^2-diffeomorphisms.
- In general, the action of the (time-dependent) change of coordinates Ψ_t on (3.52) gives a new system on $\mathbb{R}^{2n}$ which is not Hamiltonian. First of all note that the Hamiltonian $\mathscr{H} = \mathscr{H}(x,t)$ is transformed into a function which depends on $(\widetilde{x},t)$:
$$(\mathscr{H} \circ \Psi_t)(x,t) = \mathscr{H}(\widetilde{x},t) = \mathscr{H}(x(\widetilde{x},t),t),$$
so that
$$\mathsf{grad}_x \mathscr{H}(x,t) = \left(\frac{\partial \widetilde{x}}{\partial x}\right)^{\top} \mathsf{grad}_{\widetilde{x}} \mathscr{H}(\widetilde{x},t).$$
Therefore the new system of ODEs in the variables $(\widetilde{x},t)$ reads
$$\begin{aligned} \dot{\widetilde{x}} &= \frac{\partial \widetilde{x}}{\partial x}\,\dot{x} + \frac{\partial \widetilde{x}}{\partial t} \\ &= \frac{\partial \widetilde{x}}{\partial x}\,\mathbb{J}\,\mathsf{grad}_x \mathscr{H}(x,t) + \frac{\partial \widetilde{x}}{\partial t} \\ &= \frac{\partial \widetilde{x}}{\partial x}\,\mathbb{J}\left(\frac{\partial \widetilde{x}}{\partial x}\right)^{\top} \mathsf{grad}_{\widetilde{x}} \mathscr{H}(\widetilde{x},t) + \frac{\partial \widetilde{x}}{\partial t}. \end{aligned} \tag{3.53}$$
- It is evident that the r.h.s. of (3.53) does not define, in general, a canonical Hamiltonian vector field. Indeed, this happens if and only is one has
$$\dot{\widetilde{x}} = \mathbb{J}\,\mathsf{grad}_{\widetilde{x}} \widetilde{\mathscr{H}}(\widetilde{x},t),$$
for some new Hamiltonian $\widetilde{\mathscr{H}} \in C^2\left(\mathbb{R}^{2n} \times \mathbb{R}, \mathbb{R}\right)$.

- The theory of canonical transformations deals with the problem of characterizing those coordinate transformations which preserve the form of (canonical) Hamilton equations.

Example 3.15 (*Twisting canonical coordinates*)

The time-independent transformation exchanging (up to sign) the canonical coordinates $q \in \mathbb{R}^n$ with the corresponding canonical momenta $p \in \mathbb{R}^n$ preserves the form of the canonical Hamilton equations.

- Define the time-independent invertible map

$$\Psi : \mathbb{R}^{2n} \longrightarrow \mathbb{R}^{2n} : (q,p) \mapsto (\tilde{q},\tilde{p}) := (-p,q),$$

whose Jacobian is

$$\frac{\partial(\tilde{q},\tilde{p})}{\partial(q,p)} = \begin{pmatrix} \mathbb{O}_n & -\mathbb{1}_n \\ \mathbb{1}_n & \mathbb{O}_n \end{pmatrix} = -\mathbb{J} = \mathbb{J}^{-1} = \mathbb{J}^\top.$$

- If $\mathscr{H} = \mathscr{H}(q,p,t)$ is the original Hamiltonian, then

$$\mathsf{grad}_{(q,p)} \mathscr{H}(q,p,t) = \mathbb{J}\, \mathsf{grad}_{(\tilde{q},\tilde{p})} \mathscr{H}(\tilde{q},\tilde{p},t).$$

- The new system of ODEs in the variables $(\tilde{q},\tilde{p},t)$ reads

$$(\dot{\tilde{q}},\dot{\tilde{p}}) = \mathbb{J}^{-1}\mathbb{J}\,\mathbb{J}\; \mathsf{grad}_{(\tilde{q},\tilde{p})} \mathscr{H}(\tilde{q},\tilde{p},t) = \mathbb{J}\; \mathsf{grad}_{(\tilde{q},\tilde{p})} \mathscr{H}(\tilde{q},\tilde{p},t),$$

which is, in form, a canonical system of Hamilton equations.

- This transformation shows how, within the Hamiltonian formalism, there is no essential difference between the role of the canonical coordinates q and of the conjugate canonical momenta p.

Example 3.16 (*A time-dependent change of coordinates*)

In the canonical phase space $\mathbb{R}^{2n}$ define the invertible time-dependent change of coordinates

$$\Psi_t : \mathbb{R} \times \mathbb{R}^{2n} \longrightarrow \mathbb{R}^{2n} : (t,(q,p)) \mapsto (\tilde{q},\tilde{p}) := \left(a(t)q, \frac{p}{a(t)} \right),$$

where $a : \mathbb{R} \longrightarrow \mathbb{R}$ is a smooth function. Its Jacobian is

$$\frac{\partial(\tilde{q},\tilde{p})}{\partial(q,p)} = \frac{1}{a(t)} \begin{pmatrix} a^2(t)\mathbb{1}_n & \mathbb{O}_n \\ \mathbb{O}_n & \mathbb{1}_n \end{pmatrix}.$$

- If $\mathscr{H} = \mathscr{H}(q,p,t)$ is the original Hamiltonian, then

$$\mathsf{grad}_{(q,p)} \mathscr{H}(q,p,t) = \frac{1}{a(t)} \begin{pmatrix} a^2(t)\,\mathbb{1}_n & \mathbb{O}_n \\ \mathbb{O}_n & \mathbb{1}_n \end{pmatrix} \mathsf{grad}_{(\tilde{q},\tilde{p})} \mathscr{H}(\tilde{q},\tilde{p},t).$$

- Therefore the new system of ODEs in the variables $(\tilde{q},\tilde{p},t)$ reads

$$\begin{aligned} (\dot{\tilde{q}},\dot{\tilde{p}}) &= \frac{1}{a^2(t)} \begin{pmatrix} a^2(t)\,\mathbb{1}_n & \mathbb{O}_n \\ \mathbb{O}_n & \mathbb{1}_n \end{pmatrix} \mathbb{J} \begin{pmatrix} a^2(t)\,\mathbb{1}_n & \mathbb{O}_n \\ \mathbb{O}_n & \mathbb{1}_n \end{pmatrix} \mathsf{grad}_{(\tilde{q},\tilde{p})} \mathscr{H}(\tilde{q},\tilde{p},t) + \frac{\partial(\tilde{q},\tilde{p})}{\partial t} \\ &= \mathbb{J}\; \mathsf{grad}_{(\tilde{q},\tilde{p})} \mathscr{H}(\tilde{q},\tilde{p},t) + \left(\dot{a}(t)\,q, -\frac{\dot{a}(t)}{a^2(t)}\,p \right) \\ &= \mathbb{J}\; \mathsf{grad}_{(\tilde{q},\tilde{p})} \mathscr{H}(\tilde{q},\tilde{p},t) + \frac{\dot{a}(t)}{a(t)}\,(\tilde{q},-\tilde{p}) \\ &= \mathbb{J}\left(\mathsf{grad}_{(\tilde{q},\tilde{p})} \mathscr{H}(\tilde{q},\tilde{p},t) + \frac{\dot{a}(t)}{a(t)}\,(\tilde{q},\tilde{p}) \right). \end{aligned} \tag{3.54}$$

- System (3.54) is a canonical Hamiltonian system in the variables $(\widetilde{q}, \widetilde{p})$:
$$(\dot{\widetilde{q}}, \dot{\widetilde{p}}) = \mathbb{J}\, \mathsf{grad}_{(\widetilde{q},\widetilde{p})} \widetilde{\mathscr{H}}(\widetilde{q}, \widetilde{p}, t),$$
where the new Hamiltonian is:
$$\begin{aligned} \widetilde{\mathscr{H}}(\widetilde{q}, \widetilde{p}, t) &= \mathscr{H}(q(\widetilde{q}, \widetilde{p}, t), p(\widetilde{q}, \widetilde{p}, t), t) + \frac{\dot{a}(t)}{a(t)} \langle \widetilde{q}, \widetilde{p} \rangle \\ &= \mathscr{H}\left(\frac{\widetilde{q}}{a(t)}, a(t)\, \widetilde{p}, t \right) + \frac{\dot{a}(t)}{a(t)} \langle \widetilde{q}, \widetilde{p} \rangle . \end{aligned}$$

► We now give the following definition.

Definition 3.10

Let
$$\Psi_t : \mathbb{R} \times \mathbb{R}^{2n} \to \mathbb{R}^{2n} \ : \ (t, x) \mapsto \widetilde{x} := \Psi_t(x)$$
be a one-parameter family of C^2*-diffeomorphisms on the canonical Hamiltonian phase space. If*
$$\frac{\partial \widetilde{x}}{\partial x} \in \mathbf{SP}(n, \mathbb{R}) \qquad \forall\, (x, t) \in \mathbb{R}^{2n} \times \mathbb{R},$$
then Ψ_t *is a* **canonical transformation**. *If the transformation does not depend on* t *then* Ψ *is a* **symplectic transformation**.

► It is useful to introduce the set of matrices:
$$\mathbf{SP}_\alpha(n, \mathbb{R}) := \left\{ A \in \mathbf{GL}(2\,n, \mathbb{R}) \ : \ A^\top \mathbb{J}\, A = \alpha\, \mathbb{J} \right\}, \tag{3.55}$$
where $\alpha \in \mathbb{R} \setminus \{0\}$. One has the following properties:

- $\mathbf{SP}_1(n, \mathbb{R}) = \mathbf{SP}(n, \mathbb{R})$.
- The following formulas hold:
$$A^{-1} = -\frac{1}{\alpha}\, \mathbb{J}\, A^\top \mathbb{J}, \qquad A^\top = -\alpha\, \mathbb{J}\, A^{-1} \mathbb{J}, \qquad A\, \mathbb{J}\, A^\top = \alpha\, \mathbb{J},$$
for all $A \in \mathbf{SP}_\alpha(n, \mathbb{R})$.
- If $A \in C^2\left(\mathbb{R}^{2n} \times \mathbb{R}, \mathbf{SP}_\alpha(n, \mathbb{R})\right)$ then
$$\left(\frac{\partial A}{\partial t} \right) A^{-1} \in \mathfrak{sp}(n, \mathbb{R}) \qquad \forall\, (x, t) \in \mathbb{R}^{2n} \times \mathbb{R}, \tag{3.56}$$
for all $\alpha \in \mathbb{R} \setminus \{0\}$.
- If $A \in \mathbf{SP}_\alpha(1, \mathbb{R})$ then $\det A = \alpha \neq 0$.

▶ We now provide a first characterization of canonical transformations as those time-dependent transformations on the canonical Hamiltonian phase space which preserve the form of canonical Hamilton equations.

Theorem 3.15

Let $\mathscr{H} \in C^2(\mathbb{R}^{2n} \times \mathbb{R}, \mathbb{R})$ be the Hamiltonian of a mechanical system in the canonical Hamiltonian phase space.

1. *If $\Psi_t : \mathbb{R} \times \mathbb{R}^{2n} \to \mathbb{R}^{2n} : (t,x) \mapsto \widetilde{x} := \Psi_t(x)$ is a canonical transformation then the canonical Hamilton equations are form-invariant under the action of Ψ_t for all $(x,t) \in \mathbb{R}^{2n} \times \mathbb{R}$. More precisely, there exists a new Hamiltonian $\widetilde{\mathscr{H}} \in C^2(\mathbb{R}^{2n} \times \mathbb{R}, \mathbb{R})$,*
$$\widetilde{\mathscr{H}}(\widetilde{x},t) := \mathscr{H}(\widetilde{x},t) + \mathscr{K}(\widetilde{x},t),$$
where $\mathscr{H}(\widetilde{x},t) = \mathscr{H}(x(\widetilde{x},t),t)$ and $\mathscr{K}(\widetilde{x},t)$ is the Hamiltonian of the canonical Hamiltonian vector field $\partial\widetilde{x}/\partial t$, such that the new canonical Hamilton equations are
$$\dot{\widetilde{x}} = \mathbb{J}\,\mathsf{grad}_{\widetilde{x}}\widetilde{\mathscr{H}}(\widetilde{x},t).$$
2. *If $\Psi : \mathbb{R}^{2n} \to \mathbb{R}^{2n} : x \mapsto \widetilde{x} := \Psi(x)$ is a symplectic transformation then the canonical Hamilton equations are form-invariant under the action of Ψ for all $(x,t) \in \mathbb{R}^{2n} \times \mathbb{R}$. More precisely, the new canonical Hamilton equations are*
$$\dot{\widetilde{x}} = \mathbb{J}\,\mathsf{grad}_{\widetilde{x}}\mathscr{H}(\widetilde{x},t),$$
where $\mathscr{H}(\widetilde{x},t) = \mathscr{H}(x(\widetilde{x},t),t)$.

Proof. We proceed by steps. The second claim is a plane consequence of the first claim.

- We know from (3.53) that for a generic change of coordinates $\Psi_t : \mathbb{R} \times \mathbb{R}^{2n} \to \mathbb{R}^{2n} : (t,x) \mapsto \widetilde{x} := \Psi_t(x)$ Hamilton equations are transformed into the following system of $2\,n$ ODEs:
$$\dot{\widetilde{x}} = \frac{\partial\widetilde{x}}{\partial x}\,\mathbb{J}\left(\frac{\partial\widetilde{x}}{\partial x}\right)^{\top} \mathsf{grad}_{\widetilde{x}}\mathscr{H}(\widetilde{x},t) + \frac{\partial\widetilde{x}}{\partial t}.$$
- Assuming that Ψ_t is canonical we have
$$\frac{\partial\widetilde{x}}{\partial x}\,\mathbb{J}\left(\frac{\partial\widetilde{x}}{\partial x}\right)^{\top} = \mathbb{J},$$
so that it suffices to show that $\partial\widetilde{x}/\partial t$ defines a canonical Hamiltonian vector

field, i.e., that (see Theorem 3.11)

$$C := \frac{\partial}{\partial \widetilde{x}} \left(\frac{\partial \widetilde{x}(x(\widetilde{x},t),t))}{\partial t} \right) \in \mathfrak{sp}(n,\mathbb{R}) \qquad \forall\, (x,t) \in \mathbb{R}^{2n} \times \mathbb{R}.$$

- We have:

$$C_{ij} = \frac{\partial}{\partial \widetilde{x}_j} \left(\frac{\partial \widetilde{x}_i}{\partial t} \right) = \sum_{k=1}^{2n} \frac{\partial^2 \widetilde{x}_i}{\partial t\, \partial x_k} \frac{\partial x_k}{\partial \widetilde{x}_j},$$

so that

$$C = \left(\frac{\partial}{\partial t} \frac{\partial \widetilde{x}}{\partial x} \right) \left(\frac{\partial \widetilde{x}}{\partial x} \right)^{-1}.$$

From (3.56) we obtain $C \in \mathfrak{sp}(n,\mathbb{R})$ for all $(x,t) \in \mathbb{R}^{2n} \times \mathbb{R}$.

- We proved that there exists a a new Hamiltonian $\widetilde{\mathscr{H}} \in C^2\left(\mathbb{R}^{2n} \times \mathbb{R}, \mathbb{R}\right)$ such that the transformed canonical Hamilton equations take the form

$$\dot{\widetilde{x}} = \mathbb{J}\, \mathsf{grad}_{\widetilde{x}} \widetilde{\mathscr{H}}(\widetilde{x},t),$$

where

$$\widetilde{\mathscr{H}}(\widetilde{x},t) := \mathscr{H}(\widetilde{x},t) + \mathscr{K}(\widetilde{x},t).$$

Here $\mathscr{H}(\widetilde{x},t) = \mathscr{H}(x(\widetilde{x},t),t)$, $\mathscr{H}$ being the original Hamiltonian, and $\mathscr{K}(\widetilde{x},t)$ is the Hamiltonian of the canonical Hamiltonian vector field $\partial \widetilde{x}/\partial t$:

$$\frac{\partial \widetilde{x}}{\partial t} = \mathbb{J}\ \mathsf{grad}_{\widetilde{x}} \mathscr{K}(\widetilde{x},t).$$

The Theorem is proved. ∎

▶ An immediate consequence of Theorem 3.15 is the next claim.

Corollary 3.1

Let $\Psi_t : \mathbb{R} \times \mathbb{R}^{2n} \to \mathbb{R}^{2n}$ be a canonical transformation. Then Ψ_t is a canonical Hamiltonian flow.

▶ Indeed we can identify the class of canonical transformations with the class of Hamiltonian flows. The next Theorem shows that canonical transformations are all and exclusively canonical Hamiltonian flows. As a consequence, we can claim that canonical transformations are one-parameter *Lie group* of diffeomorphisms on the canonical Hamiltonian phase space.

Theorem 3.16

A flow $\Phi_t : \mathbb{R} \times \mathbb{R}^{2n} \to \mathbb{R}^{2n}$ on the canonical Hamiltonian phase space is a *canonical Hamiltonian flow if and only if Φ_t is a symplectic transformation for every fixed $t \in \mathbb{R}$.*

Proof. Corollary 3.1 assures that a canonical transformation is a canonical Hamiltonian flow. We need to prove the a canonical Hamiltonian flow defines for every fixed t a symplectic transformation. We proceed by steps.

- Assume that $\Phi_t : \mathbb{R} \times \mathbb{R}^{2n} \to \mathbb{R}^{2n}$ is a canonical Hamiltonian flow with Hamiltonian $\mathscr{H}$. We want to show that for every fixed t the map $\Phi_t : \mathbb{R}^{2n} \to \mathbb{R}^{2n}$ is a symplectic transformation, i.e. (see Definition 3.10),

$$\frac{\partial}{\partial x}\Phi_t(x) \in \mathbf{SP}(n,\mathbb{R}) \qquad \forall\,(x,t) \in \mathbb{R}^{2n} \times \mathbb{R},$$

  or, equivalently,

$$\left(\frac{\partial}{\partial x}\Phi_t(x)\right)^{\top} \mathbb{J}\,\frac{\partial}{\partial x}\Phi_t(x) = \mathbb{J} \qquad \forall\,(x,t) \in \mathbb{R}^{2n} \times \mathbb{R}.$$

- We know that Φ_t is a flow, so that $\Phi_0 = \mathbb{1}_{2n}$. Therefore it suffices to prove that

$$\frac{\mathsf{d}}{\mathsf{d}t}\left(\left(\frac{\partial}{\partial x}\Phi_t(x)\right)^{\top} \mathbb{J}\,\frac{\partial}{\partial x}\Phi_t(x)\right) = \mathbb{0}_{2n} \qquad \forall\,(x,t) \in \mathbb{R}^{2n} \times \mathbb{R}.$$

- Thanks to the group property of Φ_t we have:

$$\frac{\mathsf{d}}{\mathsf{d}t}\left(\left(\frac{\partial}{\partial x}\Phi_t(x)\right)^{\top} \mathbb{J}\,\frac{\partial}{\partial x}\Phi_t(x)\right) = \left.\frac{\mathsf{d}}{\mathsf{d}\varepsilon}\right|_{\varepsilon=0}\left(\left(\frac{\partial}{\partial x}\Phi_{t+\varepsilon}(x)\right)^{\top} \mathbb{J}\,\frac{\partial}{\partial x}\Phi_{t+\varepsilon}(x)\right)$$

$$= \left(\frac{\partial}{\partial x}\Phi_t(x)\right)^{\top} \left.\frac{\mathsf{d}}{\mathsf{d}\varepsilon}\right|_{\varepsilon=0}\left(\left(\frac{\partial}{\partial x}\Phi_\varepsilon(\Phi_t(x))\right)^{\top} \mathbb{J}\,\frac{\partial}{\partial x}\Phi_\varepsilon(\Phi_t(x))\right)\frac{\partial}{\partial x}\Phi_t(x).$$

- Therefore it suffices to prove that

$$\left.\frac{\mathsf{d}}{\mathsf{d}\varepsilon}\right|_{\varepsilon=0}\left(\left(\frac{\partial}{\partial x}\Phi_\varepsilon(x)\right)^{\top} \mathbb{J}\,\frac{\partial}{\partial x}\Phi_\varepsilon(x)\right) = \mathbb{0}_{2n} \qquad \forall\, x \in \mathbb{R}^{2n}.$$

- Recalling that

$$\begin{aligned}\frac{\partial}{\partial x}\Phi_\varepsilon(x) &= \mathbb{1}_{2n} + \varepsilon\,\frac{\partial}{\partial x}\left(\mathbb{J}\ \mathsf{grad}_x\mathscr{H}(x)\right) + O(\varepsilon^2)\\ &= \mathbb{1}_{2n} + \varepsilon\,\mathbb{J}\,\frac{\partial^2}{\partial x^2}\mathscr{H}(x) + O(\varepsilon^2),\end{aligned}$$

where $\partial^2 \mathscr{H}(x)/\partial x^2$ is the (symmetric) Hessian matrix of $\mathscr{H}$, we get

$$\begin{aligned}\frac{\mathrm{d}}{\mathrm{d}\varepsilon}\bigg|_{\varepsilon=0}&\left(\left(\frac{\partial}{\partial x}\Phi_\varepsilon(x)\right)^\top \mathbb{J}\,\frac{\partial}{\partial x}\Phi_\varepsilon(x)\right)\\ &= \frac{\mathrm{d}}{\mathrm{d}\varepsilon}\bigg|_{\varepsilon=0}\left(\mathbb{1}_{2n}-\varepsilon\frac{\partial^2}{\partial x^2}\mathscr{H}(x)\,\mathbb{J}\right)\mathbb{J}\left(\mathbb{1}_{2n}+\varepsilon\,\mathbb{J}\,\frac{\partial^2}{\partial x^2}\mathscr{H}(x)\right)\\ &= \frac{\partial^2}{\partial x^2}\mathscr{H}(x)-\frac{\partial^2}{\partial x^2}\mathscr{H}(x)=\mathbb{O}_{2n}.\end{aligned}$$

Here we used $\mathbb{J}^2 = \mathbb{1}_{2n}$ and $\mathbb{J}^\top = -\mathbb{J}$.

The Theorem is proved. ∎

Example 3.17 (*A canonical transformation on* $\mathbb{R}^2$)

Consider the one-parameter family of diffeomorphisms

$$\Psi_t : \mathbb{R}\times\mathbb{R}^2 \to \mathbb{R}^2 \,:\, (t,(q,p)) \mapsto (\widetilde{q},\widetilde{p}) := \left(q - p\,t - \frac{1}{2}\alpha\,t^2, p+\alpha\,t\right),$$

where $\alpha \in \mathbb{R}$ is a parameter.

- The inverse map is

$$\Psi_t^{-1} : \mathbb{R}\times\mathbb{R}^2 \to \mathbb{R}^2 \,:\, (t,(\widetilde{q},\widetilde{p})) \mapsto (q,p) := \left(\widetilde{q} + \widetilde{p}\,t - \frac{1}{2}\alpha\,t^2, \widetilde{p}-\alpha\,t\right),$$

- The map Ψ_t is canonical:

$$\frac{\partial(\widetilde{q},\widetilde{p})}{\partial(q,p)} = \begin{pmatrix}1 & -t\\ 0 & 1\end{pmatrix} \in \mathbf{SP}(1,\mathbb{R}) \qquad \forall\, t\in\mathbb{R}.$$

- By Theorem 3.15 the form of the canonical structure of Hamilton equation is preserved under the action of Ψ_t. The new Hamiltonian is

$$\widetilde{\mathscr{H}}(\widetilde{q},p,t) := \mathscr{H}(\widetilde{q},\widetilde{p},t) + \mathscr{K}(\widetilde{q},\widetilde{p}).$$

  where

$$\mathscr{H}(\widetilde{q},\widetilde{p},t) = \mathscr{H}\left(\widetilde{q}+\widetilde{p}\,t-\frac{1}{2}\alpha\,t^2, \widetilde{p}-\alpha\,t, t\right),$$

  and $\mathscr{K}(\widetilde{q},\widetilde{p})$ is the Hamiltonian of the canonical Hamiltonian vector field

$$\frac{\partial(\widetilde{q},\widetilde{p})}{\partial t} = (-p-\alpha\,t,\alpha) = (-\widetilde{p},\alpha) = \left(\frac{\partial\mathscr{K}}{\partial\widetilde{p}}, -\frac{\partial\mathscr{K}}{\partial\widetilde{q}}\right).$$

  Therefore we have

$$\mathscr{K}(\widetilde{q},\widetilde{p}) = -\frac{\widetilde{p}^2}{2} - \alpha\,\widetilde{q}.$$

► A more general characterization of the class of time-dependent transformations on the canonical Hamiltonian phase space which preserves the form of canonical Hamilton equations is described in the following Theorem. Theorem 3.15 is indeed a consequence of it.

Theorem 3.17

Let $\mathscr{H} \in C^2\left(\mathbb{R}^{2n} \times \mathbb{R}, \mathbb{R}\right)$ be the Hamiltonian of a mechanical system in the canonical Hamiltonian phase space. A one-parameter family of diffeomorphisms $\Psi_t : \mathbb{R} \times \mathbb{R}^{2n} \to \mathbb{R}^{2n} : (t, x) \mapsto \widetilde{x} := \Psi_t(x)$ preserves the form of the the canonical Hamilton equations if and only if

$$\frac{\partial \widetilde{x}}{\partial x} \in \mathbf{SP}_\alpha(n, \mathbb{R}) \qquad \forall\,(x, t) \in \mathbb{R}^{2n} \times \mathbb{R},$$

for some $\alpha \in \mathbb{R} \setminus \{0\}$. More precisely, there exists a new Hamiltonian $\widetilde{\mathscr{H}} \in C^2\left(\mathbb{R}^{2n} \times \mathbb{R}, \mathbb{R}\right)$,

$$\widetilde{\mathscr{H}}(\widetilde{x}, t) := \alpha\, \mathscr{H}(\widetilde{x}, t) + \mathscr{K}(\widetilde{x}, t),$$

where $\mathscr{H}(\widetilde{x}, t) = \mathscr{H}(x(\widetilde{x}, t), t)$ and $\mathscr{K}(\widetilde{x}, t)$ is the Hamiltonian of the canonical Hamiltonian vector field $\partial \widetilde{x}/\partial t$, such that the new canonical Hamilton equations are

$$\dot{\widetilde{x}} = \mathbb{J}\, \mathsf{grad}_{\widetilde{x}} \widetilde{\mathscr{H}}(\widetilde{x}, t).$$

The transformation Ψ_t is a canonical transformation if and only if $\alpha = 1$.

No Proof.

► In the next example we give an illustration of Theorem 3.17.

Example 3.18 (*A non-canonical transformation on* $\mathbb{R}^2$)

Consider the one-parameter family of diffeomorphisms $\Psi_t : \mathbb{R} \times \mathbb{R}^2 \to \mathbb{R}^2$ defined by

$$(\widetilde{q}, \widetilde{p}) := \left(-\frac{\gamma}{\beta}\, p \sin(\omega\, t) + \frac{\gamma}{\alpha}\, q \cos(\omega\, t), \alpha\, p \cos(\omega\, t) + \beta q \sin(\omega\, t)\right),$$

where $\alpha, \beta, \gamma, \omega \in \mathbb{R} \setminus \{0\}$ are parameters. By Theorem 3.17 this map preserves the form of the canonical Hamiltonian equations because

$$\det\left(\frac{\partial(\widetilde{q}, \widetilde{p})}{\partial(q, p)}\right) = \gamma,$$

but Ψ_t is a canonical transformation if and only if $\gamma = 1$.

► We now provide the second characterization of canonical transformations as those time-dependent transformations on the canonical Hamiltonian phase space which preserve the form of the canonical Poisson brackets.

- As mentioned after Definition 3.9 the canonical Poisson bracket depends on the coordinate system.

- To emphasize such a dependence we shall write $\{\,\cdot,\cdot\,\}_x$ for the canonical Poisson bracket in the coordinates $x \in \mathbb{R}^{2n}$ and $\{\,\cdot,\cdot\,\}_{\widetilde{x}}$ for the canonical Poisson bracket in the coordinates $\widetilde{x} \in \mathbb{R}^{2n}$.

- In general, if
$$\Psi_t : \mathbb{R} \times \mathbb{R}^{2n} \longrightarrow \mathbb{R}^{2n} \; : \; (t, x) \mapsto \widetilde{x} := \Psi_t(x)$$
is a one-parameter family of C^2-diffeomorphisms on the canonical Hamiltonian phase space, then
$$\{ F(x,t), G(x,t) \}_x \neq \{ F(\widetilde{x},t), G(\widetilde{x},t) \}_{\widetilde{x}}, \qquad F, G \in C^1(\mathbb{R}^{2n} \times \mathbb{R}, \mathbb{R}).$$

▶ We have the following Theorem.

Theorem 3.18

Let
$$\Psi_t : \mathbb{R} \times \mathbb{R}^{2n} \longrightarrow \mathbb{R}^{2n} \; : \; (t, x) \mapsto \widetilde{x} := \Psi_t(x)$$
be a one-parameter family of C^2-diffeomorphisms on the canonical Hamiltonian phase space. The following statements are equivalent.

1. *Ψ_t is a canonical transformation.*
2. *There holds*
$$\{ F(x,t), G(x,t) \}_x = \{ F(\widetilde{x},t), G(\widetilde{x},t) \}_{\widetilde{x}}, \qquad \forall\, F, G \in C^1(\mathbb{R}^{2n} \times \mathbb{R}, \mathbb{R}).$$
3. *Coordinates $\widetilde{x}$ have a canonical Poisson structure w.r.t. x:*
$$\{ \widetilde{x}_i, \widetilde{x}_j \}_x = \mathbb{J}_{ij}, \qquad i, j = 1, \ldots, n.$$

Proof. We prove that *1. ⇒ 2. ⇒ 3. ⇒ 1.*

- *(1. ⇒ 2.).* If Ψ_t is a canonical transformation then
$$\frac{\partial \widetilde{x}}{\partial x} \in \mathbf{SP}(n, \mathbb{R}) \qquad \forall\, (x,t) \in \mathbb{R}^{2n} \times \mathbb{R}.$$
For any $F, G \in C^1(\mathbb{R}^{2n} \times \mathbb{R}, \mathbb{R})$ we have:
$$\begin{aligned} \{ F(x,t), G(x,t) \}_x &= \langle \operatorname{grad}_x F(x,t), \mathbb{J} \operatorname{grad}_x G(x,t) \rangle \\ &= \left\langle \left(\frac{\partial \widetilde{x}}{\partial x}\right)^{\top} \operatorname{grad}_{\widetilde{x}} F(\widetilde{x},t), \mathbb{J} \left(\frac{\partial \widetilde{x}}{\partial x}\right)^{\top} \operatorname{grad}_{\widetilde{x}} G(\widetilde{x},t) \right\rangle \\ &= \left\langle \operatorname{grad}_{\widetilde{x}} F(\widetilde{x},t), \frac{\partial \widetilde{x}}{\partial x} \mathbb{J} \left(\frac{\partial \widetilde{x}}{\partial x}\right)^{\top} \operatorname{grad}_{\widetilde{x}} G(\widetilde{x},t) \right\rangle \\ &= \langle \operatorname{grad}_{\widetilde{x}} F(\widetilde{x},t), \mathbb{J} \operatorname{grad}_{\widetilde{x}} G(\widetilde{x},t) \rangle \\ &= \{ F(\widetilde{x},t), G(\widetilde{x},t) \}_{\widetilde{x}}. \end{aligned}$$

- (*2.* $\Rightarrow$ *3.*). This is obvious because *3.* is a special case of *2.*
- (*3.* $\Rightarrow$ *1.*). It is enough to prove that

$$\frac{\partial \widetilde{x}}{\partial x} \in \mathbf{SP}(n,\mathbb{R}) \qquad \forall\,(x,t) \in \mathbb{R}^{2n} \times \mathbb{R}.$$

We have:

$$\begin{aligned} \mathbb{J}_{ij} &= \{\,\widetilde{x}_i, \widetilde{x}_j\,\}_x = \big\langle\, \mathsf{grad}_x \widetilde{x}_i, \mathbb{J}\ \mathsf{grad}_x \widetilde{x}_j\,\big\rangle \\ &= \sum_{k,\ell=1}^{2n} \frac{\partial \widetilde{x}_i}{\partial x_k}\, \mathbb{J}_{k\ell}\, \frac{\partial \widetilde{x}_j}{\partial x_\ell} = \left(A\,\mathbb{J}\,A^\top\right)_{ij}, \end{aligned}$$

that is $A\,\mathbb{J}\,A^\top = \mathbb{J}$ as desired.

The Theorem is proved. ■

3.4.5 The Lie condition and the canonical symplectic 2-form

► In order to give a characterization of canonical and symplectic transformations in terms of the preservation of the so called *canonical symplectic 2-form* and of existence of *generating functions* we necessarily need some facts about *differential forms.*

- Due to the fact that we are working on Euclidean spaces our presentation will be very elementary (a more complete understanding of differential forms will be a subject of Chapter 4).
- For simplicity, we assume that all functions are smooth and globally defined. Moreover, for a reason that will be clear soon, we come back to our notation $x := (x_1, \ldots, x_n)$ for coordinates on $\mathbb{R}^n$.

► Let us recall some facts from Analysis and Linear Algebra. We shall first introduce *algebraic forms* (the dual version of constant vectors on a vector space) and then *differential forms* (the dual version of vectors fields on a vector space).

- Let V be a real n-dimensional vector space. Let $\{e_i\}_{1\leqslant i\leqslant n}$ be a basis of V.
  (a) The *dual vector space* V^* is the n-dimensional vector space of all linear maps $\omega : V \longrightarrow \mathbb{R}$. Elements of V^* are called *algebraic 1-forms* (or *covectors*).
  (b) A basis $\{e_i^*\}_{1\leqslant i\leqslant n}$ of V^* is defined uniquely by the *duality pairing*

$$\left\langle\, e_i, e_j^* \,\right\rangle = \delta_{ij}, \qquad i,j = 1,\ldots,n.$$

  Every element of V^* can be expressed as

$$\omega = \sum_{j=1}^{n} f_j\, e_j^*, \qquad f_j \in \mathbb{R}.$$

(c) If $V = \mathbb{R}^n$ is interpreted as the space of columns of n real numbers, its dual space $V^* = (\mathbb{R}^n)^*$ is typically viewed as the space of rows of n real numbers. Such a row acts on $\mathbb{R}^n$ as a linear functional by ordinary matrix multiplication.

- An *algebraic k-form* on $\mathbb{R}^n$ is a map

$$\omega : \underbrace{\mathbb{R}^n \times \cdots \times \mathbb{R}^n}_{k \text{ times}} \to \mathbb{R},$$

which satisfies the following properties:

1. *Multilinearity.* For any k vectors $(v_1, \ldots, v_k)$, a vector w and two scalars $\lambda_1, \lambda_2 \in \mathbb{R}$, there holds:

$$\omega(\lambda_1 v_1 + \lambda_2 w, v_2, \ldots, v_k) = \lambda_1 \omega(v_1, \ldots, v_k) + \lambda_2 \omega(w, \ldots, v_k).$$

2. *Skew-symmetry.* For any k vectors $(v_1, \ldots, v_k)$ there holds:

$$\omega(v_1, \ldots, v_k) = (-1)^\nu \omega(v_{i_1}, \ldots, v_{i_k}),$$

where $\nu = \pm 1$ according to the parity of the permutation $(i_1, \ldots, i_k)$ of $(1, \ldots, k)$.

- The set of all algebraic k-forms on $\mathbb{R}^n$ is a vector space denoted by $\Omega^k(\mathbb{R}^n)$, with

$$\dim\left(\Omega^k(\mathbb{R}^n)\right) = \frac{n!}{k!(n-k)!}.$$

In the simplest case, $k = 1$, we have $\Omega^1(\mathbb{R}^n) \simeq (\mathbb{R}^n)^*$.

- One can define a bilinear map which associates to two algebraic forms ω and ω' their *wedge product* $\omega \wedge \omega'$. More precisely, the wedge product of two algebraic forms $\omega \in \Omega^k(\mathbb{R}^n)$, $\omega' \in \Omega^\ell(\mathbb{R}^n)$ is the algebraic $(k+\ell)$-form defined by

$$(\omega \wedge \omega')(v_1, \ldots, v_{k+\ell}) := \sum_{\sigma \in \pi} \nu(\sigma) \omega(v_{\sigma_1}, \ldots, v_{\sigma_k}) \omega'(v_{\sigma_{k+1}}, \ldots, v_{\sigma_{k+\ell}}), \quad (3.57)$$

where $\sigma := (\sigma_1, \ldots, \sigma_{k+\ell})$, π is the set of all permutations of $(1, \ldots, k+\ell)$ and $\nu = \pm 1$ according to the parity of the permutation. This operation is associative, distributive and graded commutative, which means that

$$\omega \wedge \omega' = (-1)^{k\ell} \omega' \wedge \omega, \qquad \omega \in \Omega^k(\mathbb{R}^n),\ \omega' \in \Omega^\ell(\mathbb{R}^n).$$

- Given a collection of k algebraic 1-forms $\omega_1, \ldots, \omega_k$ one can construct an algebraic k-form whose action on k vectors is expressed by a determinant formula:

$$(\omega_1 \wedge \cdots \wedge \omega_k)(v_1, \ldots, v_k) = \det\left(\omega_i(v_j)\right),$$

where the r.h.s. is a $k \times k$ matrix with indicated (i, j)-th entry.

Example 3.19 (*Wedge product of two algebraic 1-forms*)

If $(r,s) = (1,1)$ we obtain from (3.57) the algebraic 2-form

$$\begin{aligned}(\omega \wedge \omega')(v_1, v_2) &:= \omega(v_1)\omega'(v_2) - \omega'(v_1)\omega(v_2) \\ &= \det \begin{pmatrix} \omega(v_1) & \omega'(v_1) \\ \omega(v_2) & \omega'(v_2) \end{pmatrix},\end{aligned}$$

that is the oriented area of a parallelogram with sides given by $(\omega(v_1), \omega'(v_1))$ and $(\omega(v_2), \omega'(v_2))$.

- As the relation between vectors and vector fields there exists a relation between algebraic forms and differential forms. The simplest example of differential form is a *differential* 1-*form* on $\mathbb{R}^n$. This takes the form

$$\omega = \sum_{j=1}^{n} f_j(x)\, \mathrm{d}x_j, \qquad f_j \in \mathscr{F}(\mathbb{R}^n, \mathbb{R}).$$

  Note that the differential of a smooth function (i.e., a differential 0-form) $F \in \mathscr{F}(\mathbb{R}^n, \mathbb{R})$ is a differential 1-form:

$$\mathrm{d}F = \sum_{j=1}^{n} \frac{\partial F}{\partial x_j}\, \mathrm{d}x_j.$$

- A *differential* 2-*form* on $\mathbb{R}^n$ can be written as

$$\omega = \sum_{1 \leqslant i < j \leqslant n} f_{ij}(x)\, \mathrm{d}x_i \wedge \mathrm{d}x_j, \tag{3.58}$$

  where $f_{ij} \in \mathscr{F}(\mathbb{R}^n, \mathbb{R})$.

- In general a *differential k-form* on $\mathbb{R}^n$ is specified by $n!/k!/(n-k)!$ smooth functions $f_{\ell_1 \cdots \ell_k} \in \mathscr{F}(\mathbb{R}^n, \mathbb{R})$:

$$\omega = \sum_{1 \leqslant \ell_1 < \cdots < \ell_k \leqslant n} f_{\ell_1 \cdots \ell_k}(x)\, \mathrm{d}x_{\ell_1} \wedge \cdots \wedge \mathrm{d}x_{\ell_k}. \tag{3.59}$$

  More intrinsically we say that differential k-form on $\mathbb{R}^n$ is a multilinear skew-symmetric map

$$\omega : \underbrace{\mathfrak{X}(M) \times \cdots \times \mathfrak{X}(M)}_{k \text{ times}} \to \mathscr{F}(\mathbb{R}^n, \mathbb{R}).$$

  We denote the set of all differential k-forms again by $\Omega^k(\mathbb{R}^n)$.

- A *differential n-form* on $\mathbb{R}^n$ is also called *volume form*. It is specified by a single function $f \in \mathscr{F}(\mathbb{R}^n, \mathbb{R})$:

$$\omega = f(x)\, \mathrm{d}x_1 \wedge \cdots \wedge \mathrm{d}x_n.$$

Note that ω can be used to define the measure (the volume if $f(x) \equiv 1$) of a compact subset $D \subset \mathbb{R}^n$,

$$\mathsf{Vol}(D) := \int_D \omega.$$

Example 3.20 (*Wedge product of two differential 1-forms on* $\mathbb{R}^2$)

Let $\mathrm{d}F_1, \mathrm{d}F_2$ be the differentials of two functions (i.e., 0-forms) $F_1, F_2 \in \mathscr{F}(\mathbb{R}^2, \mathbb{R})$:

$$\mathrm{d}F_1 = \frac{\partial F_1}{\partial x_1}\mathrm{d}x_1 + \frac{\partial F_1}{\partial x_2}\mathrm{d}x_2, \qquad \mathrm{d}F_2 = \frac{\partial F_2}{\partial x_1}\mathrm{d}x_1 + \frac{\partial F_2}{\partial x_2}\mathrm{d}x_2.$$

Then their wedge product is:

$$\mathrm{d}F_1 \wedge \mathrm{d}F_2 = \left(\frac{\partial F_1}{\partial x_1}\frac{\partial F_2}{\partial x_2} - \frac{\partial F_1}{\partial x_2}\frac{\partial F_2}{\partial x_1}\right)\mathrm{d}x_1 \wedge \mathrm{d}x_2.$$

- The *differential* (or *exterior derivative*) of the differential k-form (3.59) is the differential $(k+1)$-form

$$\mathrm{d}\omega := \sum_{i=1}^{n} \sum_{1 \leqslant \ell_1 < \cdots < \ell_k \leqslant n} \frac{\partial f_{\ell_1 \cdots \ell_k}}{\partial x_i}\,\mathrm{d}x_i \wedge \mathrm{d}x_{\ell_1} \wedge \cdots \wedge \mathrm{d}x_{\ell_k}. \tag{3.60}$$

It has the following properties:

1. *Linearity*. There holds:

$$\mathrm{d}(\lambda_1\,\omega + \lambda_2\,\omega') = \lambda_1\,\mathrm{d}\omega + \lambda_2\,\mathrm{d}\omega', \qquad \omega, \omega' \in \Omega^k(\mathbb{R}^n),\ \lambda_1, \lambda_2 \in \mathbb{R}.$$

2. *Anti-derivation* (graded Leibniz rule). There holds:

$$\mathrm{d}(\omega \wedge \omega') = \mathrm{d}\omega \wedge \omega' + (-1)^k \omega \wedge \mathrm{d}\omega', \qquad \omega \in \Omega^k(\mathbb{R}^n),\ \omega' \in \Omega^\ell(\mathbb{R}^n).$$

3. *Closure*. There holds:

$$\mathrm{d}(\mathrm{d}\omega) = 0, \qquad \omega \in \Omega^k(\mathbb{R}^n). \tag{3.61}$$

- The *exactness property* of differential forms means that a a *closed differential form* $\omega \in \Omega^k(\mathbb{R}^n)$, i.e., $\mathrm{d}\omega = 0$, is necessarily an *exact differential form*, meaning that there exists a $(k-1)$-form $\omega' \in \Omega^{k-1}(\mathbb{R}^n)$ such that

$$\omega = \mathrm{d}\omega'.$$

Clearly, any exact form is closed, but the converse is in general not true. However, on the local side and for special types of domains $M \subset \mathbb{R}^n$ we do have exactness. This famous result, known as "Poincaré Lemma", holds for star-shaped domains M, where "star-shaped" means that whenever $x \in M$, so is the entire line segment joining x to the origin: $\{\lambda\, x : \lambda \in [0,1]\} \subset M$.

Example 3.21 (*Differential of a differential 1-form on* $\mathbb{R}^n$)

Consider a differential 1-form on $\mathbb{R}^n$ defined as the differential of a smooth function $F \in \mathscr{F}(\mathbb{R}^n, \mathbb{R})$:

$$\mathrm{d}F = \sum_{j=1}^{n} \frac{\partial F}{\partial x_j} \,\mathrm{d}x_j.$$

The closure property (3.61) is equivalent to the equality of mixed derivatives:

$$\mathrm{d}(\mathrm{d}F) = \sum_{i,j=1}^{n} \frac{\partial^2 F}{\partial x_i \partial x_j} \mathrm{d}x_i \wedge \mathrm{d}x_j = \sum_{i<j} \left(\frac{\partial^2 F}{\partial x_i \partial x_j} - \frac{\partial^2 F}{\partial x_j \partial x_i} \right) \mathrm{d}x_i \wedge \mathrm{d}x_j = 0.$$

Example 3.22 (*Differential n- and* $(n-1)$*-forms on* $\mathbb{R}^n$)

Consider a differential n-form on $\mathbb{R}^n$:

$$\omega = f(x)\,\mathrm{d}x_1 \wedge \cdots \wedge \mathrm{d}x_n, \qquad f \in \mathscr{F}(\mathbb{R}^n, \mathbb{R}).$$

Any n-form on $\mathbb{R}^n$ is always closed, i.e., $\mathrm{d}\omega = 0$, as there are no nonzero $(n+1)$-forms.

- A differential $(n-1)$-form on $\mathbb{R}^n$ is determined by n functions $(g_1, \ldots, g_n)$, $g_i \in \mathscr{F}(\mathbb{R}^n, \mathbb{R})$:

$$\omega' = \sum_{j=1}^{n} (-1)^{j-1} g_j(x)\,\mathrm{d}x_1 \wedge \cdots \wedge \widehat{\mathrm{d}x_j} \wedge \cdots \wedge \mathrm{d}x_n.$$

- If we relate ω with ω' through a differential, i.e., $\omega = \mathrm{d}\omega'$, namely we ask for exactness of ω, we find that f is given by the divergence of $(g_1, \ldots, g_n)$:

$$f(x) = \mathrm{div}_x(g_1(x), \ldots, g_n(x)) := \sum_{j=1}^{n} \frac{\partial g_j}{\partial x_j}.$$

Example 3.23 (*Vector calculus in* $\mathbb{R}^3$)

This example shows how to use differential forms in standard vector calculus in $\mathbb{R}^3$ with coordinates $x := (x_1, x_2, x_3)$.

- Let $K \in \mathscr{F}(\mathbb{R}^3, \mathbb{R})$ be a function. Then its differential is the 1-form

$$\mathrm{d}K = \frac{\partial K}{\partial x_1} \mathrm{d}x_1 + \frac{\partial K}{\partial x_2} \mathrm{d}x_2 + \frac{\partial K}{\partial x_3} \mathrm{d}x_3 \in \Omega^1(\mathbb{R}^3),$$

which can be identified with the gradient of K:

$$\mathrm{grad}_x K(x) = \frac{\partial K}{\partial x_1} e_1 + \frac{\partial K}{\partial x_2} e_2 + \frac{\partial K}{\partial x_3} e_3,$$

if we identify the canonical basis $\{e_1, e_2, e_3\}$ of $\mathbb{R}^3$ with the standard basis of $\{\mathrm{d}x_1, \mathrm{d}x_2, \mathrm{d}x_3\}$ of $\Omega^1(\mathbb{R}^3) \simeq (\mathbb{R}^3)^*$.

- Let $\omega \in \Omega^1(\mathbb{R}^3)$ be a 1-form. Then:

$$\omega = g_1(x)\,\mathrm{d}x_1 + g_2(x)\,\mathrm{d}x_2 + g_3(x)\,\mathrm{d}x_3,$$

for some smooth functions g_1, g_2, g_3, so that

$$\begin{aligned} \mathrm{d}\omega &= \left(\frac{\partial g_3}{\partial x_2} - \frac{\partial g_2}{\partial x_3} \right) \mathrm{d}x_2 \wedge \mathrm{d}x_3 + \left(\frac{\partial g_1}{\partial x_3} - \frac{\partial g_3}{\partial x_1} \right) \mathrm{d}x_3 \wedge \mathrm{d}x_1 \\ &+ \left(\frac{\partial g_2}{\partial x_1} - \frac{\partial g_1}{\partial x_2} \right) \mathrm{d}x_1 \wedge \mathrm{d}x_2 \in \Omega^2(\mathbb{R}^3), \end{aligned}$$

which can be identified with the curl of the vector field $g(x) := (g_1(x), g_2(x), g_3(x))$:

$$\text{curl}_x g(x) = \left(\frac{\partial g_3}{\partial x_2} - \frac{\partial g_2}{\partial x_3}\right) e_1 + \left(\frac{\partial g_1}{\partial x_3} - \frac{\partial g_3}{\partial x_1}\right) e_2 + \left(\frac{\partial g_2}{\partial x_1} - \frac{\partial g_1}{\partial x_2}\right) e_3.$$

Note that $\mathrm{d}x_2 \wedge \mathrm{d}x_3 \equiv e_2 \times e_3 = e_1$ and cyclic permutations.

- Let $\omega \in \Omega^2(\mathbb{R}^3)$ be a 2-form. Then:

$$\omega = f_1(x)\,\mathrm{d}x_2 \wedge \mathrm{d}x_3 + f_2(x)\,\mathrm{d}x_3 \wedge \mathrm{d}x_1 + f_3(x)\,\mathrm{d}x_1 \wedge \mathrm{d}x_2,$$

for some smooth functions f_1, f_2, f_3, so that

$$\mathrm{d}\omega = \left(\frac{\partial f_1}{\partial x_1} + \frac{\partial f_2}{\partial x_2} + \frac{\partial f_3}{\partial x_3}\right) \mathrm{d}x_1 \wedge \mathrm{d}x_2 \wedge \mathrm{d}x_3 \in \Omega^3(\mathbb{R}^3),$$

which can be identified with the divergence of $f(x) := (f_1(x), f_2(x), f_3(x))$:

$$\text{div}_x f(x) = \frac{\partial f_1}{\partial x_1} + \frac{\partial f_2}{\partial x_2} + \frac{\partial f_3}{\partial x_3}.$$

- The vanishing of the second exterior derivative, $\mathrm{d}^2 = 0$, summarizes the vector identities $\text{curl}_x(\text{grad}_x f(x)) = 0$ and $\text{div}_x(\text{curl}_x g(x)) = 0$.

► We now come back to our canonical Hamiltonian phase space and we define a special differential 1-form.

Definition 3.11

Let $F \in C^1(\mathbb{R}^{2n} \times \mathbb{R}, \mathbb{R})$ be a function on the canonical Hamiltonian phase space with coordinates $x := (q, p)$. The **time-frozen differential** *of F is the differential 1-form on $\mathbb{R}^{2n} \times \mathbb{R}$ defined by*

$$\begin{aligned} \underline{\mathrm{d}}F &:= \mathrm{d}F - \frac{\partial F}{\partial t}\mathrm{d}t = \langle\, \text{grad}_x F(x,t), \mathrm{d}x \,\rangle \\ &= \Big\langle\, \text{grad}_q F(q,p,t), \mathrm{d}q \,\Big\rangle + \Big\langle\, \text{grad}_p F(q,p,t), \mathrm{d}p \,\Big\rangle. \end{aligned}$$

If F does not depend explicitly on t then $\underline{\mathrm{d}}F = \mathrm{d}F$.

► The time-frozen differential is used to give the following characterization of canonical transformations on the canonical Hamiltonian phase space.

Theorem 3.19

Let

$$\Psi_t : \mathbb{R} \times \mathbb{R}^{2n} \to \mathbb{R}^{2n} : (t, (q,p)) \mapsto (\widetilde{q}, \widetilde{p}) := \Psi_t(q,p)$$

be a one-parameter family of C^2-diffeomorphisms on the canonical Hamiltonian phase space. Then Ψ_t is a canonical transformation if and only if $\langle\, p, \underline{\mathrm{d}}q \,\rangle - \langle\, \widetilde{p}, \underline{\mathrm{d}}\widetilde{q} \,\rangle$ is an exact time-frozen differential, i.e.,

$$\langle\, p, \underline{\mathrm{d}}q \,\rangle - \langle\, \widetilde{p}, \underline{\mathrm{d}}\widetilde{q} \,\rangle = \underline{\mathrm{d}}F \qquad (\textbf{Lie condition}), \tag{3.62}$$

for some function $F = F(\widetilde{q}, \widetilde{p}, t)$, $F \in C^1(\mathbb{R}^{2n} \times \mathbb{R}, \mathbb{R})$.

No Proof.

► Formula (3.62) can be written more explicitly.

- Expressing q as a function of $(\widetilde{q}, \widetilde{p}, t)$ we have

$$\underline{d}q_k = \left\langle \operatorname{grad}_{\widetilde{q}} q_k(\widetilde{q}, \widetilde{p}, t), d\widetilde{q} \right\rangle + \left\langle \operatorname{grad}_{\widetilde{p}} q_k(\widetilde{q}, \widetilde{p}, t), d\widetilde{p} \right\rangle, \qquad k = 1, \ldots, n.$$

- Formula (3.62) takes the form

$$\sum_{k=1}^{n} \left(p_k \left\langle \operatorname{grad}_{\widetilde{q}} q_k(\widetilde{q}, \widetilde{p}, t), d\widetilde{q} \right\rangle + \left\langle \operatorname{grad}_{\widetilde{p}} q_k(\widetilde{q}, \widetilde{p}, t), d\widetilde{p} \right\rangle - \widetilde{p}_k \, d\widetilde{q}_k \right)$$

$$= \left\langle \operatorname{grad}_{\widetilde{q}} F(\widetilde{q}, \widetilde{p}, t), d\widetilde{q} \right\rangle + \left\langle \operatorname{grad}_{\widetilde{q}} F(\widetilde{q}, \widetilde{p}, t), d\widetilde{p} \right\rangle.$$

- The Lie condition (3.62) is equivalent to the following $2\,n$ ODEs for the function F:

$$\begin{cases} \dfrac{\partial F}{\partial \widetilde{q}_k} = -\widetilde{p}_k + \displaystyle\sum_{i=1}^{n} p_i \frac{\partial q_i}{\partial \widetilde{q}_k}, \\[2ex] \dfrac{\partial F}{\partial \widetilde{p}_k} = \displaystyle\sum_{i=1}^{n} p_i \frac{\partial q_i}{\partial \widetilde{p}_k}, \end{cases}$$

with $k = 1, \ldots, n$.

Example 3.24 (***A canonical transformation on*** $\mathbb{R}^2$)

Consider the canonical transformation of Example 3.17:

$$\Psi_t : \mathbb{R} \times \mathbb{R}^2 \to \mathbb{R}^2 \, : \, (t, (q, p)) \mapsto (\widetilde{q}, \widetilde{p}) := \left(q - p\,t - \frac{1}{2}\alpha\, t^2, p + \alpha\, t \right),$$

where $\alpha \in \mathbb{R}$ is a parameter.

- The inverse map is

$$\Psi_t^{-1} : \mathbb{R} \times \mathbb{R}^2 \to \mathbb{R}^2 \, : \, (t, (\widetilde{q}, \widetilde{p})) \mapsto (q, p) := \left(\widetilde{q} + \widetilde{p}\,t - \frac{1}{2}\alpha\, t^2, \widetilde{p} - \alpha\, t \right),$$

- Let us verify that the Lie condition is satisfied. We have:

$$\underline{d}q = \frac{\partial q}{\partial \widetilde{q}} d\widetilde{q} + \frac{\partial q}{\partial \widetilde{p}} d\widetilde{p} = d\widetilde{q} + t\, d\widetilde{p}, \qquad \underline{d}\widetilde{q} = \frac{\partial \widetilde{q}}{\partial q} dq = dq,$$

so that

$$\begin{aligned} p\,\underline{d}q - \widetilde{p}\,\underline{d}\widetilde{q} &= (\widetilde{p} - \alpha\, t)(d\widetilde{q} + t\, d\widetilde{p}) - \widetilde{p}\, d\widetilde{q} \\ &= -\alpha\, t\, d\widetilde{q} + \left(\widetilde{p}\, t - \alpha\, t^2 \right) d\widetilde{p} \\ &= \underline{d}F(\widetilde{q}, \widetilde{p}, t) \\ &= \frac{\partial F}{\partial \widetilde{q}} d\widetilde{q} + \frac{\partial F}{\partial \widetilde{p}} d\widetilde{p}. \end{aligned}$$

Therefore we obtain

$$\left(\frac{\partial F}{\partial \widetilde{q}}, \frac{\partial F}{\partial \widetilde{p}}\right) = \left(-\alpha\, t, \widetilde{p}\, t - \alpha\, t^2\right),$$

namely,

$$F(\widetilde{q}, \widetilde{p}, t) = \frac{1}{2}\, t\, \widetilde{p}^2 - \alpha\, t^2\, \widetilde{p} - \alpha\, t\, \widetilde{q} + G(t),$$

where G is an arbitrary function of time.

► We now focus our attention on symplectic transformations, namely on those canonical transformations which do not depend explicitly on time.

- By Theorem 3.19 the symplecticity of $\Psi : \mathbb{R}^{2n} \to \mathbb{R}^{2n} : (q, p) \mapsto (\widetilde{q}, \widetilde{p}) := \Psi(q, p)$ holds true if and only if $\langle\, p, \mathrm{d}q\,\rangle - \langle\, \widetilde{p}, \mathrm{d}\widetilde{q}\,\rangle$ is an exact differential.

- From this characterization we see that the differential 1-form (called *Liouville 1-form*)

$$\vartheta := \langle\, p, \mathrm{d}q\,\rangle = \sum_{i=1}^{n} p_i\, \mathrm{d}q_i \tag{3.63}$$

  plays a special role. Indeed, one can claim that Ψ is a symplectic transformation if and only if the Liouville 1-form is preserved under the action of Ψ up to a total differential of a function:

$$\vartheta - \widetilde{\vartheta} = \mathrm{d}F.$$

- We will see that one can give an equivalent formulation of this last claim in terms of a differential 2-form defined as the exterior derivative of ϑ (see Theorem 3.20 and Corollary 3.2). Such a new object has a nice geometric interpretation.

- Fix $n = 1$, so that the Hamiltonian phase space is just $\mathbb{R}^2$ with coordinates (q, p). We know that a C^2-diffeomorphism $\Psi : \mathbb{R}^2 \to \mathbb{R}^2 : (q, p) \mapsto (\widetilde{q}, \widetilde{p}) := \Psi(q, p)$ is area-preserving if and only if the Jacobian of Ψ is 1 for all $(q, p) \in \mathbb{R}^2$:

$$\frac{\partial \widetilde{q}}{\partial q}\frac{\partial \widetilde{p}}{\partial p} - \frac{\partial \widetilde{q}}{\partial p}\frac{\partial \widetilde{p}}{\partial q} = 1.$$

  The above condition can be expressed in matrix form as

$$\frac{\partial(\widetilde{q}, \widetilde{p})}{\partial(q, p)}\, \mathbb{J} \left(\frac{\partial(\widetilde{q}, \widetilde{p})}{\partial(q, p)}\right)^{\top} = \mathbb{J},$$

  for all $(q, p) \in \mathbb{R}^2$, meaning that Ψ is a symplectic transformation.

- Differential forms provide an alternative language to express the preservation of area. The differentials of $\widetilde{q}$ and $\widetilde{p}$ are given by

$$\mathrm{d}\widetilde{q} = \frac{\partial \widetilde{q}}{\partial q}\,\mathrm{d}q + \frac{\partial \widetilde{q}}{\partial p}\,\mathrm{d}p, \qquad \mathrm{d}\widetilde{p} = \frac{\partial \widetilde{p}}{\partial q}\,\mathrm{d}q + \frac{\partial \widetilde{p}}{\partial p}\,\mathrm{d}p.$$

The wedge product of $\mathrm{d}\widetilde{q}$ and $\mathrm{d}\widetilde{p}$ gives rise to the following differential 2-form

$$\mathrm{d}\widetilde{q} \wedge \mathrm{d}\widetilde{p} = \left(\frac{\partial \widetilde{q}}{\partial q}\frac{\partial \widetilde{p}}{\partial p} - \frac{\partial \widetilde{q}}{\partial p}\frac{\partial \widetilde{p}}{\partial q} \right) \mathrm{d}q \wedge \mathrm{d}p.$$

We see that the conservation of area in the canonical Hamiltonian phase space $\mathbb{R}^2$ is equivalent to the condition

$$\mathrm{d}\widetilde{q} \wedge \mathrm{d}\widetilde{p} = \mathrm{d}q \wedge \mathrm{d}p.$$

► The above characterization of preservation of area for symplectic maps admits a generalization for for $n > 1$ (see Corollary 3.2). Let us define the following differential 2-form.

Definition 3.12

The **canonical symplectic 2-form** *is (minus) the exterior derivative of the Liouville 1-form (3.63):*

$$\omega := -\mathrm{d}\vartheta = \sum_{i=1}^{n} \mathrm{d}q_i \wedge \mathrm{d}p_i. \tag{3.64}$$

► It is immediate to verify that ω is skew-symmetric, closed and non-degenerate. In particular:

- The closure condition follows from the fact that all coefficients of $\mathrm{d}q_i \wedge \mathrm{d}p_i$, $i = 1, \ldots, n$, are just constants (equal to 1) and from the closure of the differential.
- The matrix associated to (3.64) is the canonical symplectic matrix

$$\mathbb{J} := \begin{pmatrix} \mathbb{O}_n & \mathbb{1}_n \\ -\mathbb{1}_n & \mathbb{O}_n \end{pmatrix},$$

whose determinant is 1 everywhere. This proves that the canonical symplectic 2-form is non-degenerate.

Theorem 3.20

Let

$$\Psi : \mathbb{R}^{2n} \to \mathbb{R}^{2n} \;:\; (q, p) \mapsto (\widetilde{q}, \widetilde{p}) := \Psi(q, p)$$

be a C^2-diffeomorphism on the canonical Hamiltonian phase space. Then Ψ is a symplectic transformation if and only if Ψ preserves the canonical symplectic 2-form, i.e.,

$$\sum_{i=1}^{n} \mathrm{d}\widetilde{q}_i \wedge \mathrm{d}\widetilde{p}_i = \sum_{i=1}^{n} \mathrm{d}q_i \wedge \mathrm{d}p_i.$$

Proof. We prove the Theorem in both directions.

- Assuming that Ψ is a symplectic transformation we know from Theorem 3.19 that the Lie condition
$$\langle\, p, \mathrm{d}q \,\rangle - \langle\, \widetilde{p}, \mathrm{d}\widetilde{q} \,\rangle = \mathrm{d}F, \tag{3.65}$$
holds for some function $F \in C^1(\mathbb{R}^{2n}, \mathbb{R})$. Now, taking the exterior derivative of the Lie condition we get
$$\mathrm{d}\,\langle\, p, \mathrm{d}q \,\rangle = \mathrm{d}\,\langle\, \widetilde{p}, \mathrm{d}\widetilde{q} \,\rangle\,, \tag{3.66}$$
that is
$$\sum_{i,j=1}^{n} \left(\frac{\partial p_j}{\partial q_i}\, \mathrm{d}q_i \wedge \mathrm{d}q_j + \frac{\partial p_j}{\partial p_i}\, \mathrm{d}p_i \wedge \mathrm{d}q_j \right) = \sum_{i,j=1}^{n} \left(\frac{\partial \widetilde{p}_j}{\partial q_i}\, \mathrm{d}q_i \wedge \mathrm{d}\widetilde{q}_j + \frac{\partial \widetilde{p}_j}{\partial p_i}\, \mathrm{d}p_i \wedge \mathrm{d}\widetilde{q}_j \right),$$
namely
$$\sum_{j=1}^{n} \mathrm{d}p_j \wedge \mathrm{d}q_j = \sum_{j=1}^{n} \mathrm{d}\widetilde{p}_j \wedge \mathrm{d}\widetilde{q}_j,$$
which is the claim.
- Conversely, assuming (3.66) we conclude that (3.65) holds for some function $F \in C^1(\mathbb{R}^{2n}, \mathbb{R})$ because every closed form in $\mathbb{R}^{2n}$ is exact.

The Theorem is proved. ■

► From Theorem 3.20 we deduce the following claim.

Corollary 3.2

Let

$$\Psi : \mathbb{R}^{2n} \to \mathbb{R}^{2n} \;:\; (q, p) \mapsto (\widetilde{q}, \widetilde{p}) := \Psi(q, p)$$

be a symplectic transformation on the canonical Hamiltonian phase space. Then Ψ preserves the differential $2\,k$-forms

$$\omega^{2k} := \sum_{1 \leqslant i_1 < \cdots < i_k \leqslant n} \mathrm{d}q_{i_1} \wedge \cdots \wedge \mathrm{d}q_{i_k} \wedge \mathrm{d}p_{i_1} \wedge \cdots \wedge \mathrm{d}p_{i_k}, \tag{3.67}$$

with $k = 1, \dots, n$. In particular Ψ preserves the volume form ($k = n$)

$$\omega^{2n} := \mathrm{d}q_1 \wedge \cdots \wedge \mathrm{d}q_n \wedge \mathrm{d}p_1 \wedge \cdots \wedge \mathrm{d}p_n.$$

Proof. From Theorem 3.20 we know that Ψ preserves the canonical symplectic 2-form ω. Therefore Ψ preserves also the wedge product of ω with itself k times:

$$\underbrace{\omega \wedge \cdots \wedge \omega}_{k\,\text{times}} = \sum_{i_1,\dots,i_k} \mathrm{d}q_{i_1} \wedge \cdots \wedge \mathrm{d}q_{i_k} \wedge \mathrm{d}p_{i_1} \wedge \cdots \wedge \mathrm{d}p_{i_k},$$

but $\omega \wedge \cdots \wedge \omega$ is indeed proportional to ω^{2k} defined in (3.67)

$$\underbrace{\omega \wedge \cdots \wedge \omega}_{k\,\text{times}} = (-1)^{k-1} k!\, \omega^{2k}.$$

The claim is proved. ■

► The $2k$-forms ω^{2k} have an important geometrical interpretation.

- If $k = 1$ the integral of the form $\omega^2 \equiv \omega$ on a submanifold M of $\mathbb{R}^{2n}$ is equal to the sum of the areas (the sign keeps track of the orientation) of the projections of M onto the planes (q_i, p_i).
- For arbitrary k, the integral of ω^{2k} is equal to the sum of the measures (with sign) of the projections of M onto all hyperplanes $(q_{i_1}, \dots, q_{i_k}, p_{i_1}, \dots, p_{i_k})$, with $1 \leqslant i_1 < \cdots < i_k \leqslant n$.

Corollary 3.2 claims that a symplectic transformation preserves the sum of the measures of the projections onto all coordinate hyperplanes $(q_{i_1}, \dots, q_{i_k}, p_{i_1}, \dots, p_{i_k})$ with $1 \leqslant i_1 < \cdots < i_k \leqslant n$.

3.4.6 Generating functions of canonical transformations

► We now present another characterization of canonical transformations, which is indeed an immediate consequence of the Lie condition (3.62).

Theorem 3.21

Let $F = F(q, \tilde{q}, t)$, $F \in C^2(\mathbb{R}^{2n} \times \mathbb{R}, \mathbb{R})$, be a function such that

$$\det \left(\frac{\partial^2 F}{\partial q_i \partial \tilde{q}_j} \right)_{1 \leqslant i,j \leqslant n} \neq 0. \tag{3.68}$$

The one-parameter family of C^2-diffeomorphisms

$$\Psi_t : \mathbb{R} \times \mathbb{R}^{2n} \to \mathbb{R}^{2n} \; : \; (t, (q, p)) \mapsto (\widetilde{q}, \widetilde{p}) := \Psi_t(q, p)$$

defined implicitly by

$$\begin{cases} p = \operatorname{grad}_q F(q, \widetilde{q}, t), \\ \widetilde{p} = -\operatorname{grad}_{\widetilde{q}} F(q, \widetilde{q}, t), \end{cases} \tag{3.69}$$

is a canonical transformation. The function F is called **generating function of the first kind**.

Proof. Condition (3.68) allows us to apply the "Implicit function Theorem" to invert the first equation in (3.69) to express $\widetilde{q} = \widetilde{q}(q, p, t)$. Introducing this expression in the second equation in (3.69) we find $\widetilde{p} = \widetilde{p}(q, p, t)$. Thus we have a one-parameter family of C^2-diffeomorphisms $(t, (q, p)) \mapsto (\widetilde{q}, \widetilde{p}) := \Psi_t(q, p)$. Now the Lie condition is satisfied:

$$\langle p, \underline{d}q \rangle - \langle \widetilde{p}, \underline{d}\widetilde{q} \rangle = \left\langle \operatorname{grad}_q F(q, \widetilde{q}, t), \underline{d}q \right\rangle + \left\langle \operatorname{grad}_{\widetilde{q}} F(q, \widetilde{q}, t), \underline{d}\widetilde{q} \right\rangle = \underline{d}F(q, \widetilde{q}, t),$$

from which the claim follows. ■

► In a similar way it can be proved the following claim.

Theorem 3.22

1. *Let $F = F(q, \widetilde{p}, t)$, $F \in C^2(\mathbb{R}^{2n} \times \mathbb{R}, \mathbb{R})$, be a function such that*

$$\det \left(\frac{\partial^2 F}{\partial q_i \partial \widetilde{p}_j} \right)_{1 \leqslant i,j \leqslant n} \neq 0.$$

The one-parameter family of C^2-diffeomorphisms

$$\Psi_t : \mathbb{R} \times \mathbb{R}^{2n} \to \mathbb{R}^{2n} \; : \; (t, (q, p)) \mapsto (\widetilde{q}, \widetilde{p}) := \Psi_t(q, p)$$

defined implicitly by

$$\begin{cases} \widetilde{q} = \operatorname{grad}_{\widetilde{p}} F(q, \widetilde{p}, t), \\ p = \operatorname{grad}_q F(q, \widetilde{p}, t), \end{cases}$$

is a canonical transformation. The function F is called **generating function of the second kind**.

2. *Let $F = F(\widetilde{q}, p, t)$, $F \in C^2(\mathbb{R}^{2n} \times \mathbb{R}, \mathbb{R})$, be a function such that*

$$\det \left(\frac{\partial^2 F}{\partial \widetilde{q}_i \partial p_j} \right)_{1 \leqslant i,j \leqslant n} \neq 0.$$

The one-parameter family of C^2-diffeomorphisms

$$\Psi_t : \mathbb{R} \times \mathbb{R}^{2n} \to \mathbb{R}^{2n} \; : \; (t,(q,p)) \mapsto (\widetilde{q},\widetilde{p}) := \Psi_t(q,p)$$

defined implicitly by

$$\begin{cases} q = \operatorname{grad}_p F(\widetilde{q},p,t), \\ \widetilde{p} = -\operatorname{grad}_{\widetilde{q}} F(\widetilde{q},p,t), \end{cases}$$

is a canonical transformation. The function F is called **generating function of the third kind**.

3. *Let $F = F(p,\widetilde{p},t)$, $F \in C^2(\mathbb{R}^{2n} \times \mathbb{R},\mathbb{R})$, be a function such that*

$$\det \left(\frac{\partial^2 F}{\partial p_i \partial \widetilde{p}_j} \right)_{1 \leqslant i,j \leqslant n} \neq 0.$$

The one-parameter family of C^2-diffeomorphisms

$$\Psi_t : \mathbb{R} \times \mathbb{R}^{2n} \to \mathbb{R}^{2n} \; : \; (t,(q,p)) \mapsto (\widetilde{q},\widetilde{p}) := \Psi_t(q,p)$$

defined implicitly by

$$\begin{cases} q = \operatorname{grad}_p F(p,\widetilde{p},t), \\ \widetilde{q} = \operatorname{grad}_{\widetilde{q}} F(p,\widetilde{p},t), \end{cases}$$

is a canonical transformation. The function F is called **generating function of the fourth kind**.

No Proof.

▶ Remarks:

- It can be shown that if a canonical transformation admits more that one generating functions, these are related by a Legendre transformation. For instance, if $F_1(q,\widetilde{q},t)$ and $F_2(q,\widetilde{p},t)$ are generating functions of first and second kind, then

$$F_2(q,\widetilde{p},t) = F_1(q,\widetilde{q},t) + \langle \widetilde{q},\widetilde{p} \rangle .$$

Similarly, if $F_3(\widetilde{q},p,t)$ and $F_4(p,\widetilde{p},t)$ are generating functions of third and fourth kind, one can prove that

$$\begin{aligned} F_3(\widetilde{q},p,t) &= F_1(q,\widetilde{q},t) - \langle q,p \rangle , \\ F_4(\widetilde{q},p,t) &= F_1(q,\widetilde{q},t) - \langle q,p \rangle + \langle \widetilde{q},\widetilde{p} \rangle . \end{aligned}$$

- By Theorem 3.15 we can associate to each canonical transformation a new Hamiltonian. For instance, one can prove that, in the case of generating functions of the first kind, $F = F(q, \tilde{q}, t)$, $F \in C^2(\mathbb{R}^{2n} \times \mathbb{R}, \mathbb{R})$, the new Hamiltonian (in the variables $(\tilde{q}, \tilde{p})$) is

$$\widetilde{\mathscr{H}}(\tilde{q}, \tilde{p}, t) = \mathscr{H}(\tilde{q}, \tilde{p}, t) + \frac{\partial F}{\partial t},$$

where $\mathscr{H}(\tilde{q}, \tilde{p}, t) = \mathscr{H}(q(\tilde{q}, \tilde{p}, t), p(\tilde{q}, \tilde{p}, t), t)$ is the original Hamiltonian and $F = F(q(\tilde{q}, \tilde{p}, t), \tilde{q}, t)$.

- A generating function is defined up to an arbitrary additive term which is a function of time. This term does not change the transformation, but it modifies the corresponding Hamiltonian.

- The transformations associated with generating functions exhaust all canonical transformations.

▶ The *identity transformation* $(q, p) \mapsto (\tilde{q}, \tilde{p}) = (q, p)$ is a canonical transformation associated with a generating function of second kind, $F(q, \tilde{p}, t) = \langle q, \tilde{p} \rangle$:

$$\begin{cases} \tilde{q} = \mathsf{grad}_{\tilde{p}} \langle q, \tilde{p} \rangle = q, \\ p = \mathsf{grad}_q \langle q, \tilde{p} \rangle = \tilde{p}, \end{cases}$$

The identity transformation plays a special role.

- Suppose we have a canonical Hamiltonian system with Hamiltonian $\mathscr{H} \in C^2(\mathbb{R}^{2n} \times \mathbb{R}, \mathbb{R})$, for which we know the solutions of canonical Hamilton equations.

- Consider a *perturbation* of $\mathscr{H}$ given by $\mathscr{H} + \varepsilon \mathscr{K}$ for $|\varepsilon| \ll 1$. One may ask about the existence of a canonical transformation, at least of class C in ε, which maps the solutions of the unperturbed system to the solutions of the perturbed system.

- If such a transformation exists, then it must be a perturbation of the identity transformation. Its generating function must be of the form $\langle q, \tilde{p} \rangle + G_\varepsilon(q, \tilde{p}, t)$, with $\lim_{\varepsilon \to 0} G_\varepsilon(q, \tilde{p}, t) = 0$.

▶ It is worthwhile to mention the existence of the so called *Hamilton-Jacobi method* to integrate canonical Hamilton equations by using canonical transformations. Even if we do not give any detail about this technique, which is indeed a quite important and developed part of Hamiltonian Theory, we just say that it is based on the following observation.

- Consider a canonical Hamiltonian system defined by a given Hamiltonian $\mathscr{H} \in C^2\left(\mathbb{R}^{2n}, \mathbb{R}\right)$.
- If $\Psi_t : \mathbb{R} \times \mathbb{R}^{2n} \to \mathbb{R}^{2n} : (t,(q,p)) \mapsto (\widetilde{q},\widetilde{p}) := \Psi_t(q,p)$ is a canonical transformation such that the in the variables $(\widetilde{q},\widetilde{p})$ the Hamiltonian $\mathscr{H}$ takes the simple form $\widetilde{\mathscr{H}} = \widetilde{\mathscr{H}}(\widetilde{q},t)$ then the canonical Hamilton equations in the in the variables $(\widetilde{q},\widetilde{p})$ read
$$\begin{cases} \dot{\widetilde{q}} = \mathsf{grad}_{\widetilde{p}} \widetilde{\mathscr{H}}(\widetilde{q},t) = 0, \\ \dot{\widetilde{p}} = -\,\mathsf{grad}_{\widetilde{q}} \widetilde{\mathscr{H}}(\widetilde{q},t), \end{cases}$$
which are easily integrated to give
$$\widetilde{q}(t) = \widetilde{q}(0), \qquad \widetilde{p}(t) = \widetilde{p}(0) - \int_0^t \mathsf{grad}_{\widetilde{q}} \widetilde{\mathscr{H}}(\widetilde{q},s)\big|_{\widetilde{q}(0)}\, \mathsf{ds}.$$

3.5 Exercises

Ch3.E1 In $\mathbb{R}^n$ consider the system of n Newton equations

$$\sum_{i=1}^{n} A_{ik}(q)\,\ddot{q}_i = \sum_{i,j=1}^{n} \left(\frac{1}{2}\frac{\partial A_{ij}}{\partial q_k} - \frac{\partial A_{ik}}{\partial q_j} \right) \dot{q}_i\,\dot{q}_j - \frac{\partial U}{\partial q_k},$$

where $k = 1, \ldots, n$, $A := (A_{ij})_{1 \leqslant i,j \leqslant n} \in C^2(\mathbb{R}^n, \mathbf{GL}(n, \mathbb{R}))$ is symmetric and positive definite and $U \in C^2(\mathbb{R}^n, \mathbb{R})$ is the potential energy. Prove that the total energy

$$E(q, \dot{q}) := \frac{1}{2} \langle \dot{q}, A(q)\,\dot{q} \rangle + U(q),$$

is an integral of motion.

~~~~~~~~~~~~~~~~~~~~

Ch3.E2 Consider two point masses $m_1, m_2 > 0$ in $\mathbb{R}^3$ with time-dependent coordinates

$$q^{(j)} := \left(q_1^{(j)}, q_2^{(j)}, q_3^{(j)} \right) \in \mathbb{R}^3, \qquad j = 1, 2.$$

Define

$$Q := \left(q^{(1)}, q^{(2)} \right).$$

The total coordinate space is $M = \mathbb{R}^6$. Newton equations of the system form a system of 6 ODEs of the form

$$\ddot{q}^{(1)} = G\,m_2 \frac{q^{(2)} - q^{(1)}}{\|q^{(2)} - q^{(1)}\|^3}, \qquad \ddot{q}^{(2)} = G\,m_1 \frac{q^{(1)} - q^{(2)}}{\|q^{(1)} - q^{(2)}\|^3},$$

where G is the gravitational constant. Set $G \equiv 1$.

(a) Define the vectors

$$q := q^{(1)} - q^{(2)}, \qquad q_{CM} := \frac{m_1\,q^{(1)} + m_2\,q^{(2)}}{m_1 + m_2}.$$

Prove that q and q_{CM} obey the ODEs

$$\frac{1}{m_1 + m_2}\ddot{q} = \mathsf{grad}_q \left(\frac{1}{\|q\|} \right), \qquad \ddot{q}_{CM} = 0.$$

(b) Prove that the total linear momentum,

$$P\left(\dot{Q}\right) := m_1\,\dot{q}^{(1)} + m_2\,\dot{q}^{(2)},$$

is an integral of motion.

(c) Prove that the total angular momentum,

$$\ell\left(Q, \dot{Q}\right) := m_1\,q^{(1)} \times \dot{q}^{(1)} + m_2\,q^{(2)} \times \dot{q}^{(2)},$$

is an integral of motion.

(d) Prove that Newton equations are invariant under the change of coordinates

$$q^{(1)} \mapsto \widetilde{q}^{(1)} := A\,q^{(1)} + a, \qquad q^{(2)} \mapsto \widetilde{q}^{(2)} := A\,q^{(2)} + a,$$

where $A \in \mathbf{GL}(3, \mathbb{R})$, $a \in \mathbb{R}^3$.
~~~~~~~~~~~~~~~~~~~~

Ch3.E3 In $\mathbb{R}$ consider the system of N Newton equations

$$\ddot{q}_k = e^{q_{k+1}-q_k} - e^{q_k - q_{k-1}},$$

where $k = 1, \dots, N$ and $q_{N+k} \equiv q_k$ (mod. N). Prove that the total energy

$$E(q_1, \dots, q_N, \dot{q}_1, \dots, \dot{q}_N) := \sum_{k=1}^{N} \left(\frac{\dot{q}_k^2}{2} + e^{q_{k+1}-q_k} \right)$$

is an integral of motion.

Ch3.E4 Find the shortest path joining two fixed points on $\mathbb{R}^2$.

Ch3.E5 Consider a string of length ℓ with one end fixed at the origin of the (x, y)-plane, and the other end on the x-axis. Find the shape of the string maximizing the area enclosed between the string and the x-axis.

Ch3.E6 Fix $T, \alpha > 0$. Consider the functional $\psi : K \longrightarrow \mathbb{R}$ defined by

$$\psi(\gamma) := \int_0^T \dot{q}^2 \, \mathrm{d}t,$$

where K is the space of all C^1-curves

$$\gamma := \left\{ (t, q) \, : \, q = q(t), \, q \in C^1([0, T], \mathbb{R}), \, q(0) = 0, \, q(T) = \alpha \right\}.$$

Find an extremal point of ψ. Is this extremal point a candidate to be a maximum or a minimum?

Ch3.E7 Consider the functional $\psi : K \longrightarrow \mathbb{R}$ defined by

$$\psi(\gamma) := \int_0^1 \sqrt{q^2 + \dot{q}^2} \, \mathrm{d}t,$$

where K is the space of all C^1-curves

$$\gamma := \left\{ (t, q) \, : \, q = q(t), \, q \in C^1([0, 1], \mathbb{R}), \, q(0) = 0, \, q(1) = 1 \right\}.$$

Prove that $\psi(\gamma) > 1$ for all $\gamma \in K$.

Ch3.E8 Consider the functional $\psi : K \to \mathbb{R}$ defined by

$$\psi(\gamma) := \int_1^2 \frac{1}{t}\sqrt{1+\dot{q}^2}\,\mathrm{d}t,$$

where K is the space of all C^1-curves

$$\gamma := \left\{(t,q) \,:\, q = q(t),\, q \in C^1([1,2],\mathbb{R}),\, q(1) = 0,\, q(2) = 1\right\}.$$

Find an extremal point of ψ.

Ch3.E9 Consider a function $\mathscr{L} \in C^2(\mathbb{R}^2 \times \mathbb{R}, \mathbb{R})$ and a functional $\psi : K \to \mathbb{R}$ defined by

$$\psi(\gamma) := \int_{t_1}^{t_2} \mathscr{L}(q,\dot{q},t)\,\mathrm{d}t,$$

where K is the space of all C^1-curves

$$\gamma := \left\{(t,q) \,:\, q = q(t),\, q \in C^1([t_1,t_2],\mathbb{R}),\, q(t_1) = q_1,\, q(t_2) = q_2\right\}.$$

Assume that the corresponding Euler-Lagrange equation is identically satisfied for any γ. Prove that ψ does not depend on γ but only on (t_1,q_1) and (t_2,q_2).

Ch3.E10 We look for the optimal shape of a wire that connects two fixed points A and B on a vertical plane. A bead of unit mass falls along this wire, without friction, under the influence of gravity. The shape of the wire is defined to be optimal if the bead falls from A to B in as short a time as possible.

Let $y = y(x)$ be the function which describes the shape of the wire on the (x,y)-plane, connecting $A := (0,0)$ and $B := (a,b)$ with $a > 0$ and $b \geqslant 0$. We assume that the positive y-axis is pointing downward. The associated falling time follows from elementary mechanics:

$$T(y) = \frac{1}{\sqrt{2g}}\int_0^a \sqrt{\frac{1+(y')^2}{y}}\,\mathrm{d}x, \qquad y' := \frac{\mathrm{d}y}{\mathrm{d}x}. \tag{3.70}$$

Here g is the constant gravitational acceleration (fix $g \equiv 1/2$). To solve the problem one has to minimize the functional T over the set of all functions $y \in C^1([0,a],(0,\infty))$ with $(y(0),y(a)) = (0,b)$.

(a) Justify, by elementary mechanics, formula (3.70).

(b) Construct the Euler-Lagrange equation of the problem.

(c) Reduce the second-order ODE obtained in (b) to a first-order ODE of the form

$$y(1+(y')^2) = c,$$

where $c \in \mathbb{R}$ is a constant of integration.

(d) Introduce the angular variable φ, which measures the angle that the tangent to the curve makes with the vertical. Find the family of parametric equations for the plane curve $y = y(x)$ which minimizes T (φ is the parameter).

Ch3.E11 Consider a heavy flexible cable of length L hanging between two points $(\pm\ell, h)$, $h > 0$, $L > 2\ell$, on a vertical (x, y)-plane (one can think at telephone poles and hanging telephone lines). It hangs in a curve that looks like a parabola, but, in fact, it is not.

It turns out that the cable has a potential energy which is given by the functional

$$U(y) := \rho g \int_{-\ell}^{\ell} y\sqrt{1 + (y')^2}\, dx, \qquad y' := \frac{dy}{dx}.$$

Here ρ and g are respectively the lineal density of the cable and the constant gravitational acceleration (fix $\rho g = 1$). To solve the problem one has to minimize the functional U over the set of all functions $y \in C^1(\mathbb{R}, \mathbb{R})$ with $(y(-\ell), y(\ell)) = (h, h)$ under the *constraint* that the length of the cable is the constant L.

(a) Introduce a *Lagrange multiplier* $\lambda \in \mathbb{R}$ due to the constraint and write down the Euler-Lagrange equation of the problem.

(b) Reduce the ODE obtained in (a) to a first-order ODE of the form

$$(y')^2 = \frac{(y + \lambda)^2}{c} - 1,$$

where $c \in \mathbb{R}$ is a constant of integration.

(c) Integrate the ODE obtained in (b). What is the curve describing the shape of the cable?

~~~~~~~~~~~~~~~~~~~~~~~~

Ch3.E12 In $\mathbb{R}^3$ consider two point-like masses $m_1 = m_2 = 2$ which interact through a central force field whose potential energy is

$$U(r) := r - \frac{r^4}{4},$$

where $r := \|q^{(1)} - q^{(2)}\|$, $q^{(j)} := \left(q_1^{(j)}, q_2^{(j)}, q_3^{(j)}\right) \in \mathbb{R}^3$, $j = 1, 2$, being the positions of the two masses.

(a) Describe the dynamics in the center of mass frame. Use polar coordinates (r, θ).

(b) Discuss the time evolution of the variable r and describe the orbits in the $(r, \dot{r})$-plane when the angular momentum is varied. Find the fixed points and discuss their stability. Find the condition that the angular momentum must satisfy in order to have periodic orbits.

(c) Discuss the time evolution of the variable θ in terms of r.

(d) Find a periodic orbit and determine its period.

~~~~~~~~~~~~~~~~~~~~~~~~

Ch3.E13 Consider a one-dimensional mechanical system describing the motion of a point (mass $m = 1$) under the influence of a potential energy

$$U(q) := \frac{\alpha}{q} + \log\left(\frac{q^2}{1 + q^2}\right),$$

where $\alpha > 0$ is a parameter.

(a) Write down Newton equations and the corresponding dynamical system defined through a system of ODEs in $\mathbb{R}^2$.

(b) Find the fixed points and investigate their stability. Draw the graph of the potential energy.

(c) Make a qualitative analysis of the motion in the phase space $(q, \dot{q})$. Find the value(s) of α such that there exist periodic orbits.

Ch3.E14 Consider a one-dimensional mechanical system describing the motion of a point (mass $m = 1$) under the influence of a potential energy

$$U(q) := \frac{q^4}{\alpha + \beta q^2},$$

where $\alpha, \beta \in \mathbb{R} \setminus \{0\}$ are parameters.

(a) Determine for which values of α and of β the origin $q = 0$ is a stable fixed point.

(b) Linearize the dynamical system around this fixed point and determine the frequency of small oscillations.

Ch3.E15 Consider a particle of mass $m > 0$ in $\mathbb{R}^n$ with Lagrangian

$$\mathscr{L}(q, \dot{q}, t) := e^{\alpha t/m} \left(\frac{1}{2} m \langle \dot{q}, \dot{q} \rangle - U(q) \right), \qquad \alpha > 0.$$

(a) Find the Euler-Lagrange equations.

(b) Fix $n = 1$ and $m = 1$ and consider $U(q) := -\beta q$, $\beta > 0$. Solve the Euler-Lagrange equation with $(q(0), \dot{q}(0)) = (q_0, 0)$, $q_0 > 0$. Compute $\lim_{t \to +\infty} \dot{q}$.

Ch3.E16 Consider a particle of mass $m > 0$ in $\mathbb{R}^3$ moving under the influence of a central potential energy $U : (0, \infty) \to \mathbb{R}$,

$$U(r) := -\alpha \frac{e^{-kr}}{r}, \qquad \alpha, k > 0.$$

Prove that for sufficiently small values of the angular momentum there exists a stable closed orbit.

Ch3.E17 Consider a point with mass $m = 1$ moving in $\mathbb{R}^3$ with Lagrangian

$$\mathscr{L}(q, \dot{q}) := \frac{1}{2} \left(\dot{q}_1^2 + \dot{q}_2^2 + \dot{q}_3^2 \right) + \alpha (q_1 \dot{q}_2 - q_2 \dot{q}_1), \qquad \alpha > 0.$$

(a) Write down the Euler-Lagrange equations.

(b) Show that the system is invariant under rotations about the q_3-axis.

(c) Use Noether Theorem to find the integral of motion corresponding to the above symmetry. Verify that the integral of motion is an actual conserved quantity.

Ch3.E18 Consider a point with mass $m = 1$ moving in $\mathbb{R}^2$ with Lagrangian

$$\mathscr{L}(q, \dot{q}) := \frac{1}{2}\left(\dot{q}_1^2 + \dot{q}_2^2\right) + 2\alpha^2(q_1\dot{q}_2 + q_2\dot{q}_1)^2 - \alpha\beta\left(q_1^2 + q_2^2\right), \qquad \alpha, \beta > 0.$$

(a) Use Noether Theorem to prove that the third component of the angular momentum, $\ell_3 := q_1\dot{q}_2 - q_2\dot{q}_1$ is an integral of motion.

(b) Find the value of ℓ_3 for which the circle $q_1^2 + q_2^2 = c$, $c > 0$, is an orbit.

Ch3.E19 Let $F \in C^2(\mathbb{R}^n, \mathbb{R})$ be a convex function, i.e., $F(\lambda q_1 + (1-\lambda)q_2) \leqslant \lambda F(q_1) + (1-\lambda)F(q_2)$ for all $q_1, q_2 \in \mathbb{R}^n$ and $\lambda \in [0,1]$. The *Legendre transformation* is by definition

$$F^*(p) := \sup_{q\in\mathbb{R}^n}\left(\langle q, p\rangle - F(q)\right).$$

(a) Show that $F^*(p)$ is a convex function.

(b) Show that the Legendre transformation is an involution, i.e., $(F^*(p))^*(q) = F(q)$.

Ch3.E20 (a) Compute the Legendre transformation of the function

$$F(q) := \frac{1}{2}q^2 + \frac{1}{3}q^3, \qquad q \in \mathbb{R}.$$

(b) Compute the Legendre transformation of the function

$$F(q) := \frac{1}{2}\langle q, Aq\rangle + c, \qquad q \in \mathbb{R}^n,$$

where $A \in \mathbf{GL}(n, \mathbb{R})$ is symmetric and positive definite and $c \in \mathbb{R}^n$ is a constant vector.

(c) Let $F^*(p)$ be the Legendre transformation of a convex function $F \in C^2(\mathbb{R}, \mathbb{R})$.

(1) Compute the Legendre transformation of $G(q) := F(q) + \alpha q$, $\alpha > 0$.

(2) Compute the Legendre transformation of $G(q) := \alpha F(q) + \beta$, $\alpha, \beta > 0$.

Ch3.E21 Let $\mathscr{L} = \mathscr{L}(q, \dot{q}, t)$, $\mathscr{L} \in C^2\left(\mathbb{R}^{2n} \times \mathbb{R}, \mathbb{R}\right)$ be a Lagrangian. Consider a diffeomorphism $\Psi : q \mapsto \widetilde{q} = \widetilde{q}(q)$. Denote by $\mathbf{J} = \mathbf{J}(q)$ the Jacobian of Ψ and by $\mathscr{L}(\widetilde{q}, \dot{\widetilde{q}}, t)$ the transformed Lagrangian.

(a) Prove that Ψ induces the momenta transformation

$$p \mapsto \widetilde{p} = \left(\mathbf{J}^\top\right)^{-1} p.$$

(b) Prove that $\langle p, \dot{q}\rangle = \langle \widetilde{p}, \dot{\widetilde{q}}\rangle$.

Ch3.E22 In the canonical Hamiltonian phase space $\mathbb{R}^6$ consider a point of unit mass with Hamiltonian

$$\mathscr{H}(q,p) := \sqrt{\alpha^2 + \langle p,p \rangle} + \langle a,q \rangle, \qquad \alpha > 0,$$

where $a \in \mathbb{R}^3$ is a constant vector.

(a) Derive the canonical Hamiltonian equations.

(b) Construct the Lagrangian.

(c) Derive the Euler-Lagrange equations.

〰〰〰

Ch3.E23 In the canonical Hamiltonian phase space $\mathbb{R}^6$ consider a point of unit mass with Hamiltonian

$$\mathscr{H}(q,p) := \frac{1}{2}\langle p - A(q), p - A(q) \rangle.$$

where

$$A(q) := \frac{\alpha}{2}(-q_2, q_1, 0), \qquad \alpha > 0.$$

Derive and solve the canonical Hamilton equations.

〰〰〰

Ch3.E24 In $\mathbb{R}$ consider the system of N Newton equations

$$\ddot{q}_k = e^{q_{k+1}-q_k} - e^{q_k - q_{k-1}},$$

where $k = 1, \dots, N$ and $q_{N+k} \equiv q_k$ (mod. N).

(a) Prove that the above equations of motion are Euler-Lagrange equations for the Lagrangian

$$\mathscr{L}(q_1, \dots, q_N, \dot{q}_1, \dots, \dot{q}_N) := \sum_{k=1}^{N} \left(\frac{\dot{q}_k^2}{2} - e^{q_{k+1}-q_k} \right).$$

(b) Construct the Hamiltonian of the system and write down Hamilton equations.

(c) Prove that the total linear momentum

$$P(p_1, \dots, p_N) := \sum_{k=1}^{N} p_k$$

is an integral of motion.

(d) Prove that the function

$$F(q_1, \dots, q_N, p_1, \dots, p_N) := \sum_{k=1}^{N} \left(\frac{p_k^3}{3} + e^{q_{k+1}-q_k}(p_{k+1} + p_k) \right)$$

is an integral of motion functionally independent on the Hamiltonian and P.

〰〰〰

Ch3.E25 The *symplectic Lie group* $\mathbf{SP}(n,\mathbb{R})$ is defined as the following group of matrices:

$$\mathbf{SP}(n,\mathbb{R}) := \left\{ A \in \mathbf{GL}(2n,\mathbb{R}) \,:\, A^\top \mathbb{J}\, A = \mathbb{J} \right\}.$$

Here $\mathbb{J}$ is the canonical symplectic matrix

$$\mathbb{J} := \begin{pmatrix} \mathbb{O}_n & \mathbb{1}_n \\ -\mathbb{1}_n & \mathbb{O}_n \end{pmatrix},$$

where $\mathbb{1}_n$ is the $n \times n$ identity matrix and $\mathbb{O}_n$ is the $n \times n$ zero matrix. The *symplectic Lie algebra* $\mathfrak{sp}(n,\mathbb{R})$ (which is the Lie algebra of $\mathbf{SP}(n,\mathbb{R})$) is defined as the following vector space:

$$\mathfrak{sp}(n,\mathbb{R}) := \left\{ A \in \mathfrak{gl}(2n,\mathbb{R}) \,:\, \mathbb{J}\, A^\top \mathbb{J} = A \right\}.$$

Prove that if $A \in \mathfrak{sp}(n,\mathbb{R})$ then $\mathrm{e}^{tA} \in \mathbf{SP}(n,\mathbb{R})$ for all $t \in \mathbb{R}$.

Ch3.E26 Let $F, G \in C^1(\mathbb{R}^{2n} \times \mathbb{R}, \mathbb{R})$ be two functions defined on the canonical Hamiltonian phase space with coordinates $x := (q, p)$. The *canonical Poisson bracket* of F and G is a function that we denote by $\{ F, G \} \in C^1(\mathbb{R}^{2n} \times \mathbb{R}, \mathbb{R})$ defined by the canonical symplectic product of the gradients w.r.t. x of F and G:

$$\begin{aligned} \{ F, G \} &:= \langle\, \mathsf{grad}_x F(x,t), \mathsf{grad}_x G(x,t) \,\rangle_{\mathbb{J}} \\ &= \langle\, \mathsf{grad}_x F(x,t), \mathbb{J}\ \mathsf{grad}_x G(x,t) \,\rangle . \end{aligned}$$

Here $\mathbb{J}$ is the canonical symplectic matrix

$$\mathbb{J} := \begin{pmatrix} \mathbb{O}_n & \mathbb{1}_n \\ -\mathbb{1}_n & \mathbb{O}_n \end{pmatrix},$$

where $\mathbb{1}_n$ is the $n \times n$ identity matrix and $\mathbb{O}_n$ is the $n \times n$ zero matrix.

Prove the following properties:

(a) *(Bi)linearity*: $\{ \lambda_1 F + \lambda_2 G, K \} = \lambda_1 \{ F, K \} + \lambda_2 \{ G, K \}$,

(b) *Skew-symmetry*: $\{ F, G \} = -\{ G, F \}$,

(c) *Leibniz rule*: $\{ F\, G, K \} = F\{ G, K \} + G\{ F, K \}$,

(d) *Jacobi identity*: $\{ F, \{ G, K \} \} + \{ K, \{ F, G \} \} + \{ G, \{ K, F \} \} = 0$,

for all $F, G, K \in C^1\left(\mathbb{R}^{2n} \times \mathbb{R}, \mathbb{R}\right)$ and $\lambda_1, \lambda_2 \in \mathbb{R}$.

Ch3.E27 In the canonical Hamiltonian phase space $\mathbb{R}^6$ consider a point of mass $m > 0$ with Hamiltonian

$$\mathscr{H}(p) := \frac{\langle p, p \rangle}{2m}.$$

(a) By using the notion of Poisson brackets prove that the angular momentum, $\ell(q,p) := q \times p$, is an integral of motion. Here $\times$ is the cross product in $\mathbb{R}^3$.

(b) Compute the Poisson brackets $\{ \ell_i, \ell_j \}$, $i, j = 1, 2, 3$.

(c) Compute the Poisson brackets $\{ L^2, \ell_i \}$, $i = 1, 2, 3$, where $L^2 := \ell_1^2 + \ell_2^2 + \ell_3^2$.

Ch3.E28 In the canonical Hamiltonian phase space $\mathbb{R}^6$ consider a point of mass $m = 1$ with Hamiltonian

$$\mathscr{H}(q,p) := \frac{\langle p, p \rangle}{2} + U(\|q\|).$$

where $U = U(\|q\|)$ is a central potential. The *Runge-Lenz vector* is defined by the formula

$$A(q,p) := p \times \ell(q,p) + U(\|q\|)\, q,$$

where $\ell(q,p) := q \times p$ is the angular momentum

(a) Prove that ℓ is an integral of motion.

(b) Prove the following formula:

$$\{ A_i(q,p), \mathscr{H}(q,p) \} = p_i\left(\|q\| U'(\|q\|) + U(\|q\|)\right), \qquad i = 1,2,3.$$

(c) Consider the *Kepler potential*

$$U(\|q\|) := -\frac{\alpha}{\|q\|}, \qquad \alpha > 0.$$

Prove that A is an integral of motion.

Ch3.E29 In the canonical Hamiltonian phase space $\mathbb{R}^2$ consider a point of mass $m > 0$ with Hamiltonian

$$\mathscr{H}(q,p) := \frac{p^2}{2m} + \frac{1}{2} m\, \omega^2 q^2, \qquad \omega > 0.$$

Introduce the complex variables

$$a_{\pm}(q,p) := \sqrt{\frac{m\,\omega}{2}} \left(q \pm \frac{\mathrm{i}\, p}{m\,\omega} \right), \qquad \mathrm{i} := \sqrt{-1}.$$

(a) Write the Hamiltonian in terms of $a_{\pm}$.

(b) Compute the Poisson brackets $\{ a_+, a_- \}$ and $\{ a_{\pm}, \mathscr{H}(a_+, a_-) \}$.

(c) Find and solve the equations of motion for the variables $a_{\pm}$. Determine the solution of the equations of motion in terms of the variables (q,p).

Ch3.E30 Consider N point particles with masses $m_i > 0$ and corresponding position vectors $q^{(i)} := \left(q_1^{(i)}, q_2^{(i)}, q_3^{(i)}\right) \in \mathbb{R}^3, i = 1, \ldots, N$. The system is described by the Hamiltonian

$$\mathscr{H}(q,p) := \sum_{i=1}^{N} \frac{\|p^{(i)}\|^2}{2\,m_i} - \alpha \sum_{1 \leqslant i < j \leqslant N} \frac{m_i\, m_j}{\|q^{(i)} - q^{(j)}\|}, \qquad \alpha > 0.$$

Prove that the the corresponding dynamical system has no fixed points.

Ch3.E31 Consider the following parametric family of ODEs in the canonical Hamiltonian phase space $\mathbb{R}^2$:

$$\begin{cases} \dot{q} = -8\,p^3 - \gamma\, p\, q^2 + 6\,p, \\ \dot{p} = 5\,p^2\, q + \alpha\, q^3 - \beta\, q, \end{cases}$$

with $\alpha, \beta, \gamma > 0$.

(a) Determine α, β, γ in such a way that the above system of ODEs is Hamiltonian and compute the corresponding Hamiltonian.

(b) Set $\alpha = 0$. Determine β and γ in such a way that the system is Hamiltonian and compute the corresponding Hamiltonian. Determine $\delta, \varepsilon \in \mathbb{R}$ so that the two families of curves $4\,p^2 + 5\,q^2 + \delta = 0, 2\,p^2 + \varepsilon = 0$, are invariant under the Hamiltonian flow.

Ch3.E32 In the canonical Hamiltonian phase space $\mathbb{R}^4$ consider the Hamiltonian

$$\mathscr{H}(q,p) := \frac{p_1^2}{q_1^2 - 1}\left(1 + q_2^2 + p_2^2\right).$$

(a) Derive the canonical Hamilton equations.

(b) Determine two functionally independent and Poisson involutive integrals of motion.

Ch3.E33 In the canonical Hamiltonian phase space $\mathbb{R}^2$ consider the the parametric family of vector fields

$$f(q,p) := \left(p^\alpha q^\beta, -p^{\alpha+1} q^\delta\right),$$

with $\alpha, \beta, \gamma, \delta \in \mathbb{R}$.

(a) Compute for which values of $\alpha, \beta, \gamma, \delta$ the vector field f is a canonical Hamiltonian vector field.

(b) Compute the corresponding Hamiltonians.

(c) Solve the canonical Hamilton equations for $\alpha \neq -1$.

Ch3.E34 Consider the following IVP in $\mathbb{R}^2$:

$$\begin{cases} \dot{x} = \mathsf{grad}_x G(x), \\ x(0) \in \mathbb{R}^2, \end{cases}$$

where $G \in \mathscr{F}(\mathbb{R}^2, \mathbb{R})$.

(a) In which case is the flow of this IVP a canonical Hamiltonian flow?

(b) In such a case find the Hamiltonian.

Ch3.E35 In the canonical Hamiltonian phase space $\mathbb{R}^{2n}$ consider the transformation

$$(q,p) \mapsto (\widetilde{q}, \widetilde{p}) := (q, f(q,p)),$$

for some smooth function f.

(a) Determine the structure that f must have for the transformation to be symplectic.

(b) Find a generating function for the transformation.

Ch3.E36 In the canonical Hamiltonian phase space $\mathbb{R}^4$ consider the Hamiltonian

$$\mathscr{H}(q_1,q_2,p_1,p_2) := \frac{1}{2}\left(p_1^2 + q_1^2\, q_2\, p_2\right).$$

(a) Find a one-parameter group of symplectic transformations

$$(q_1,q_2,p_1,p_2) \mapsto (\widetilde{q}_1,\widetilde{q}_2,\widetilde{p}_1,\widetilde{p}_2) = \Psi_s(q_1,q_2,p_1,p_2), \qquad s \in \mathbb{R},$$

which preserves the form of the function $\mathscr{H}$ for all $s \in \mathbb{R}$, i.e.,

$$\mathscr{H}\left(\widetilde{q}_1,\widetilde{q}_2,\widetilde{p}_1,\widetilde{p}_2\right) := \frac{1}{2}\left(\widetilde{p}_1^2 + \widetilde{q}_1^2\, \widetilde{q}_2\, \widetilde{p}_2\right).$$

(b) Construct the infinitesimal generator $\mathbf{v}$ of Ψ_s.

(c) Find the Hamiltonian of the vector field $\mathbf{v}$ and verify that such function is an integral of motion of the canonical Hamiltonian flow of $\mathscr{H}$.

Ch3.E37 In the canonical Hamiltonian phase space $\mathbb{R}^{2n}$ consider the transformation

$$(q_1,\ldots,q_n,p_1,\ldots,p_n) \mapsto (\alpha_1\, q_1,\ldots,\alpha_n\, q_n,\beta_1\, p_1,\ldots,\beta_n\, p_n),$$

with $\alpha_i,\beta_i \in \mathbb{R}\setminus\{0\}$ and $\alpha_i\beta_i = \lambda \in \mathbb{R}$ for all $i = 1,\ldots,n$.

(a) Prove the above transformation preserves the canonical structure of the Hamiltonian equations.

(b) For which value(s) of λ are these transformations symplectic?

Ch3.E38 In the canonical Hamiltonian phase space $\mathbb{R}^2$ consider the transformation

$$(q,p) \mapsto (\widetilde{q},\widetilde{p}) := \left(q\sqrt{1+q^2\,p^2}, \frac{p}{\sqrt{1+q^2\,p^2}}\right).$$

Show that this transformation is symplectic by proving that the canonical symplectic 2-form is preserved.

Ch3.E39 In the canonical Hamiltonian phase space $\mathbb{R}^4$ consider a discrete dynamical system defined by iterations of the map $G : \mathbb{R}^4 \setminus \{q_1q_2 = 1\} \to \mathbb{R}^4$ given by:

$$\begin{cases} \widetilde{q}_1 = -p_2 + \dfrac{2\,q_1}{1-q_1\,q_2}, \\[2ex] \widetilde{q}_2 = -p_1 + \dfrac{2\,q_2}{1-q_1\,q_2}, \\[2ex] \widetilde{p}_1 = q_2, \\ \widetilde{p}_2 = q_1. \end{cases}$$

(a) Prove that G is a symplectic transformation.

(b) Prove that the functions

$$F_1(q_1,q_2,p_1,p_2) := q_1\,p_1 - q_2\,p_2,$$
$$F_2(q_1,q_2,p_1,p_2) := q_1\,q_2 + p_1\,p_2 - q_1\,q_2\,p_1\,p_2 - q_1\,p_1 - q_2\,p_2,$$

are two functionally independent integrals of motion of G.

(c) Prove that F_1 and F_2 are in involution.

Ch3.E40 In the canonical Hamiltonian phase space $\mathbb{R}^4$ consider the discrete dynamical system defined by iterations of the map $G : \mathbb{R}^4 \setminus \{q_1^2 = q_2^2\} \to \mathbb{R}^4$ given by:

$$\begin{cases} \widetilde{q}_1 = p_1(q_1^2+q_2^2) + 2\,q_1\,q_2\,p_2, \\ \widetilde{q}_2 = p_2(q_1^2+q_2^2) + 2\,q_1\,q_2\,p_1, \\ \widetilde{p}_1 = \dfrac{q_1}{q_1^2-q_2^2}, \\ \widetilde{p}_2 = -\dfrac{q_2}{q_1^2-q_2^2}. \end{cases}$$

It can be proved that the map G preserves the canonical Poisson brackets.

(a) Prove that the functions

$$F_1(q_1,q_2,p_1,p_2) := q_1\,p_1 + q_2\,p_2,$$
$$F_2(q_1,q_2,p_1,p_2) := q_1\,p_2 + q_2\,p_1,$$

are two functionally independent integrals of motion of G.

(b) Prove that F_1 and F_2 are in involution.

Consider the change of coordinates $W : (q_1,q_2,p_1,p_2) \mapsto (Q_1,Q_2,P_1,P_2)$ defined by:

$$\begin{cases} Q_1 = \dfrac{1}{2}\big(\ln(q_1+q_2) + \ln(q_1-q_2)\big), \\ Q_2 = \dfrac{1}{2}\big(\ln(q_1+q_2) - \ln(q_1-q_2)\big), \\ P_1 = q_1\,p_1 + q_2\,p_2, \\ P_2 = q_1\,p_2 + q_2\,p_1. \end{cases}$$

It can be proved that the map W preserves the canonical Poisson brackets.

(c) Prove that W provides the following *linearization* of G:

$$\begin{cases} \widetilde{Q}_1 = Q_1 + \nu_1(I_1,I_2), \\ \widetilde{Q}_2 = Q_2 + \nu_2(I_1,I_2), \\ \widetilde{P}_1 = P_1, \\ \widetilde{P}_2 = P_2, \end{cases}$$

where ν_1 and ν_2 are two functions of the integrals of motion to be determined.

Ch3.E41 (a) In the canonical Hamiltonian phase space $\mathbb{R}^2$ consider the following parametric family of maps which approximates the identity transformation:

$$(q,p) \mapsto (\widetilde{q},\widetilde{p}) := \big(q + \varepsilon\, f_1(q,p), p + \varepsilon\, f_2(q,p)\big), \qquad |\varepsilon| \ll 1,$$

with

$$f_i(q,p) := \alpha_i\, q^2 + 2\,\beta_i\, q\, p + \gamma_i\, p^2, \qquad i = 1,2,$$

where $\alpha_i, \beta_i, \gamma_i \in \mathbb{R}$ are parameters.

Determine the conditions on $\alpha_i, \beta_i, \gamma_i$ such that the above transformation is symplectic up to second order terms in ε.

(b) In the system of canonical coordinates $(\widetilde{q},\widetilde{p}) \in \mathbb{R}^2$, determined in (a), consider the Hamiltonian of the following perturbation of a one-dimensional harmonic oscillator of mass $m > 0$:

$$\mathscr{H}(\widetilde{q},\widetilde{p}) := \frac{\widetilde{p}^2}{2\,m} + \frac{1}{2}m\,\omega^2\,\widetilde{q}^2 + \varepsilon\,\widetilde{q}^3, \qquad \omega > 0.$$

(i) Perform the inverse symplectic transformation found in (a) and adjust the parameters $\alpha_i, \beta_i, \gamma_i$ to kill the anharmonic term of the Hamiltonian $\mathscr{H}(\widetilde{q},\widetilde{p})$, so that the new Hamiltonian will be of the form

$$\mathscr{H}(q,p) = \frac{p^2}{2\,m} + \frac{1}{2}m\,\omega^2\,q^2 + O(\varepsilon^2).$$

(ii) Construct a first order approximation of the solutions of the canonical Hamilton equations of the anharmonic oscillator starting from the solutions of the harmonic oscillator.

Ch3.E42 In the canonical Hamiltonian phase space $\mathbb{R}^2$ consider the parametric transformation

$$(q,p) \mapsto (\widetilde{q},\widetilde{p}) := \big(-p, q + \alpha\,p^2\big), \qquad \alpha \in \mathbb{R}.$$

(a) Prove that this is a symplectic transformation by verifying the Lie condition corresponding to a generating function of the first kind, $F_1 = F_1(q,\widetilde{q})$.

(b) Find a generating function of the second kind, $F_2 = F_2(q,\widetilde{p})$, by computing the Legendre transformation of F_1. Verify that F_2 generates the symplectic transformation $(q,p) \mapsto (\widetilde{q},\widetilde{p})$.

Ch3.E43 In the canonical Hamiltonian phase space $\mathbb{R}^2$ consider the parametric transformation

$$(q,p) \mapsto (\widetilde{q},\widetilde{p}) := \left(q\cos\theta - p\,\frac{\sin\theta}{m\,\omega}, q\,m\,\omega\,\sin\theta + p\cos\theta\right), \qquad \theta \in [0, 2\,\pi).$$

Show that it is a symplectic transformation.

(a) By computing the Poisson bracket $\{\widetilde{q},\widetilde{p}\}$.

(b) By finding a generating function of the transformation equations to express $(p,\widetilde{p})$ in terms of $(q,\widetilde{q})$.

Assume that (q, p) are the canonical coordinates for a one-dimensional harmonic oscillator of mass $m > 0$ with Hamiltonian

$$\mathscr{H}(q,p) := \frac{1}{2}\left(\frac{p^2}{m} + m\,\omega^2\,q^2\right), \qquad \omega > 0.$$

(c) Find the Hamiltonian $\widetilde{\mathscr{H}}(\widetilde{q}, \widetilde{p}, t)$ assuming that $\theta = \theta(t)$.

(d) Show that it is possible to choose $\theta(t)$ so that $\widetilde{\mathscr{H}}(\widetilde{q}, \widetilde{p}, t) = 0$. With this choice of $\theta(t)$ solve the canonical Hamilton equations to find the time evolution of the original variables (q, p).

4

Introduction to Hamiltonian Mechanics on Poisson Manifolds

4.1 Introduction

► Roughly speaking, a manifold is a (hyper)surface that locally looks like a region of $\mathbb{R}^n$. More formally a smooth manifold is a Hausdorff topological space that is locally diffeomorphic to the Euclidean space. So, from the geometric point of view, a smooth n-dimensional manifold is the natural generalization of the Euclidean space $\mathbb{R}^n$, which is flat and admits a global system of coordinates $q := (q_1, \ldots, q_n)$.

► There are several reasons to generalize the Lagrangian and Hamiltonian formalism on linear spaces (Chapter 3) to smooth manifolds. We will restrict our generalization only to the Hamiltonian formulation of mechanics, even though the Lagrangian side can be generalized as well.

► As a matter of fact the canonical Hamiltonian phase space $\mathbb{R}^{2n}$, with global coordinates (q, p), has two evident and severe limitations:

1. In Chapter 3 our *configuration (vector) space*, where coordinates q live, was $\mathbb{R}^n$. Many mechanical systems have a configuration space which can be described in terms of Euclidean coordinates $q := (q_1, \ldots, q_n)$ only at a *local* level. For instance, the position of a particle moving on a smooth two-dimensional closed surface embedded in the ambient space $\mathbb{R}^3$ admits a natural parametrization in terms of the coordinates which parametrize the surface. In other words, in this case, the configuration space is a surface, which is smooth manifold. To fix ideas: if the surface is the (unit) sphere $\mathbb{S}^2$ the position of the particle is completely determined in terms of two angles $(\theta, \phi) \in [0, \pi) \times [0, 2\pi)$.

2. The canonical Hamiltonian phase space is by construction even-dimensional. In other words, a canonical Hamiltonian system is always described in terms of an IVP consisting of an even number of ODEs. But there are systems whose description leads naturally to an odd-dimensional phase space. In other words, the mechanics is described by an IVP consisting of an odd number of ODEs. A natural question is to understand if such systems admit some Hamiltonian formulation. Furthermore, the even-dimensional nature of the canonical Hamiltonian phase space does not really fit with the description of systems depending on parameters. Suppose to have a canonical Hamiltonian system on $\mathbb{R}^{2n}$ which is characterized by some odd number, say s (maybe $s = 1$), of parameters that are constant in time. Then even though for a fixed value of the parameter(s), we can describe the dynamics in terms of the canonical formalism, it is useful to envisage the $(2n + s)$-dimensional space in order to keep track of how the

behavior of the system depends on the parameters. For example, this may be very useful to investigate stability properties and occurrence of bifurcations.

► Indeed we already considered some Hamiltonian systems whose configuration space is not $\mathbb{R}^n$ but a *submanifold of* $\mathbb{R}^n$. In fact, as anticipated in the Introduction of Chapter 3, the canonical Hamiltonian formulation of mechanics given in Chapter 3 *is* the correct *local* formulation of Hamiltonian mechanics when the phase space is a *symplectic manifold* (which is, maybe not surprisingly, always even-dimensional). The standard way to obtain a symplectic manifold as phase space is to start with a n-dimensional smooth manifold as configuration space. Then the $2n$-dimensional *cotangent bundle* of the manifold, to be defined later, can be locally equipped with a symplectic structure induced by the canonical symplectic 2-form (see Definition 3.12).

Example 4.1 (***Planar and spherical penduli***)

1. The planar pendulum considered several times in the previous Chapters admits both a Lagrangian and a Hamiltonian formulation. In particular its Hamiltonian

$$\mathscr{H}(q,p) := \frac{p^2}{2} - \cos q,$$

leads to the canonical Hamilton equations

$$\begin{cases} \dot{q} = \dfrac{\partial \mathscr{H}}{\partial p} = p, \\ \dot{p} = -\dfrac{\partial \mathscr{H}}{\partial q} = -\sin q. \end{cases}$$

The configuration space is diffeomorphic to the circle S^1, which is a one-dimensional closed curve embedded in $\mathbb{R}^2$. Therefore, even if the ambient space of the system is two-dimensional, say the plane $\mathbb{R}^2$, the system has only one *degree of freedom*. In other words, the configuration is completely described in terms of the angle of rotation $q \in [0, 2\pi)$.

2. A *spherical pendulum* is a mass attached to a fixed centre by a rigid rod, free to swing in any direction in $\mathbb{R}^3$. The state of the pendulum is entirely determined by the position of the mass, which is constrained to move on the surface of a sphere. The configuration space is diffeomorphic to the sphere S^2. The number of degrees of freedom of the system is two.

► From the above examples, we argue that a possible manifestation of the first limitation is when we consider systems where the spatial configurations are subject to some geometric constraints, as for instance a rigid rod for a pendulum. One of the effects of these constraints is that the dimension of the configuration space, the so called *number of degrees of freedom*, is less than the dimension of the ambient space. Let us illustrate the meaning of such constraints.

- Consider a mechanical system in $\mathbb{R}^3$ consisting of N points. If all configurations are possible, the system is *free*. It can be described by a global system of coordinates in $\mathbb{R}^{3N}$. Then one can define a canonical Hamiltonian phase space as a subset of $\mathbb{R}^{3N} \times \mathbb{R}^{3N} \simeq \mathbb{R}^{6N}$. The dimension of the ambient space, $3N$, coincides with the dimension of the configuration space.

- Nevertheless, it happens quite often that there are some geometric *constraints* on the allowed configurations of the system. If such constraints do not depend on time one says that the system admis *holonomic constraints*. For example, if $N = 1$, we can require that the point lies on a given regular surface $F(x_1, x_2, x_3) = 0$ in $\mathbb{R}^3$. In such a case it is possible to introduce a local parametrization of the surface, $x_i = x_i(q_1, q_2)$, $i = 1, 2, 3$, with the property that the Jacobian matrix has maximum rank

$$\operatorname{rank}\begin{pmatrix} \frac{\partial x_1}{\partial q_1} & \frac{\partial x_1}{\partial q_2} \\ \frac{\partial x_2}{\partial q_1} & \frac{\partial x_2}{\partial q_2} \\ \frac{\partial x_3}{\partial q_1} & \frac{\partial x_3}{\partial q_2} \end{pmatrix} = 2,$$

where (q_1, q_2) vary in an certain open subset of $\mathbb{R}^2$. The vectors $\partial x/\partial q_1$ and $\partial x/\partial q_2$ are then linearly independent and form a basis in the *tangent plane* to the surface at x. Recall that a vector $v \in \mathbb{R}^3$ is tangent to a surface $S := \{x \in \mathbb{R}^3 : F(x_1, x_2, x_3) = 0\}$ at $p \in S$ if there exists a differentiable curve $\gamma : (-\varepsilon, \varepsilon) \to S$ such that $\gamma(0) = p$ and $\dot{\gamma}(0) = v$.

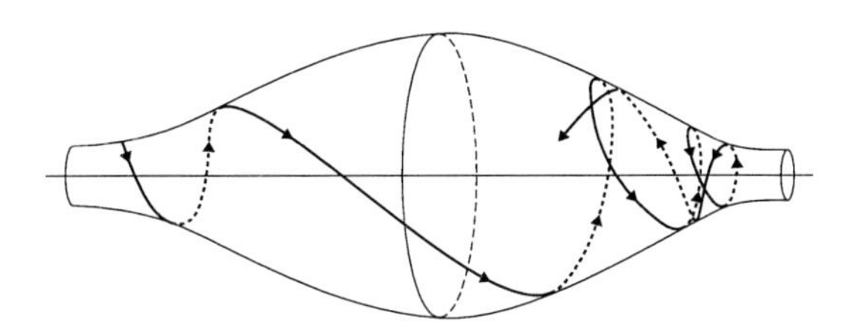

Fig. 4.1. A particle moving on a surface ([FaMa]).

The geometric and hence the kinematic description of the system becomes much more complicated when the system contains two or more points (i.e., $N > 1$), mutually constrained. An example is the case when the distance between each pair of points in the system is fixed.

► Concerning the second limitation, a standard example of a mechanical system governed by an odd number of first-order ODEs is the three-dimensional rigid body with a fixed point (the so called *Euler top*). An elementary mechanical analysis leads to a description of the body by the three components of the angular momentum (relative to body coordinates, i.e., coordinates fixed in the body): these components, say x_1, x_2, x_3, evolve according to the following system of ODEs

$$\begin{cases} \dot{x}_1 = (a_2 - a_3)\, x_2\, x_3, \\ \dot{x}_2 = (a_3 - a_1)\, x_3\, x_1, \\ \dot{x}_3 = (a_1 - a_2)\, x_1\, x_2, \end{cases} \tag{4.1}$$

where a_1, a_2, a_3 are real parameters. This situation prompts two fundamental questions.

1. We note that a configuration of the body is given by three real numbers (to specify the rotation required to rotate the body into the given configuration). A canonical Hamiltonian description of the rigid body would use six first-order equations (for example, three ODEs for the evolution of three angles and three ODEs for the evolution of the corresponding momenta). However many questions are easier to study using the three ODEs (4.1). So how is the description of the Euler top related to a six-dimensional Hamiltonian description?

2. Is the dynamics Hamiltonian is some sense? We anticipate that the answer is yes. Indeed, the set of the three coordinates x_1, x_2, x_3 is a prototype example of a *Poisson manifold* and the ODEs (4.1) are *already* in Hamiltonian form.

► We shall see that the natural framework to develop Hamiltonian mechanics is a *Poisson manifold,* that is a smooth manifold, not necessarily of even dimension, equipped with a Poisson structure. Such a Poisson structure will allow us to properly define Hamiltonian vector fields. Remarkably a Poisson bracket on a Poisson manifold may be degenerate, unlike the canonical Poisson bracket in Definition 3.9. But, if the bracket is non-degenerate it allows to define locally a closed and non-degenerate symplectic 2-form. This leads to the concept of *symplectic manifold,* whose dimension is necessarily even due to the non-degeneracy condition. It turns out that every symplectic manifold is locally diffeomorphic to the canonical symplectic Hamiltonian phase space $\mathbb{R}^{2n}$. This last claim is the content of the famous "Darboux Theorem" 4.1.

4.2 Basic facts on smooth manifolds

► In order to define Hamiltonian dynamics on a smooth manifold we first need a formal definition of this object supplemented with the definition of the main geometrical structures one can define on it. Before starting let us note that:

- The main geometrical features of a smooth manifold will be independent of any particular coordinate system on the open subset which might be used to define them. Therefore, it becomes of great importance to free ourselves from the dependence on particular local coordinates. From this point of view, manifolds provide the natural setting for studying objects that do not depend on coordinates.

- We will not be concerned about the degree of differentiability of all objects we are going to define: everything will be *smooth*, namely C^∞. Furthermore we will be interested only in *real* manifolds.
- We will provide only those fundamental facts and notions necessary for the our purposes. All theorems in this Section will be given without proof.

4.2.1 Definition of a smooth manifold

► We start with the following definition (refinements of it such as maximality conditions and equivalence classes of charts are ignored).

Definition 4.1

A real n-dimensional **smooth manifold** *M is a set of points together with a countable collection (called* **atlas***) of open sets $U_\alpha \subset M$ (called* **coordinate charts***), and one-to-one mappings $\chi_\alpha : U_\alpha \to \mathbb{R}^n$ (called* **local coordinate functions***) which satisfy the following properties:*

1. *The coordinate charts cover M:*

$$\bigcup_\alpha \chi_\alpha (U_\alpha) = M.$$

2. *For each pair of indices α, β such that $W := \chi_\alpha(U_\alpha) \cap \chi_\beta(U_\beta) \neq \emptyset$, the one-to-one overlap functions $\chi_\beta^{-1} \circ \chi_\alpha$ and $\chi_\alpha^{-1} \circ \chi_\beta$ are smooth.*

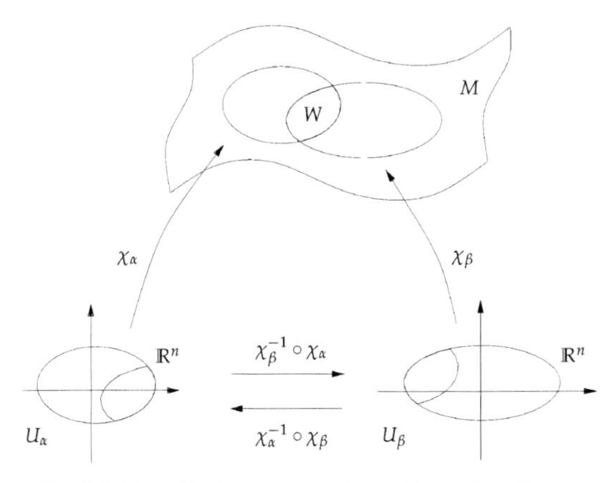

Fig. 4.2. Coordinate charts and coordinate functions.

► Let us give some examples of smooth manifolds.

Example 4.2 (*The Euclidean space* $\mathbb{R}^n$)

1. The simplest n-dimensional smooth manifold is just the Euclidean space $\mathbb{R}^n$. There is a single ($\alpha = 1$) coordinate chart $U = \mathbb{R}^n$, with a global coordinate function given by the identity $\chi = \mathbb{1}_n : \mathbb{R}^n \to \mathbb{R}^n$.
2. Any open subset $M \subset \mathbb{R}^n$ is a n-dimensional smooth manifold with a single coordinate chart given by $U = M$, with global coordinate function the identity again. Conversely, if M is any smooth manifold with a single global coordinate chart $\chi : M \to \mathbb{R}^n$, we can identify M with its image $\chi(M)$, which is an open subset of $\mathbb{R}^n$.

Example 4.3 (*The unit sphere* $\mathbb{S}^2$)

The unit sphere

$$\mathbb{S}^2 := \left\{(x_1, x_2, x_3) \in \mathbb{R}^3 \,:\, x_1^2 + x_2^2 + x_3^2 = 1\right\}$$

is an example of a two-dimensional smooth manifold realized as a surface embedded in $\mathbb{R}^3$.

- Let
$$U_1 := \mathbb{S}^2 \setminus \{(0,0,1)\}, \qquad U_2 := \mathbb{S}^2 \setminus \{(0,0,-1)\},$$
be the subsets obtained by deleting the north and south poles respectively.
- Let $\chi_\alpha : U_\alpha \to \mathbb{R}^2$, $\alpha = 1, 2$ be *stereographic projections* form the respective poles:
$$\chi_1(x_1, x_2, x_3) := \left(\frac{x_1}{1 - x_3}, \frac{x_2}{1 - x_3}\right), \quad \chi_2(x_1, x_2, x_3) := \left(\frac{x_1}{1 + x_3}, \frac{x_2}{1 + x_3}\right).$$
- It can be verified that the overlap function
$$\chi_1 \circ \chi_2^{-1} : \mathbb{R}^2 \setminus \{(0,0)\} \to \mathbb{R}^2 \setminus \{(0,0)\}$$
is a smooth diffeomorphism given by the inversion
$$\left(\chi_1 \circ \chi_2^{-1}\right)(x_1, x_2) = \left(\frac{x_1}{x_1^2 + x_2^2}, \frac{x_2}{x_1^2 + x_2^2}\right).$$

The unit sphere is a particular case of the general concept of a surface in $\mathbb{R}^3$, which historically provided the motivating example for the development of the general theory of manifolds.

Example 4.4 (*The torus* $\mathbb{T}^2$)

If we start with two identical unit circles $\mathbb{S}^1$ in the (x_1, x_3)-plane centered at $x_1 = \pm 2$, then rotate them round the x_3-axis in $\mathbb{R}^3$, the result is a *torus* $\mathbb{T}^2$. It is a two-dimensional smooth manifold.

Example 4.5 (*Implicit submanifolds of* $\mathbb{R}^n$)

Smooth manifolds often arise as zero-level sets

$$M := \{x \in \mathbb{R}^n \,:\, F_i(x) = 0,\, i = 1, \ldots, k\},$$

for a given set of k smooth functions $F_i : \mathbb{R}^n \to \mathbb{R}$. If the Jacobian matrix of $F := (F_1, \ldots, F_k)$ is a constant $\ell \leqslant k < n$ for all x, then M is a smooth manifold of dimension $n - \ell$. In this situation, the set M is called an *implicit submanifolds of* $\mathbb{R}^n$.

4.2.2 1-forms and vector fields

► Let M be a real n-dimensional smooth manifold and $p \in M$ be a point on M.

- The *algebra of smooth functions* on M will be denoted by $\mathscr{F}(M) \equiv C^\infty(M, \mathbb{R})$.
- The *tangent space* to M at p is denoted by T_pM and its dual space, the *cotangent space* to M at p, is denoted by T_p^*M. The tangent and cotangent spaces to M, which are n-dimensional vector spaces, form the *fibers* of the *tangent bundle* TM, resp. the *cotangent bundle* T^*M:

$$TM := \bigcup_{p \in M} \{p\} \times T_pM, \qquad T^*M := \bigcup_{p \in M} \{p\} \times T_p^*M.$$

(a) The bundles TM and T^*M carry a natural structure of real smooth manifolds of dimension $2\,n$.

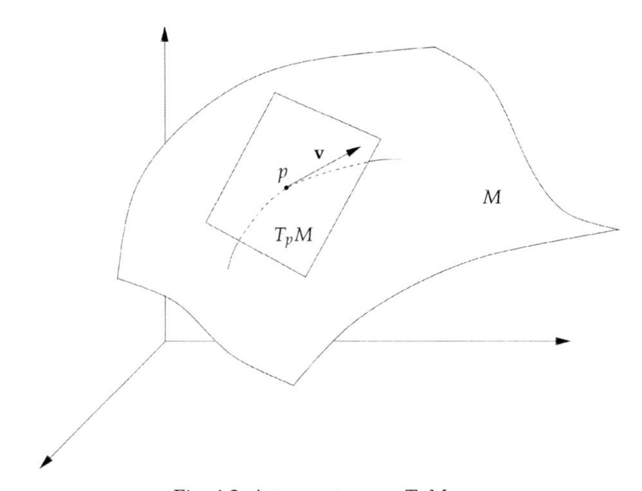

Fig. 4.2. A tangent space T_pM.

(b) A (smooth) *vector field* $\mathbf{v}$ on M is a smooth section of the tangent bundle (i.e., a smooth map $\mathbf{v} : M \to TM$ that assigns a vector $\mathbf{v}_p \in T_pM$ at each point $p \in M$). The $\mathscr{F}(M)$-module of vector fields on M is denoted by $\mathfrak{X}(M)$.

(c) A *1-form* ω (also called *covector field*) is a smooth section of the cotangent bundle (i.e., a smooth map $\omega : M \to T^*M$ that assigns a 1-form $\omega_p \in T_p^*M$ at each point $p \in M$). The $\mathscr{F}(M)$-module of 1-forms on M is denoted by $\Omega(M)$.

(d) We denote the pairing between a vector space and its dual by $\langle \cdot, \cdot \rangle$. Thus if $\mathbf{v}_p \in T_pM$ we may define a function $\omega(\mathbf{v}) \in \mathscr{F}(M)$ at p by setting

$$(\omega(\mathbf{v}))_p := \langle \omega_p, \mathbf{v}_p \rangle \qquad \forall\, p \in M. \tag{4.2}$$

Formula (4.2) expresses the *evalutation* at p of a 1-form $\omega \in \Omega(M)$ on a smooth vector field $\mathbf{v} \in \mathfrak{X}(M)$. The result is a smooth function evaluated at p, namely a real number.

Example 4.6 (*The tangent space to implicit submanifolds of* $\mathbb{R}^n$)

Let

$$M := \{x \in \mathbb{R}^n \, : \, F_i(x) = 0, \, i = 1, \dots, k\},$$

be an implicit submanifold of $\mathbb{R}^n$. The tangent space to M at $p \in M$ is

$$T_pM := \{y \in \mathbb{R}^n \, : \, \langle \, \mathrm{grad}_x F_i(x), y \, \rangle = 0, \, i = 1, \dots, k\} \, .$$

- If $M = S^2$, so that $F(x_1, x_2, x_3) := x_1^2 + x_2^2 + x_3^2 - 1$, we get

$$T_pS^2 := \left\{ (y_1, y_2, y_3) \in \mathbb{R}^3 \, : \, x_1 y_1 + x_2 y_2 + x_3 y_3 = 0 \right\} .$$

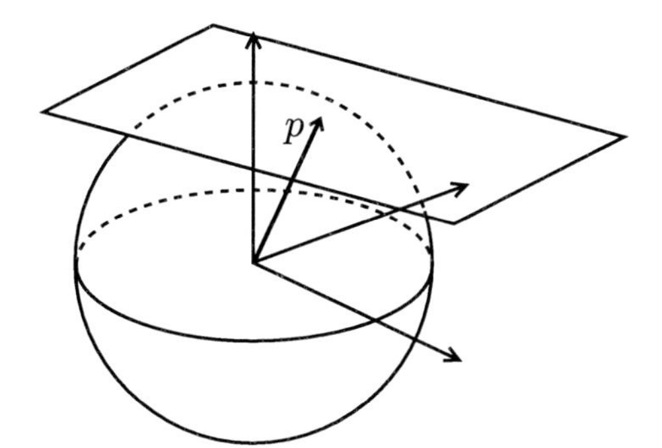

Fig. 4.3. A tangent space T_pS^2.

- The tangent bundle TS^2 is by definition

$$TS^2 := \bigcup_{p \in S^2} \{p\} \times T_pS^2,$$

a four-dimensional smooth manifold which is not possible to visualize.

- A simpler example is the unit circle S^1, whose tangent bundle is isomorphic to $S^1 \times \mathbb{R}$. Geometrically, this is a cylinder of infinite height.

- To a function $F \in \mathscr{F}(M)$ we may associate its *differential* $\mathrm{d}F \in \Omega(M)$, which is a 1-form, that can be applied to vector fields on M.

  (a) The notion of differential is used to associate to every vector field $\mathbf{v}$ on M a *derivation* on $\mathscr{F}(M)$. For $F \in \mathscr{F}(M)$ we define $\mathbf{v}[F] \in \mathscr{F}(M)$ by

$$\mathbf{v}[F] := \mathrm{d}F(\mathbf{v}) \tag{4.3}$$

  or, equivalently,

$$(\mathbf{v}[F])_p := \left\langle \, \mathrm{d}F_p, \mathbf{v}_p \, \right\rangle \qquad \forall \, p \in M. \tag{4.4}$$

Note that $(\mathbf{v}[F])_p$ is the (directional) Lie derivative of F along the direction of $\mathbf{v}$ at p.

(b) Saying that $\mathbf{v}$ is a *derivation* on $\mathscr{F}(M)$ means

$$\mathbf{v}[F\,G] = \mathbf{v}[F]G + F\,\mathbf{v}[G] \qquad \forall\, F, G \in \mathscr{F}(M),$$

which is a consequence of (4.4) and the *Leibniz rule* for differentials.

- It is a fundamental fact that every derivation on M corresponds to a unique vector field on M and that every derivation at p corresponds to a unique tangent vector at p. As a consequence, since the commutator of two derivations is a derivation we may define the *Lie bracket* between two vector fields on M as the vector field (see (2.23) for the coordinate representation)

$$[\mathbf{v}_1, \mathbf{v}_2][F] := \mathbf{v}_1[\mathbf{v}_2[F]] - \mathbf{v}_2[\mathbf{v}_1[F]] \qquad \forall\, \mathbf{v}_1, \mathbf{v}_2 \in \mathfrak{X}(M),\ F \in \mathscr{F}(M). \tag{4.5}$$

In this way $\mathfrak{X}(M)$ becomes the infinite-dimensional *Lie algebra of smooth vector fields*. In particular, for $F \in \mathscr{F}(M)$ and $\mathbf{v}_1, \mathbf{v}_2 \in \mathfrak{X}(M)$, one has

$$[F\,\mathbf{v}_1, \mathbf{v}_2] = F[\mathbf{v}_1, \mathbf{v}_2] - \mathbf{v}_2[F]\mathbf{v}_1.$$

► We now provide a local coordinate description. Let $U \subset \mathbb{R}^n$ be a coordinate chart coordinatized by $x := (x_1, \ldots, x_n)$ around a point $p \in M$. Then a derivation on $\mathscr{F}(U)$ is completely determined once its effect on all coordinates x_i.

- The differential of $F \in \mathscr{F}(U)$ can be written as

$$\mathrm{d}F = \sum_{i=1}^{n} \frac{\partial F}{\partial x_i} \mathrm{d}x_i,$$

so that, in view of (4.4), we obtain

$$\mathbf{v}[F(x)] = \sum_{i=1}^{n} \frac{\partial F}{\partial x_i} \mathbf{v}[x_i], \tag{4.6}$$

which expresses the action of a smooth vector field on a smooth function.

- T_pM has basis $\{\partial/\partial x_1, \ldots, \partial/\partial x_n\}$, while T_p^*M has basis $\{\mathrm{d}x_1, \ldots, \mathrm{d}x_n\}$. In other words, a vector field on M has a local coordinate form

$$\mathbf{v} = \sum_{i=1}^{n} f_i(x) \frac{\partial}{\partial x_i}, \qquad f_i \in \mathscr{F}(U), \tag{4.7}$$

while a 1-form on M can be represented as

$$\omega = \sum_{i=1}^{n} g_i(x)\, \mathrm{d}x_i, \qquad g_i \in \mathscr{F}(U).$$

There holds the *duality pairing*

$$\left\langle \mathrm{d}x_i, \frac{\partial}{\partial x_j} \right\rangle = \delta_{ij}, \qquad i,j = 1,\dots,n.$$

- A local formulation of Theorem 2.2 assures that there is a one-to-one correspondence between smooth vector fields on U and systems of ODEs on U of the form

$$\frac{\mathrm{d}x_i}{\mathrm{d}t} = f_i(x_1,\dots,x_n), \qquad i = 1,\dots,n, \tag{4.8}$$

where $f_i \in \mathscr{F}(U)$. Indeed, given a vector field $\mathbf{v}$ in the form (4.7) we have $f_i := \mathbf{v}[x_i]$. Conversely, given the functions $f_i \in \mathscr{F}(U)$ we can define $\mathbf{v}[x_i] := f_i$ and extend $\mathbf{v}$ to a derivation on $\mathscr{F}(U)$ by using (4.6).

(a) Solutions to (4.8) define parametrized curves in U, called *integral curves*, whose tangent vector at each point coincides with the value of $\mathbf{v}$ at that point.

(b) The local existence and uniqueness of integral curves is guaranteed by the local formulation of Theorem 1.1. More precisely, Theorem 1.1 implies that given a smooth vector field $\mathbf{v}$ on an n-dimensional manifold M we can find for any $p \in M$ a coordinate chart U of p, with coordinates $(x_1,\dots,x_n)$, an open subset $U' \subseteq U$ and $\varepsilon > 0$, such that the solution $(t,p) \mapsto x(t,p)$ of (4.8) is defined for $(t,p) \in I_\varepsilon \times U'$, where $I_\varepsilon := \{t \in \mathbb{R} : |t| < \varepsilon\}$. The map

$$\Phi_t : I_\varepsilon \times U' \to U : (t,p) \mapsto \Phi_t(p) := x(t,p)$$

is the *local flow* of $\mathbf{v}$. For a fixed $t \in I_\varepsilon$ the map Φ_t is a local diffeomorphism from U' to $\Phi_t(U')$. Flows are *local one-parameter Lie groups of diffeomorphisms* and to describe them we can apply the theory and machinery developed in Chapter 2 (see Section 2.3). Let us recall that if M is compact, then all smooth vector fields on M are *complete*, so that Φ_t exists for all times t.

4.2.3 Maps between manifolds

► Let M, N be two real smooth manifolds of dimension n and ℓ respectively.

- A map $F : M \to N$ is *differentiable* (resp. *smooth*) if for each point $p \in M$, parametrized by a local coordinate function $\chi_\alpha : U_\alpha \to \mathbb{R}^n$ there exists a coordinate chart V_β on N and a local coordinate function $\psi_\beta : V_\beta \to \mathbb{R}^\ell$ with $F(\chi_\alpha(U)) \subset \psi_\beta(V_\beta)$ such that the composite function $\psi_\beta^{-1} \circ F \circ \chi_\alpha : U \subset \mathbb{R}^n \to \mathbb{R}^\ell$, called *local representation of* F, is a differentiable (resp. *smooth*) function.

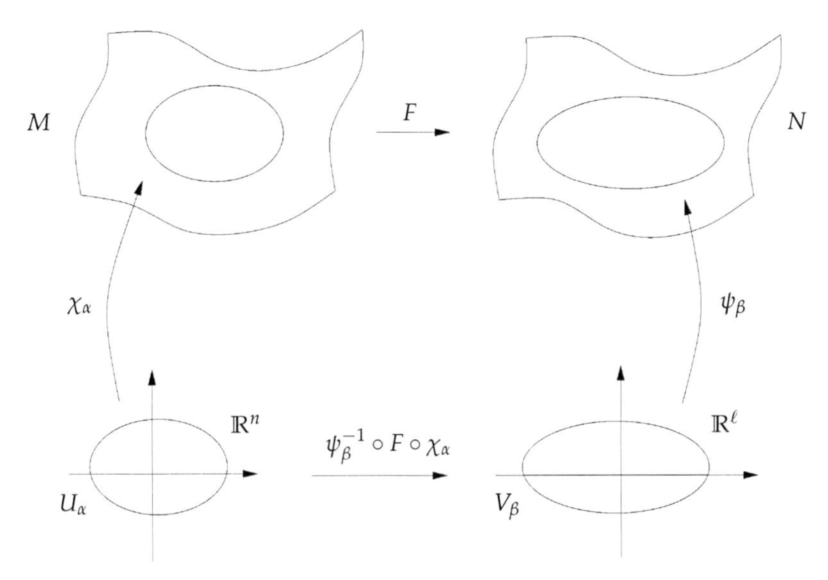

Fig. 4.4. A map $F : M \to N$.

- The *differential* of a differentiable map $F : M \to N$ at $p \in M$ is the linear map

$$\mathsf{d}F_p : T_pM \to T_{F(p)}N.$$

It is constructed as follows. Fix $p \in M$ with coordinates $x := (x_1, \ldots, x_n)$. For $\mathbf{v} \in T_pM$ choose a curve γ that maps an open interval $I_\varepsilon := \{t \in \mathbb{R} \, : \, |t| < \varepsilon\}$ to M with $\gamma(0) = x$. The *velocity vector* is

$$\left.\frac{\mathsf{d}}{\mathsf{d}t}\right|_{t=0} \gamma(t) = \mathbf{v}_p.$$

Then $\mathsf{d}F_p(\mathbf{v})$ is the velocity vector at $t = 0$ of the curve $F \circ \gamma : \mathbb{R} \to N$:

$$\mathsf{d}F_p(\mathbf{v}) := \left.\frac{\mathsf{d}}{\mathsf{d}t}\right|_{t=0} F(\gamma(t)).$$

- The *rank* of a differentiable map $F : M \to N$ at $p \in M$ is the rank of $\mathsf{d}F_p$, i.e., the dimension of the image of $\mathsf{d}F_p$, that is at most $\min(n, \ell)$. One says that F is of *maximal rank* on a subset $S \subset M$ if for each $p \in S$ the rank of F is $\min(n, \ell)$. When $n < \ell$, the best we can hope for is that $\mathsf{d}F_p$ is injective.

► Let us define some important smooth maps $F : M \to N$.

- The map F is a *diffeomorphism* if it is bijective and F^{-1} is differentiable. In such a case the manifolds M and N are *diffeomorphic*.

- The map F is an *immersion at a point* $p \in M$ if $\mathrm{d}F_p$ is injective. It is an *immersion* if it is a immersion at all points $p \in M$. Equivalently, F is an immersion if
$$\mathsf{rank}(\mathrm{d}F_p) = n \qquad \forall\, p \in M.$$
- The map F is a *submersion at a point* $p \in M$ if $\mathrm{d}F_p$ is surjective. Note that we necessarily have $n \geqslant \ell$. It is a *submersion* if it is a submersion at all points $p \in M$. Equivalently, F is a submersion if
$$\mathsf{rank}(\mathrm{d}F_p) = \ell \qquad \forall\, p \in M.$$
- The map F is an *embedding* if it is an (injective) immersion which is also a homeomorphism onto its image $F(M)$.

► A point $q \in N$ is called *regular value* of F if $\mathrm{d}F_p$ is surjective for all $p \in F^{-1}(q)$, i.e.,
$$\mathsf{rank}(\mathrm{d}F_p) = \ell \qquad \forall\, p \in F^{-1}(q).$$
Otherwise it is called a *singular value*.

► Let $F : M \to N$ be a smooth map.

- Assume that F is an immersion. Then $F(M)$ is an *immersed manifold* of N. In general this image will not be a submanifold as a subset because an immersion need not to be injective. By requiring that the immersion map F be injective one has an injective immersion and one can define an *immersed submanifold* of N as the image subset $F(M)$ together with a topology and differential structure such that $F(M)$ is a manifold and the injective immersion is a diffeomorphism.
- Assume that F be an embedding. Then $F(M)$ is an *embedded submanifold* (or *regular submanifold*) of N. An embedded submanifold of N is such that the image $F(M)$ with the subspace topology is homeomorphic to M under F.

► The following results hold true:

- $F : M \to N$ is a (local) diffeomorphism if and only if $\mathrm{d}F_p$ is an isomorphism for all $p \in M$.
- "Maximal rank Theorem". *Let $F : M \to N$ be a smooth map with maximal rank at $p \in M$, i.e., an immersion at p. Then there exist coordinates $(x_1, \dots, x_n)$ in a neighborhood of p and coordinates $(y_1, \dots, y_\ell)$ in a neighborhood of $F(p)$ where F has the following coordinate representation:*
$$F(x_1, \dots, x_n) = \begin{cases} (x_1, \dots, x_n, 0, \dots, 0) & \text{if } \ell > n, \\ (x_1, \dots, x_\ell) & \text{if } \ell \leqslant n. \end{cases}$$
The above map defines a *canonical immersion*.

- "Regular value Theorem". *Let $F : M \to N$ be a smooth map with $\ell < n$. Let $q \in N$ be a regular value of F, with $F^{-1}(q) \neq \emptyset$. Then:*

  1. *$F^{-1}(q)$ is a smooth manifold of dimension $n - \ell$.*
  2. *The tangent space at $p \in F^{-1}(q)$ is*

$$T_p F^{-1}(q) = \mathsf{kernel}(\mathsf{d}F_p).$$

- "Whitney Theorem". *Any n-dimensional smooth manifold can be embedded in $\mathbb{R}^{2n}$.* As a consequence any n-dimensional smooth manifold is diffeomorphic to a submanifold of $\mathbb{R}^{2n}$.

Example 4.7 (***The unit hypersphere*** $\mathbb{S}^n$)

The fact that the hypersphere $\mathbb{S}^n$ is a regular n-dimensional submanifold of $\mathbb{R}^{n+1}$ follows easily from the "Regular value Theorem".

- $\mathbb{S}^n$ is the regular level set $F^{-1}(1)$ of the smooth function $F : \mathbb{R}^{n+1} \to [0, \infty)$ defined by

$$F(x) := \sum_{i=1}^{n+1} x_i^2.$$

  Note that

$$\mathsf{d}F = 2 \sum_{i=1}^{n+1} x_i \, \mathsf{d}x_i$$

  vanishes only at $x = 0$.

- Moreover, given $p \in \mathbb{S}^n$, with coordinates $(x_1, \ldots, x_{n+1})$, we have

$$T_p F^{-1}(1) = T_p \mathbb{S}^n = \mathsf{kernel}(\mathsf{d}F_p) = \left\{ y \in \mathbb{R}^{n+1} : \sum_{i=1}^{n+1} x_i \, y_i = 0 \right\},$$

  which is a hyperplane tangent to $\mathbb{S}^n$ at p.

Example 4.8 (***A two-dimensional submanifold of*** $\mathbb{R}^4$)

Define a map $F : \mathbb{R}^4 \to \mathbb{R}^2$ by

$$F : (x_1, x_2, x_3, x_2) \mapsto \left(x_1^2 + x_2^2, x_1 \, x_3 + x_2 \, x_4 \right).$$

- We see that the differential is

$$\mathsf{d}F = \begin{pmatrix} 2\,x_1 & 2\,x_2 & 0 & 0 \\ x_3 & x_4 & x_1 & x_2 \end{pmatrix}.$$

  If $(x_1, x_2) = (0, 0)$ then $\mathsf{d}F$ has rank 1 and $F(x) = 0$. So $(0, 0)$ is the only singular value of F.

- Define $M := F^{-1}(q)$, $q \in \mathbb{R}^2 \setminus \{0\}$. It is clear that q is a regular value of F. By the "Regular value Theorem" M is a regular submanifold of $\mathbb{R}^4$ with dimension $4 - 2 = 2$.
- The tangent space $T_p M$ with $p := (1, 0, 0, 0) \in \mathbb{R}^4$ is:

$$T_p M = \mathsf{kernel}(\mathsf{d}F_p) = \mathsf{kernel} \begin{pmatrix} 2 & 0 & 0 & 0 \\ 0 & 0 & 1 & 0 \end{pmatrix} = \left\{ y \in \mathbb{R}^4 : y_1 = y_3 = 0 \right\}.$$

Example 4.9 (***Maps between manifolds***)

- The image of $\mathbb{R}$ under the smooth map

$$F : \mathbb{R} \to \mathbb{R}^2 : t \mapsto \left(t^2, t^3\right)$$

is the planar curve $x_1^2 = x_2^3$, which has a cusp at $(0,0)$. The differential (Jacobian) of F is $(2\,t, 3\,t^2)$, which is not of maximal rank at $t = 0$, indicating the appearance of a singularity in the image of F. Therefore F is not an immersion and $F(\mathbb{R})$ is not an immersed submanifold of $\mathbb{R}^2$.

- The image of $\mathbb{R}$ under the smooth map

$$F : \mathbb{R} \to \mathbb{R}^2 : t \mapsto (\sin t, 2\,\sin(2\,t))$$

is the "figure-eight", which is a curve with self-intersections, namely $F(t) = (0,0)$ whenever t is an integral multiple of π. A slight modification of F,

$$\widetilde{F} : \mathbb{R} \to \mathbb{R}^2 : t \mapsto (\sin(2\,\arctan t), 2\,\sin(4\,\arctan t))$$

parametrizes the same curve of F, but provides a one-to-one parametrization, i.e., the curve passes through $(0,0)$ just once. The maximal rank condition holds everywhere. Therefore $\widetilde{F}$ provides a parametrization of a submanifold of $\mathbb{R}^2$ with apparent self-intersections, even though the immersion $\widetilde{F}$ is one-to-one.

- The image of $\mathbb{R}$ under the smooth map

$$F : \mathbb{R} \to \mathbb{R}^3 : t \mapsto (\cos t, \sin t, t)$$

is a circular helix spiraling up the x_3-axis. Here F is clearly one-to-one and the differential (Jacobian) of F, $(-\sin t, \cos t, 1)$, never vanishes, so the maximal rank condition holds everywhere. Therefore F is an injective immersion and $F(\mathbb{R})$ is an immersed submanifold of $\mathbb{R}^3$.

4.2.4 Distributions

► Instead of having a vector at every point of a manifold M, as is the case of a vector field on M, one may have a one-dimensional subspace of the tangent space to M, at every point of M. This is called a *1-dimensional smooth distribution* on M.

► A *k-dimensional smooth distribution* Δ on a smooth manifold M of dimension n $(k < n)$ is the datum of a k-dimensional subspace Δ_p of T_pM for every $p \in M$. One says that the Δ is smooth if there exist smooth vector fields $\mathbf{v}_1, \ldots, \mathbf{v}_k$ on a open neighborhood U of p, such that

$$\Delta_p = \mathsf{span}\{(\mathbf{v}_1)_p, \ldots, (\mathbf{v}_k)_p\} \qquad \forall\, p \in U.$$

- The notion of an integral curve is easily adapted to the case of a k-dimensional distribution: a k-dimensional (connected) immersed submanifold N of M is called an *integral manifold* of Δ if $T_pN = \Delta_p$ for any $p \in N$.

- In contrast to the case of integral curves, integral manifolds need not to exist in general, even locally. One obstruction comes from the following facts.

(a) If $\mathbf{v}_1$ and $\mathbf{v}_2$ are two vector fields on M which are tangent to some submanifold N (such as the candidate integral manifold) then their Lie bracket $[\mathbf{v}_1, \mathbf{v}_2]$ is also tangent to N.

(b) In order to rephrase this in the language of distributions, we say that a vector field $\mathbf{v}$ on $U \subset M$ is adapted to Δ on U if $\mathbf{v}_p \in \Delta_p$ for every $p \in U$.

(c) In these terms the obstruction reads: if $\mathbf{v}_1$ and $\mathbf{v}_2$ are adapted to Δ on U then $[\mathbf{v}_1, \mathbf{v}_2]$ is also adapted to Δ on U.

(d) One says that Δ is an *integrable distribution* if for any $\mathbf{v}_1$ and $\mathbf{v}_2$ that are adapted to Δ on an open subset U, their Lie bracket $[\mathbf{v}_1, \mathbf{v}_2]$ is also adapted to Δ on U.

► The "Frobenius Theorem" says that the above obstruction to the existence of integral manifolds is the only one.

- "Frobenius Theorem" (first formulation). *Let Δ be a smooth k-dimensional distribution on M. If Δ is integrable then there exists through any point $p \in M$ a unique maximal integral manifold for Δ.*

- "Frobenius Theorem" (second formulation). *Let Δ be an integrable smooth k-dimensional distribution on M. Then we can choose coordinates $(x_1, \ldots, x_n)$ on a neighborhood U of any $p \in M$ such that*

$$\Delta_q = \mathsf{span}\left\{ \left(\frac{\partial}{\partial x_1}\right)_q, \ldots, \left(\frac{\partial}{\partial x_k}\right)_q \right\}$$

for any $q \in U$. In terms of these coordinates the integral manifold of Δ restricted to U and through p is given by the connected component of the set $\{q \in U \,:\, x_i(q) = x_i(p),\ i = k+1, \ldots, n\}$ that contains p.

Note that the first formulation is the analog of Theorem 1.1 about existence and uniqueness of integral curves for IVPs.

► Remarks:

- A collection of smooth vector fields $\mathbf{v}_1, \ldots, \mathbf{v}_r$ defined in a chart U of $p \in M$ is *involutive* if there exist functions $F_{ij}^k \in \mathscr{F}(U)$, $i, j, k = 1, \ldots, r$, such that

$$[\mathbf{v}_i, \mathbf{v}_j] = \sum_{k=1}^{r} F_{ij}^k(x)\, \mathbf{v}_k,$$

for all $i, j = 1, \ldots, r$. In terms of this definition one can reformulate "Frobenius Theorem" saying that *a collection of smooth vector fields $\mathbf{v}_1, \ldots, \mathbf{v}_r$ is integrable (i.e., through every point $p \in M$ there passes an integral submanifold) if and only if it is involutive.*

- "Frobenius Theorem" implies that the maximal integral manifolds of an integrable distribution on M form the leaves of a *foliation* on M (see Section 4.5).

Example 4.10 (*An integrable distribution on* $\mathbb{R}^3$)

Consider on $M = \mathbb{R}^3$ the vector fields

$$\mathbf{v}_1 := -x_2 \frac{\partial}{\partial x_1} + x_1 \frac{\partial}{\partial x_2},$$

and

$$\mathbf{v}_2 := 2\,x_1 x_3 \frac{\partial}{\partial x_1} + 2\,x_2 x_3 \frac{\partial}{\partial x_2} + \left(x_3^2 - x_1^2 - x_2^2 + 1\right) \frac{\partial}{\partial x_3}.$$

- One can verify that $[\mathbf{v}_1, \mathbf{v}_2] = 0$, so that "Frobenius Theorem" implies that the pair $\mathbf{v}_1, \mathbf{v}_2$ is integrable.
- The space $T\,\mathbb{R}^3$ spanned by $\mathbf{v}_1$ and $\mathbf{v}_2$ is two-dimensional except on the x_3-axis and on the circle $\left\{x \in \mathbb{R}^3 \,:\, x_3 = 0,\, x_1^2 + x_2^2 = 1\right\}$, where its dimension is one.
- One can check that both the x_3-axis and $\left\{x \in \mathbb{R}^3 \,:\, x_3 = 0,\, x_1^2 + x_2^2 = 1\right\}$ are one-dimensional integral submanifolds of $\mathbf{v}_1, \mathbf{v}_2$. All other integral submanifolds are two-dimensional tori

$$\left\{ x \in \mathbb{R}^3 : F(x) := \frac{x_1^2 + x_2^2 + x_3^2 + 1}{\sqrt{x_1^2 + x_2^2}} = c,\ c > 2 \right\}.$$

Indeed one has $\mathrm{d}F(\mathbf{v}_1) = \mathbf{v}_1[F] = 0$ and $\mathrm{d}F(\mathbf{v}_2) = \mathbf{v}_2[F] = 0$ everywhere. By the "Regular value Theorem" both $\mathbf{v}_1$ and $\mathbf{v}_2$ are tangent to each level set of F where $\mathrm{grad}_x F(x) \neq 0$.

4.2.5 k-forms and k-vector fields

► In the formulation of Hamiltonian mechanics on smooth manifolds we will be interested in differential 2-forms and their dual version, namely 2-vector fields (also called *bivector fields*). In order to understand the main features of these objects in a general geometric framework we give a short introductory presentation of k-forms and k-vector fields.

► The natural generalization of a 1-form on a smooth manifold M is given by a differential k-form on M, whose local definition is nothing but the definition of a differential k-form on $\mathbb{R}^n$ given in Chapter 3 (Subsection 3.4.5). The main difference is that the coordinate expression of a differential k-form on M is given in a local coordinate chart U. To simplify our presentation we omit the coordinate-free description of k-forms and we refer to Chapter 3 (Subsection 3.4.5) for any coordinate-dependent concept, as for instance, wedge product and exterior derivative.

Example 4.11 (*A differential 2-form on M*)

A differential 2-form on M admits the following local expression in a chart U (cf. formula (3.58):

$$\omega = \sum_{1 \leqslant i < j \leqslant n} f_{ij}(x)\, \mathrm{d}x_i \wedge \mathrm{d}x_j,$$

where $f_{ij} \in \mathscr{F}(U)$. Its exterior derivative gives a 3-form on M whose local expression can be computed by using (3.60). A more intrinsic definition is the following

$$d\omega(\mathbf{v}_1,\mathbf{v}_2,\mathbf{v}_3) := \mathbf{v}_1[\omega(\mathbf{v}_2,\mathbf{v}_3)] + \omega(\mathbf{v}_1,[\mathbf{v}_2,\mathbf{v}_3]) + \circlearrowleft (\mathbf{v}_1,\mathbf{v}_2,\mathbf{v}_3). \tag{4.9}$$

► For $k \in \mathbb{N}$ we denote the $\mathscr{F}(M)$-module of k-forms on a manifold M by $\Omega^k(M)$. In particular, $\Omega^0(M) = \mathscr{F}(M)$ and $\Omega^1(M) = \Omega(M)$.

- We let
$$\Omega^*(M) := \bigoplus_{k=0}^{n} \Omega^k(M), \qquad \dim\left(\Omega^k(M)\right) = \frac{n!}{k!(n-k)!}.$$
Any element of $\Omega^*(M)$ is a *differential form*.
- A *differential k-form* is an element of $\Omega^k(M)$, which is the set of all k-linear alternating maps
$$\omega : \underbrace{T_pM \times \cdots \times T_pM}_{k \text{ times}} \to \mathscr{F}(M).$$
Then the basic requirements are:
  1. *Multilinearity.* For any k smooth vectors fields $(\mathbf{v}_1,\ldots,\mathbf{v}_k)$, a smooth vector field $\mathbf{w}$ and two scalars $\lambda_1,\lambda_2 \in \mathbb{R}$, there holds:
  $$\omega(\lambda_1\,\mathbf{v}_1 + \lambda_2\,\mathbf{w},\mathbf{v}_2,\ldots,\mathbf{v}_k) = \lambda_1\,\omega(\mathbf{v}_1,\ldots,\mathbf{v}_k) + \lambda_2\,\omega(\mathbf{w},\ldots,\mathbf{v}_k).$$
  2. *Skew-symmetry.* For any k smooth vector fields $(\mathbf{v}_1,\ldots,\mathbf{v}_k)$ there holds:
  $$\omega(\mathbf{v}_1,\ldots,\mathbf{v}_k) = (-1)^\nu \omega(\mathbf{v}_{i_1},\ldots,\mathbf{v}_{i_k}),$$
  where $\nu = \pm 1$ according to the parity of the permutation $(i_1,\ldots,i_k)$ of $(1,\ldots,k)$.

► The fact that a 1-form can be evaluated on a vector field to produce an element of $\mathscr{F}(M)$ (see (4.2)) generalizes in two ways to a k-form ω, where $k > 1$.

1. We can evaluate $\omega \in \Omega^k(M)$ on k vector fields $\mathbf{v}_1,\ldots,\mathbf{v}_k$ obtaining
$$\omega(\mathbf{v}_1,\ldots,\mathbf{v}_k) \in \mathscr{F}(M).$$
From this point of view a k-form is an $\mathscr{F}(M)$-k-linear map $\mathfrak{X}(M) \to \mathscr{F}(M)$.
2. We can insert one vector field $\mathbf{v}$ as the first argument to $\omega \in \Omega^k(M)$ yielding a $(k-1)$-form which is denoted by $\mathbf{v} \lrcorner \omega$ and called *interior product* (or *contraction*):
$$\mathbf{v} \lrcorner \omega \in \Omega^{k-1}(M).$$
From this point of view a k-form is an $\mathscr{F}(M)$-linear map $\mathfrak{X}(M) \to \Omega^{k-1}(M)$. The definition of the map $\mathbf{v}\lrcorner$ can be extended to all of $\Omega^*(M)$ by defining $\mathbf{v} \lrcorner \omega = 0$ for all 0-forms on M.

Example 4.12 (*A contraction of a 3-form with a vector field*)

On $\mathbb{R}^3$ define a 3-form

$$\omega := \mathrm{d}x_1 \wedge \mathrm{d}x_2 \wedge \mathrm{d}x_3,$$

and a vector field

$$\mathbf{v} := x_1 \frac{\partial}{\partial x_1} + x_2 \frac{\partial}{\partial x_2}.$$

Then

$$\mathbf{v} \lrcorner \omega = x_1 \, \mathrm{d}x_2 \wedge \mathrm{d}x_3 - x_2 \, \mathrm{d}x_1 \wedge \mathrm{d}x_3.$$

► Let $F : M \to N$ be a smooth map between two smooth manifolds. Since $\mathrm{d}F_p$, $p \in M$, is an isomorphism between T_pM and $T_{F(p)}N$ we can dualize this map, thus getting a linear map on cotangent spaces going in the opposite direction. More precisely, we can associate to $\omega \in \Omega^k(N)$ a differential k-form on $\Omega^k(M)$, denoted by $F^*(\omega)$ and called *pull-back* of ω, by setting

$$\left((F^*\omega)\,(\mathbf{v}_1, \ldots, \mathbf{v}_k)\right)_p := \omega_{F(p)}(\mathrm{d}F_p(\mathbf{v}_1), \ldots, \mathrm{d}F_p(\mathbf{v}_k)),$$

where $p \in M$ and $\mathbf{v}_1, \ldots, \mathbf{v}_k \in T_pM$.

- In the particular case $k = 0$, i.e., ω is a smooth function, say $G \in \mathscr{F}(M)$, we simply recover the pull-back of a function, i.e., $F^*G := G \circ F$.
- Note that the pull-back of ω under F is a linear operator from $\Omega^k(N)$ to $\Omega^k(M)$, and the word "pull-back" reminds one of the fact that the direction of the arrow is reversed compared to the arrow of the map $F : M \to N$, i.e., $F^*\omega : \Omega^k(N) \to \Omega^k(M)$.
- It can be proved that the pull-back commutes with the differential, i.e.,

$$F^*(\mathrm{d}\omega) = \mathrm{d}(F^*\omega).$$

Example 4.13 (*Pull-back of a 2-form on* $\mathbb{R}^2$)

Consider the map

$$F : \mathbb{R}^2 \to \mathbb{R}^2 : (x_1, x_2) \mapsto (\widetilde{x}_1, \widetilde{x}_2) := \left(\frac{1}{2}\left(x_1^2 - x_2^2\right), x_1 \, x_2\right).$$

Then

$$\begin{aligned} \mathrm{d}\widetilde{x}_1 \wedge \mathrm{d}\widetilde{x}_2 &= (x_1 \, \mathrm{d}x_1 - x_2 \, \mathrm{d}x_2) \wedge (x_2 \, \mathrm{d}x_1 + x_1 \, \mathrm{d}x_2) \\ &= \left(x_1^2 + x_2^2\right) \mathrm{d}x_1 \wedge \mathrm{d}x_2. \end{aligned}$$

Therefore $(x_1^2 + x_2^2) \, \mathrm{d}x_1 \wedge \mathrm{d}x_2$ is the pull-back of $\mathrm{d}\widetilde{x}_1 \wedge \mathrm{d}\widetilde{x}_2$ via F:

$$F^*(\mathrm{d}\widetilde{x}_1 \wedge \mathrm{d}\widetilde{x}_2) = \left(x_1^2 + x_2^2\right) \mathrm{d}x_1 \wedge \mathrm{d}x_2.$$

► Remarks:

- The space $\Omega^n(M)$ of n-forms on M (which is n-dimensional) is one-dimensional. Thus all n-forms on M can be locally written in a coordinate chart U as

$$\omega = f(x)\,\mathrm{d}x_1 \wedge \cdots \wedge \mathrm{d}x_n, \tag{4.10}$$

for some $f \in \mathscr{F}(U)$. If f is thought of as a measure on U, then ω can be used to define the measure of a compact subset $U' \subset U$,

$$\mathsf{Vol}\,(U') := \int_{U'} \omega.$$

Indeed, ω is also called *measure form* (*volume form* if $f \equiv 1$).

- Given a n-form ω as in (4.10) we say that a diffeomorphism $F : M \to M$ is *volume preserving* if $F^*\omega = \omega$ when $f \equiv 1$ and *measure preserving* if $F^*\omega = \omega$ with $f \neq 1$. The distinction is slight since F depends on the Jacobian of any coordinate transformation.

► We now introduce k-vector fields on M as those objects which are dual to k-forms. For $k \in \mathbb{N}$ we denote the $\mathscr{F}(M)$-module of k-vector fields on a manifold M by $\mathfrak{X}^k(M)$. In particular, $\mathfrak{X}^0(M) = \mathscr{F}(M)$ and $\mathfrak{X}^1(M) = \mathfrak{X}(M)$.

- We let

$$\mathfrak{X}^*(M) := \bigoplus_{k=0}^{m} \mathfrak{X}^k(M), \qquad \mathsf{dim}\left(\mathfrak{X}^k(M)\right) = \frac{n!}{k!(n-k)!}.$$

An element of $\mathfrak{X}^*(M)$ is a *multi-vector field*.

- A *k-vector field* is an element of $\mathfrak{X}^k(M)$, which is the set of all k-linear alternating maps

$$\mathbf{V} : \underbrace{T_p^*M \times \cdots \times T_p^*M}_{k \text{ times}} \to \mathscr{F}(M).$$

Therefore $\mathbf{V}$ defines a skew-symmetric k-derivation.

- As in the case of differential forms we can define a *wedge product*:

$$\wedge : \mathfrak{X}^*(M) \times \mathfrak{X}^*(M) \to \mathfrak{X}^*(M),$$

which makes $\mathfrak{X}^*(M)$ into a graded associative algebra.

- We can evaluate any k-form ω on any k-vector field $\mathbf{V} := \mathbf{v}_1 \wedge \cdots \wedge \mathbf{v}_k$ by letting $\omega(\mathbf{V}) := \omega(\mathbf{v}_1, \ldots, \mathbf{v}_k)$.

- If $\mathbf{V}$ is a k-vector field the value of the corresponding skew-symmetric k-derivation on k functions $F_1, \ldots, F_k \in \mathscr{F}(M)$ is denoted by

$$\mathbf{V}\left[F_1, \ldots, F_k\right] := (\mathrm{d}F_1 \wedge \cdots \wedge \mathrm{d}F_k)\,(\mathbf{V}). \tag{4.11}$$

Note that (4.11) is the generalization of (4.3) to k-vector fields.

- As in the case of smooth vector fields (cf. formula (4.6)) we have that a k-vector field $\mathbf{V}$ is completely specified on a coordinate neighborhood U once it is known on all k-tuples $(x_{i_1}, \ldots, x_{i_k})$, with $1 \leqslant i_1 < i_2 < \cdots < i_k \leqslant n$, where the k-tuples are taken from any chosen system of coordinates $(x_1, \ldots, x_n)$ on U. Explicitly (4.6) is generalized to

$$\mathbf{V}\left[F_1, \ldots, F_k\right] = \sum_{i_1, \ldots, i_k = 1}^{n} \frac{\partial F_1}{\partial x_{i_1}} \cdots \frac{\partial F_k}{\partial x_{i_k}} \mathbf{V}[x_{i_1}, \ldots, x_{i_k}].$$

Example 4.14 (*A 2-vector field on M*)

A 2-vector field on M admits the following local expression in a chart U:

$$\mathbf{V} = \sum_{1 \leqslant i < j \leqslant n} f_{ij}(x) \frac{\partial}{\partial x_i} \wedge \frac{\partial}{\partial x_j},$$

where $f_{ij} \in \mathscr{F}(U)$.

Example 4.15 (*A 2-vector field on* $\mathbb{R}^3$)

Let U be a local chart on a three-dimensional manifold M.

- From Example (3.23) we know that a basis of $\Omega^2(M)$ is

$$\{\mathrm{d}x_2 \wedge \mathrm{d}x_3, \mathrm{d}x_3 \wedge \mathrm{d}x_1, \mathrm{d}x_1 \wedge \mathrm{d}x_2\}.$$

Indeed, a 2-form on M is expressed as

$$\omega = f_1(x)\,\mathrm{d}x_2 \wedge \mathrm{d}x_3 + f_2(x)\,\mathrm{d}x_3 \wedge \mathrm{d}x_1 + f_3(x)\,\mathrm{d}x_1 \wedge \mathrm{d}x_2,$$

for some functions $f_i \in \mathscr{F}(U)$, $i = 1, 2, 3$.

- A basis of $\mathfrak{X}^2(M)$ is

$$\left\{ \frac{\partial}{\partial x_2} \wedge \frac{\partial}{\partial x_3}, \frac{\partial}{\partial x_3} \wedge \frac{\partial}{\partial x_1}, \frac{\partial}{\partial x_1} \wedge \frac{\partial}{\partial x_2} \right\}.$$

A 2-vector field on M is expressed as

$$\mathbf{V} = g_1(x) \frac{\partial}{\partial x_2} \wedge \frac{\partial}{\partial x_3} + g_2(x) \frac{\partial}{\partial x_3} \wedge \frac{\partial}{\partial x_1} + g_3(x) \frac{\partial}{\partial x_1} \wedge \frac{\partial}{\partial x_2},$$

for some functions $g_i \in \mathscr{F}(U)$, $i = 1, 2, 3$.

4.2.6 Lie derivatives

► In general, Lie derivatives are important derivation operations which measure how a given object, as a k-form or a k-vector field, changes in the direction of a given vector field $\mathbf{v}$. Here are the main definitions (in the case of k-vector fields we consider only $k = 1, 2$).

- The *Lie derivative of a k-form* $\omega \in \Omega^k(M)$ along a smooth vector field $\mathbf{v} \in \mathfrak{X}(M)$ is

$$\mathfrak{L}_{\mathbf{v}}\omega := \left.\frac{\mathrm{d}}{\mathrm{d}t}\right|_{t=0} \Phi_t^* \omega,$$

where Φ_t is the local flow of $\mathbf{v}$ on M. Therefore $\mathfrak{L}_{\mathbf{v}}\omega = 0$ if and only if ω is constant on the integral curves of $\mathbf{v}$. This is a consequence of the formula

$$\Phi_t^* \left(\mathfrak{L}_{\mathbf{v}}\omega\right) = \frac{\mathrm{d}}{\mathrm{d}t} \left(\Phi_t^* \,\omega\right).$$

In the particular case $k = 0$, i.e., ω is a smooth function, we get the *Lie derivative of a smooth function* $F \in \mathscr{F}(M)$ along a smooth vector field $\mathbf{v} \in \mathfrak{X}(M)$ (cf. formula (2.17))

$$\mathfrak{L}_{\mathbf{v}}F := \mathbf{v}[F] = \mathbf{v} \lrcorner \mathrm{d}F = \mathrm{d}F(\mathbf{v}) = \left.\frac{\mathrm{d}}{\mathrm{d}t}\right|_{t=0} \Phi_t^* F.$$

- The *Lie derivative of a 2-vector field* $\mathbf{W} \in \mathfrak{X}^2(M)$ along $\mathbf{v} \in \mathfrak{X}(M)$ is the 2-vector field whose value on $F_1, F_2 \in \mathscr{F}(M)$ is given by

$$\mathfrak{L}_{\mathbf{v}}\mathbf{W}\left[F_1, F_2\right] := \mathbf{v}\left[\mathbf{W}\left[F_1, F_2\right]\right] - \mathbf{W}\left[\mathbf{v}[F_1], F_2\right] - \mathbf{W}\left[F_1, \mathbf{v}[F_2]\right]. \tag{4.12}$$

Therefore $\mathfrak{L}_{\mathbf{v}}\mathbf{W} = 0$ if and only if $\mathbf{W}$ is constant on the integral curves of $\mathbf{v}$. The *Lie derivative of a smooth vector field* $\mathbf{w} \in \mathfrak{X}(M)$ along $\mathbf{v} \in \mathfrak{X}(M)$ is defined in terms of the Lie bracket between $\mathbf{v}$ and $\mathbf{w}$:

$$\mathfrak{L}_{\mathbf{v}}\mathbf{w} := \left[\,\mathbf{v}, \mathbf{w}\,\right].$$

Therefore $\mathfrak{L}_{\mathbf{v}}\mathbf{w} = 0$ if and only if $\mathbf{w}$ is constant on the integral curves of $\mathbf{v}$.

► There hold the following useful formulas:

$$\mathfrak{L}_{\mathbf{v}}\mathrm{d}\omega = \mathrm{d}(\mathfrak{L}_{\mathbf{v}}\omega),$$

$$\mathfrak{L}_{\mathbf{v}}\omega = \mathrm{d}(\mathbf{v} \lrcorner \omega) + \mathbf{v} \lrcorner \mathrm{d}\omega \qquad (\textit{Cartan formula}), \tag{4.13}$$

and

$$\left[\,\mathbf{v}_1, \mathbf{v}_2\,\right] \lrcorner \omega = \mathfrak{L}_{\mathbf{v}_1}(\mathbf{v}_2 \lrcorner \omega) - \mathbf{v}_2 \lrcorner \mathfrak{L}_{\mathbf{v}_1}\omega, \tag{4.14}$$

for all $\mathbf{v}, \mathbf{v}_1, \mathbf{v}_2 \in \mathfrak{X}(M)$ and $\omega \in \Omega^k(M)$. Combining (4.13) and (4.14) we also get

$$\left[\,\mathbf{v}_1, \mathbf{v}_2\,\right] \lrcorner \omega = \mathrm{d}(\mathbf{v}_1 \lrcorner \mathbf{v}_2 \lrcorner \omega) + \mathbf{v}_1 \lrcorner \mathrm{d}(\mathbf{v}_2 \lrcorner \omega) - \mathbf{v}_2 \lrcorner \mathrm{d}(\mathbf{v}_1 \lrcorner \omega) - \mathbf{v}_2 \lrcorner \mathbf{v}_1 \lrcorner \mathrm{d}\omega. \tag{4.15}$$

Formula (4.13) can be considered as an alternative definition of Lie derivative of a differential k-form along a smooth vector field.

Example 4.16 (*Cartan formula*)

On $\mathbb{R}^2$ define a 1-form

$$\omega := -x_2 \,\mathrm{d}x_1 + x_1 \,\mathrm{d}x_2,$$

and a vector field

$$\mathbf{v} := -x_2 \frac{\partial}{\partial x_1} + x_1 \frac{\partial}{\partial x_2}.$$

- We have:
$$\mathfrak{L}_{\mathbf{v}}\, x_1 = \mathbf{v}[x_1] = -x_2, \qquad \mathfrak{L}_{\mathbf{v}}\, x_2 = \mathbf{v}[x_1] = x_1.$$
Then, since $\mathfrak{L}$ is a derivation and it commutes with d,
$$\begin{aligned} \mathfrak{L}_{\mathbf{v}}(-x_2\,\mathrm{d}x_1) &= -(\mathfrak{L}_{\mathbf{v}}x_2)\mathrm{d}x_1 - x_2\mathfrak{L}_{\mathbf{v}}\mathrm{d}x_1 \\ &= -(\mathfrak{L}_{\mathbf{v}}x_2)\mathrm{d}x_1 - x_2\,\mathrm{d}(\mathfrak{L}_{\mathbf{v}}x_1) \\ &= -x_1\mathrm{d}x_1 + x_2\,\mathrm{d}x_2. \end{aligned}$$
Similarly
$$\mathfrak{L}_{\mathbf{v}}(x_1\,\mathrm{d}x_2) = x_1\mathrm{d}x_1 - x_2\,\mathrm{d}x_2.$$
Therefore $\mathfrak{L}_{\mathbf{v}}\omega = 0$.
- Let us prove that $\mathfrak{L}_{\mathbf{v}}\omega = 0$. by using formula (4.13). Note that
$$\mathbf{v} \lrcorner\, \omega = x_1^2 + x_2^2,$$
and
$$\mathrm{d}(\mathbf{v} \lrcorner\, \omega) = 2\,x_1\,\mathrm{d}x_1 + 2\,x_2\,\mathrm{d}x_2.$$
Then,
$$\mathrm{d}\omega = 2\,\mathrm{d}x_1 \wedge \mathrm{d}x_2.$$
and
$$\mathbf{v} \lrcorner\, \mathrm{d}\omega = -2\,x_2\,\mathrm{d}x_2 - 2\,x_1\,\mathrm{d}x_1.$$
From formula (4.13) we have:
$$\mathfrak{L}_{\mathbf{v}}\omega = \mathrm{d}(\mathbf{v} \lrcorner\, \omega) + \mathbf{v} \lrcorner\, \mathrm{d}\omega = 0.$$

4.2.7 *Matrix Lie groups and matrix Lie algebras*

► In a purely algebraic language, a finite-dimensional real *Lie algebra* $\mathfrak{g}$ is a finite-dimensional real vector space equipped with a binary mapping $[\cdot,\cdot] : \mathfrak{g} \times \mathfrak{g} \to \mathfrak{g}$, called *Lie bracket*, satisfying the following axioms:

1. *(Bi)linearity*: $[\,\lambda_1 X + \lambda_2 Y, Z\,] = \lambda_1[\,X, Z\,] + \lambda_2[\,Y, Z\,]$,
2. *Alternating*: $[\,X, X\,] = 0$,
3. *Jacobi identity*: $[\,X, [\,Y, Z\,]\,] + \circlearrowleft (X, Y, Z) = 0$,

for all $X, Y, Z \in \mathfrak{g}$ and $\lambda_1, \lambda_2 \in \mathbb{R}$.

- Bilinearity and alternating properties imply skew-symmetry, $[\,X, Y\,] = -[\,Y, X\,]$, while skew-symmetry only implies the alternating property. Furthermore the Jacobi identity implies that the map $Y \mapsto [\,X, Y\,]$ is a *derivation*, i.e.,
$$[\,XY, Z\,] = X[\,Y, Z\,] + [\,X, Z\,]Y \qquad \forall\, X, Y, Z \in \mathfrak{g},$$
which is also called *Leibniz rule*.

- Let n be the dimension of $\mathfrak{g}$ and $\{X_1, \ldots, X_n\}$ be a basis of $\mathfrak{g}$. Then there are certain constant $c_{ij}^k \in \mathbb{R}$, $i, j, k = 1, \ldots, n$, called *structure constants* of $\mathfrak{g}$, such that

$$[\, X_i, X_j \,] = \sum_{k=1}^{n} c_{ij}^k \, X_k. \tag{4.16}$$

Formula (4.16) clearly encodes the linear structure of $\mathfrak{g}$. The structure constants of $\mathfrak{g}$ define the Lie algebra itself. Evidently one has

$$c_{ij}^k = -c_{ji}^k,$$

and

$$\sum_{k=1}^{n} \left(c_{ij}^k c_{k\ell}^r + c_{\ell i}^k c_{kj}^r + c_{j\ell}^k c_{ki}^r \right) = 0,$$

for $i, j, k, \ell, r = 1, \ldots, n$. The last formula is nothing but the Jacobi identity for the structure constants.

► In a geometric framework, Lie algebras naturally arise as tangent spaces at the identity element of Lie groups. Informally, Lie groups are smooth manifold with an additional group structure. It turns out that to study the local structure of a Lie group, it is enough to examine a neighborhood of the identity element. Therefore, it is not surprising that the tangent space at the identity of a Lie group (called *Lie algebra of the Lie group*) plays a key role.

► More formally, a n-dimensional real *Lie group* $\mathbf{G}$ is a n-dimensional real smooth manifold equipped with a smooth group structure: the group operations of multiplication, $* : \mathbf{G} \times \mathbf{G} \to \mathbf{G}$, and inversion are smooth. The *Lie algebra* $\mathfrak{g}$ of $\mathbf{G}$ is the n-dimensional real vector space defined as the tangent space $T_e\mathbf{G}$ where $e \in \mathbf{G}$ is the identity element in $\mathbf{G}$.

- The main examples of Lie groups include linear groups (*matrix Lie groups*), i.e., Lie subgroups of the *general linear Lie group*

$$\mathbf{GL}(n, \mathbb{R}) := \{A \in \mathfrak{gl}(n, \mathbb{R}) \; : \; \det A \neq 0\},$$

the non-commutative group of all invertible $n \times n$ matrices (with coefficients in $\mathbb{R}$), where the group operation is given by the usual product of matrices.

  (a) $\mathbf{GL}(n, \mathbb{R})$ is a real smooth manifold of dimension n^2 since it is an open subset of the vector space of all linear transformations (not necessarily invertible) from $\mathbb{R}^n$ to $\mathbb{R}^n$.

  (b) In fact, $\mathbf{GL}(n, \mathbb{R})$ is the inverse image of $\mathbb{R} \setminus \{0\}$ under the submersion $A \mapsto \det A$. This is true thanks to the "Regular value Theorem".

- The main examples of Lie algebras include *matrix Lie algebras*, i.e., Lie subalgebras of *general linear Lie algebra* $\mathfrak{gl}(n, \mathbb{R})$, the vector space of all linear maps (not necessarily invertible) from $\mathbb{R}^n$ to $\mathbb{R}^n$. In $\mathfrak{gl}(n, \mathbb{R})$ the Lie bracket is given by the commutator of matrices.

- $\mathfrak{gl}(n, \mathbb{R})$ is the Lie algebra of $\mathbf{GL}(n, \mathbb{R})$. Such a connection is established in terms of the *exponential map*, which is, in this case, the usual exponential of matrices,

$$\exp(A) = \mathbb{1}_n + A + \frac{1}{2!}A^2 + \frac{1}{3!}A^3 + \ldots, \qquad A \in \mathfrak{gl}(n, \mathbb{R}).$$

  Here $\mathbb{1}_n$ is the identity element in $\mathbf{GL}(n, \mathbb{R})$.

- According to "Ado Theorem", every finite-dimensional Lie algebra is isomorphic to a matrix Lie algebra (for every finite-dimensional matrix Lie algebra, there is a matrix Lie group with this algebra as its Lie algebra). The corresponding theorem does not hold true for (finite-dimensional) Lie groups: not every Lie group is isomorphic to a subgroup of $\mathbf{GL}(n, \mathbb{R})$.

- A consequence of "Ado Theorem" establishes the fundamental correspondence between Lie groups and Lie algebras: if $\mathfrak{g}$ is a finite-dimensional Lie algebra then there exists a unique (connected) Lie group $\mathbf{G}$ having $\mathfrak{g}$ as its Lie algebra.

Example 4.17 (*The orthogonal Lie group and its Lie algebra*)

The *orthogonal Lie group*

$$\mathbf{O}(n) := \left\{ A \in \mathbf{GL}(n, \mathbb{R}) \,:\, A\,A^\top = A^\top A = \mathbb{1}_n \right\}$$

is a smooth manifold of dimension $n(n-1)/2$. It is a subgroup of $\mathbf{GL}(n, \mathbb{R})$.

- Let us construct the corresponding Lie algebra, the *orthogonal Lie algebra*. For $B \in \mathfrak{gl}(n, \mathbb{R})$ and for small $|t|$, consider the invertible matrix

$$\exp(t\,B) = \mathbb{1}_n + t\,B + O(t^2).$$

  Orthogonality of $\exp(t\,B)$ yields

$$\begin{aligned} \mathbb{1}_n &= \exp(t\,B)(\exp(t\,B))^\top \\ &= \left(\mathbb{1}_n + t\,B + O(t^2)\right)\left(\mathbb{1}_n + t\,B^\top + O(t^2)\right) \\ &= \mathbb{1}_n + t\left(B + B^\top\right) + O(t^2), \end{aligned}$$

  which gives

$$T_{\mathbb{1}_n}\mathbf{O}(n) = \mathfrak{o}(n) = \left\{ B \in \mathfrak{gl}(n, \mathbb{R}) \,:\, B + B^\top = \mathbb{O}_n \right\}.$$

- Fix $n = 3$. A possible choice of a basis in $\mathfrak{o}(3)$ is

$$B_1 := \begin{pmatrix} 0 & 0 & 0 \\ 0 & 0 & -1 \\ 0 & 1 & 0 \end{pmatrix}, \quad B_2 := \begin{pmatrix} 0 & 0 & 1 \\ 0 & 0 & 0 \\ -1 & 0 & 0 \end{pmatrix}, \quad B_3 := \begin{pmatrix} 0 & -1 & 0 \\ 1 & 0 & 0 \\ 0 & 0 & 0 \end{pmatrix}.$$

  Note that

$$[\,B_i, B_j\,] = \varepsilon_{ijk} B_k, \tag{4.17}$$

where ε_{ijk} is a totally skew-symmetric tensor with $\varepsilon_{123} = 1$. Here ε_{ijk} defines the structure constants of $\mathfrak{o}(3)$.

- We can explicitly describe the corresponding subgroups in $\mathbf{O}(3)$. For instance,

$$\exp(t\,B_1) = \begin{pmatrix} 1 & 0 & 0 \\ 0 & \cos t & -\sin t \\ 0 & \sin t & \cos t \end{pmatrix},$$

which is a rotation around the axis 1 by angle t. Similarly, B_2 and B_3 generate rotations around axes 2 and 3.

Example 4.18 (*The special linear Lie group and its Lie algebra*)

The real *special linear Lie group*

$$\mathbf{SL}(n,\mathbb{R}) := \{A \in \mathbf{GL}(n,\mathbb{R}) \,:\, \det A = 1\} = \det{}^{-1}(1)$$

is a smooth manifold of dimension $n^2 - 1$. It is a subgroup of $\mathbf{GL}(n,\mathbb{R})$.

- It can be proved that the differential of the submersion $A \mapsto \det A$ is

$$(\mathrm{d}\det A)(B) = \det A\ \mathrm{Trace}\left(A^{-1}B\right).$$

- The Lie algebra of $\mathbf{SL}(n,\mathbb{R})$ is the (n^2-1)-dimensional vector space

$$\begin{aligned} T_{1_n}\mathbf{SL}(n,\mathbb{R}) &= \mathfrak{sl}(n,\mathbb{R}) = \mathrm{kernel}(\mathrm{d}\det A) \\ &= \{B \in \mathfrak{gl}(n,\mathbb{R}) \,:\, \mathrm{Trace}\, B = 0\}. \end{aligned}$$

This is called *special linear Lie algebra*.

► Lie groups appear most often through their action (say the left action) on a manifold M. Let M be a real smooth manifold and $\mathbf{G}$ be a real Lie group.

- The diffeomorphism $\Phi : \mathbf{G} \times M \to M$ is an *action of* $\mathbf{G}$ *on* M if

$$\Phi(e,p) = p$$

for all $p \in M$ and

$$\Phi(g,\Phi(h,p)) = \Phi(g * h, p)$$

for all $g, h \in \mathbf{G}$ and for all $p \in M$. The triple $(\Phi, \mathbf{G}, M)$ is called *Lie group of diffeomorphisms* on M.

- The action allows to associate to each element A in the Lie algebra $\mathfrak{g}$ of $\mathbf{G}$ a vector field $\mathbf{v}$ on M, whose value at $p \in M$, is the derivation on $\mathscr{F}(M)$ at p given by

$$\mathbf{v}(p) := \left.\frac{\mathrm{d}}{\mathrm{d}t}\right|_{t=0} \exp(t\,A)p.$$

The vector field $\mathbf{v}$ is called *fundamental vector field* corresponding to $A \in \mathfrak{g}$ (or *infinitesimal generator of* $\mathbf{G}$). Its flow is given by the action of the one-parameter group $\exp(t\,A)$. Under some quite general conditions (that we are not going to investigate) one is allowed to not distinguish between the elements of the Lie algebra $\mathfrak{g}$ and the infinitesimal generators of $\mathbf{G}$. In other words, the two vector spaces are isomorphic.

Example 4.19 (*Lie group actions*)

1. The Lie group $\mathbf{O}(3)$ acts on $\mathbb{R}^3$ by $(A, x) \mapsto A\,x$, $A \in \mathbf{O}(3)$, $x \in \mathbb{R}^3$.
2. A global flow $\Phi_t : \mathbb{R} \times M \to M$ generated by a compactly supported vector field is an action of $\mathbb{R}$ on M. It defines a global Lie group of diffeomorphisms A local flow $\Phi_t : (-\varepsilon, \varepsilon) \times M \to M$ generated by a vector field is a local action of $(\varepsilon, \varepsilon)$ on M. It defines a local Lie group of transformations (indeed a continuous dynamical system!).

► The next example shows how a Lie group naturally arises in the framework of dynamics.

Example 4.20 (*A rigid body in* $\mathbb{R}^3$)

Consider a free (i.e., no external forces) rigid (i.e., the distance between any two points of the body is unchanged during the motion) body rotating about its center of mass, taken to be the origin.

- Let $x \in \mathbb{R}^3$ be a point of the body at time $t = 0$ and $f(x, t)$ be its position at time t.
- Rigidity of the body and the assumption of a smooth motion imply that $f(x, t) = A(t)x$, where $A(t)$ is a proper rotation, that is, $A(t) \in \mathbf{SO}(3)$. Here $\mathbf{SO}(3)$ is a subgroup of the orthogonal group $\mathbf{O}(3)$ characterized by orthogonal matrices with determinant one.
- The Lie group $\mathbf{SO}(3)$, which is three-dimensional, defines the configuration manifold of the mechanical system. In fact it also plays a dual role of a symmetry group, since the same physical motion is described if we rotate our coordinate axes. Used as a symmetry group, $\mathbf{SO}(3)$ leads to conservation of angular momentum.

4.3 Hamiltonian mechanics on Poisson manifolds

► We now have all necessary ingredients to define those smooth manifolds where Hamiltonian dynamics can be defined in a natural way. Our strategy will be to define the smooth manifolds on which Hamiltonian vector fields can be constructed and then to derive consequences in a systematic way.

Definition 4.2

Let M be real n-dimensional smooth manifold.

1. *A* **Poisson bracket** *(or* **Poisson structure***) on M is a Lie algebra structure $\{\cdot, \cdot\}$ on $\mathscr{F}(M)$, which is a biderivation on $\mathscr{F}(M)$. Equivalently, for any $H \in \mathscr{F}(M)$ the linear map*

$$\mathscr{F}(M) \to \mathscr{F}(M) \, : \, F \mapsto \{F, H\},$$

is a derivation on $\mathscr{F}(M)$, i.e., it defines a smooth vector field on M.

2. *The pair* $(M, \{\cdot,\cdot\})$ *is called* **Poisson manifold**.
3. *Let* $(M, \{\cdot,\cdot\})$ *be a Poisson manifold and let* $H \in \mathscr{F}(M)$. *The vector field*

$$\mathbf{v}_H := \{\cdot, H\} \tag{4.18}$$

is called the **Hamiltonian vector field** *associated with the Hamiltonian* H. *Its local flow* $\Phi_t : I_\varepsilon \times M \to M$, $I_\varepsilon := \{t \in \mathbb{R} : |t| < \varepsilon\}$, *is a* **Hamiltonian flow** *on* M. *We write*

$$\mathsf{Ham}(M) := \{\{\cdot, H\} : H \in \mathscr{F}(M)\}$$

for the vector space of Hamiltonian vector fields. The function H *is a* **Casimir function** *if its associated vector field is (identically) zero,* $\mathbf{v}_H \equiv 0$, *i.e.,* $\{F, H\} = 0$ *for all* $F \in \mathscr{F}(M)$. *We write*

$$\mathsf{Cas}(M) := \{H \in \mathscr{F}(M) : \mathbf{v}_H \equiv 0\}$$

for the vector space of Casimir functions.

▶ From the above definition we can systematically derive the general features of Hamiltonian dynamics on Poisson manifolds.

- Item *1.* of Definition 4.2 says that a Poisson bracket on M is a bilinear operation $\{\cdot,\cdot\} : \mathscr{F}(M) \times \mathscr{F}(M) \to \mathscr{F}(M)$ possessing the following properties:
  1. *Skew-symmetry*: $\{F, H\} = -\{H, F\}$,
  2. *Leibniz rule*: $\{FG, H\} = F\{G, H\} + G\{F, H\}$,
  3. *Jacobi identity*: $\{F, \{G, H\}\} + \circlearrowleft (F, G, H) = 0$,

  for all functions $F, G, H \in \mathscr{F}(M)$.
- Since the Poisson bracket is a biderivation, it vanishes whenever one of its arguments is constant, so that we can associate to a Poisson bracket $\{\cdot,\cdot\}$ a $\mathscr{F}(M)$-bilinear map

  $$\mathbf{X} : \Omega(M) \times \Omega(M) \to \mathscr{F}(M) : (\mathrm{d}F, \mathrm{d}H) \mapsto \mathbf{X}(\mathrm{d}F, \mathrm{d}H) := \{F, H\},$$

  for all $F, H \in \mathscr{F}(M)$. $\mathbf{X}$ is a 2-vector field on M called *Poisson 2-vector field* associated with the Poisson structure $\{\cdot,\cdot\}$.

  (a) As one does for a differential 2-form (which can be evaluated either on two vector fields to give a smooth function or on a single vector field to give a differential 1-form), one can contract $\mathbf{X}$ with a single 1-form to get a vector field. In this way we derive from $\mathbf{X}$ a map

  $$\widetilde{\mathbf{X}} : \Omega(M) \to \mathfrak{X}(M),$$

simply by setting

$$\widetilde{\mathbf{X}}(\mathrm{d}H) := \mathbf{X}(\,\cdot\,, \mathrm{d}H) = \{\,\cdot, H\,\} = \mathbf{v}_H \qquad \forall\, H \in \mathscr{F}(M). \tag{4.19}$$

Formula (4.19) is an alternative definition of a Hamiltonian vector field.

(b) We see from (4.19) that a Casimir function $H \in \mathscr{F}(M)$ is such that $\widetilde{\mathbf{X}}$ annihilates its differential $\mathrm{d}H$, i.e., $\widetilde{\mathbf{X}}(\mathrm{d}H) \equiv 0$. In other words, Casimir functions are constant along the integral curves of any Hamiltonian vector field.

- The Leibniz property for $\{\,\cdot,\cdot\,\}$ implies on the one hand that $\mathsf{Cas}(M)$ is a subalgebra of $\mathscr{F}(M)$ (for the ordinary multiplication of functions) and on the other hand that $\mathsf{Ham}(M)$ is a subalgebra of $\mathfrak{X}(M)$.
- The Jacobi identity for $\{\,\cdot,\cdot\,\}$ has two fundamental consequences:

(a) The map $\mathscr{F}(M) \to \mathsf{Ham}(M) : F \mapsto \mathbf{v}_F$ is a Lie algebra anti-homomorphism, (cf. Theorem 3.12):

$$[\,\mathbf{v}_F, \mathbf{v}_G\,] = -\mathbf{v}_{\{\,F,G\,\}} \qquad \forall\, F, G \in \mathscr{F}(M). \tag{4.20}$$

It turns out (cf. Theorem 3.13) that two local Hamiltonian flows, whose Hamiltonian vector fields are $\mathbf{v}_F := \{\,\cdot, F\,\}$ and $\mathbf{v}_G := \{\,\cdot, G\,\}$, commute if and only if the Poisson bracket $\{\,F, G\,\}$ is locally constant.

(b) Any Hamiltonian vector field $\mathbf{v}_H \in \mathsf{Ham}(M)$ leaves $\{\,\cdot,\cdot\,\}$ invariant. Equivalently, if $\mathbf{X}$ is the Poisson 2-vector field associated with the Poisson structure $\{\,\cdot,\cdot\,\}$ then $\mathfrak{L}_{\mathbf{v}_H}\mathbf{X} = 0$. Indeed, identifying $\mathbf{X}[F,G] \equiv \mathbf{X}(\mathrm{d}F, \mathrm{d}G)$ and using (4.12) we find that for any $F, G \in \mathscr{F}(M)$ there holds

$$\begin{aligned}
\mathfrak{L}_{\mathbf{v}_H}\mathbf{X}[F,G] &= \mathbf{v}_H[\mathbf{X}[F,G]] - \mathbf{X}[\mathbf{v}_H[F],G] - \mathbf{X}[F,\mathbf{v}_H[G]] \\
&= \mathbf{v}_H[\{F,G\}] - \mathbf{X}[\{F,H\},G] - \mathbf{X}[F,\{G,H\}] \\
&= \{\{F,G\},H\} - \{\{F,H\},G\} - \{F,\{G,H\}\} \\
&= 0.
\end{aligned}$$

An equivalent way to claim that any $\mathbf{v}_H \in \mathsf{Ham}(M)$ leaves $\{\,\cdot,\cdot\,\}$ invariant is to say that any Hamiltonian flow Φ_t is a *Poisson morphism* (cf. Theorems 3.16 and 3.18).

► Let us consider a coordinate chart U of M where a point $p \in M$ can be expressed by local coordinates $x := (x_1, \ldots, x_n)$.

- A Poisson structure on M is locally defined in terms of its *structure functions*

$$x_{ij} := \{\,x_i, x_j\,\} \in \mathscr{F}(M), \qquad i,j = 1, \ldots, n,$$

which define a skew-symmetric matrix $X := \left(x_{ij}\right)_{1\leqslant i,j\leqslant n}$, called *Poisson matrix* associated with $\{\,\cdot,\cdot\,\}$. The structure functions satisfy the equations

$$x_{ij} = -x_{ji},, \qquad i,j = 1,\ldots,n, \tag{4.21}$$

and

$$\sum_{\ell=1}^{n}\left(\frac{\partial x_{ij}}{\partial x_\ell}x_{\ell k} + \frac{\partial x_{jk}}{\partial x_\ell}x_{\ell i} + \frac{\partial x_{ki}}{\partial x_\ell}x_{\ell j}\right) = 0, \qquad i,j,k = 1,\ldots,n. \tag{4.22}$$

Formula (4.22) is the coordinate representation of the Jacobi identity. Note that the skew-symmetry of X implies that its rank is always an even number.

- Conversely, every set of functions $x_{ij} \in \mathscr{F}(M)$ satisfying conditions (4.21-4.22) locally defines a Poisson bracket on M between functions $F, H \in \mathscr{F}(M)$ by setting

$$\{\,F(x), H(x)\,\} := \sum_{1\leqslant i<j\leqslant n} x_{ij}\left(\frac{\partial F}{\partial x_i}\frac{\partial H}{\partial x_j} - \frac{\partial H}{\partial x_i}\frac{\partial F}{\partial x_j}\right) = [\mathrm{d}F]^\top X\,[\mathrm{d}H], \tag{4.23}$$

where $[\mathrm{d}F]$ is the column vector which represents $\mathrm{d}F$ in the basis $\{\mathrm{d}x_1,\ldots,\mathrm{d}x_n\}$, i.e., the i-th component of $[\mathrm{d}F]$ is $\partial F/\partial x_i$.

- For $H \in \mathscr{F}(M)$ the Hamiltonian vector field $\mathbf{v}_H$, which is the infinitesimal generator of a Hamiltonian flow Φ_t, can be expressed as

$$\mathbf{v}_H = \sum_{j=1}^{n}\{\,x_j, H(x)\,\}\frac{\partial}{\partial x_j}.$$

- The local form of *Hamilton equations* expressing the time evolution of coordinates $x := (x_1,\ldots,x_n)$, is given by (cf. formula (3.46))

$$\mathbf{v}_H\,[x_i] \equiv \dot{x}_i = \{\,x_i, H(x)\,\} = \sum_{j=1}^{n} x_{ij}\frac{\partial H}{\partial x_j}, \qquad i = 1,\ldots,n. \tag{4.24}$$

- If $F \in \mathscr{F}(M)$ we obtain (cf. formula (3.48))

$$\begin{aligned}\mathbf{v}_H[F(x)] &= (\mathfrak{L}_{\mathbf{v}_H}F)(x) = \sum_{i=1}^{n}\{x_i, H(x)\,\}\frac{\partial F}{\partial x_i}\\ &= \sum_{1\leqslant i<j\leqslant n} x_{ij}\left(\frac{\partial F}{\partial x_i}\frac{\partial H}{\partial x_j} - \frac{\partial H}{\partial x_i}\frac{\partial F}{\partial x_j}\right) = \{\,F(x), H(x)\,\}.\end{aligned} \tag{4.25}$$

- We say that F is an *integral of motion* of Φ_t if it is in *Poisson involution* with H. In particular H is an invariant function for Φ_t, i.e., $\mathbf{v}_H[H(x)] = 0$. The Jacobi identity for $\{\cdot,\cdot\}$ implies that the set of all integrals of motion of Φ_t is an Abelian subalgebra of $\mathscr{F}(M)$ ("Poisson Theorem").

- In terms of the structure functions x_{ij} the Poisson 2-vector field $\mathbf{X}$ can be written as

$$\mathbf{X} = \sum_{1\leqslant i<j\leqslant n} x_{ij} \frac{\partial}{\partial x_i} \wedge \frac{\partial}{\partial x_j}. \tag{4.26}$$

► Given a Poisson manifold $(M, \{\cdot,\cdot\})$ the existence of Casimirs is related to a degeneracy of the the Poisson structure. Thus, we need to give a proper definition of the rank of the Poisson structure. To do so we first introduce a particular distribution on M. If we specialize the construction of Hamiltonian vector fields to a point $p \in M$ we find a linear space whose dimension varies in general with p, thereby defining a (generalized) distribution on M:

$$\mathsf{Ham}_p(M) = \mathsf{span}\left\{(\mathbf{v}_{x_1})_p, \ldots, (\mathbf{v}_{x_n})_p\right\} \subseteq T_pM,$$

where $(x_1, \ldots, x_n)$ is any system of local coordinates, defined on a coordinate chart containing p.

► We now give the following definition.

Definition 4.3

Let $(M, \{\cdot,\cdot\})$ be a n-dimensional Poisson manifold.

1. *For $p \in M$, the dimension of $\mathsf{Ham}_p(M)$ is called the* **rank** *of $\{\cdot,\cdot\}$ at p, denoted by $\mathsf{rank}_p\{\cdot,\cdot\}$. The* **rank of the Poisson manifold** *$(M, \{\cdot,\cdot\})$ is*

$$\mathsf{rank}\{\cdot,\cdot\} := \max_{p\in M}\left(\mathsf{rank}_p\{\cdot,\cdot\}\right).$$

2. *$(M, \{\cdot,\cdot\})$ is a* **regular Poisson manifold** *when $\mathsf{rank}_p\{\cdot,\cdot\} = \mathsf{rank}\{\cdot,\cdot\}$ independently of $p \in M$.*

3. *The Poisson structure $\{\cdot,\cdot\}$ has* **maximal rank at** *$p \in M$ when $\mathsf{rank}_p\{\cdot,\cdot\} = n$. The Poisson structure $\{\cdot,\cdot\}$ has* **maximal rank on** *M when $\mathsf{rank}_p\{\cdot,\cdot\} = n$ for all $p \in M$, i.e., $\mathsf{Ham}_p(M) = T_pM$ for all $p \in M$.*

► Remarks:

- In a coordinate chart U of M where a point $p \in M$ can be expressed by local coordinates $x := (x_1, \ldots, x_n)$ we have

$$\mathsf{rank}_p\{\,\cdot,\cdot\,\} = \mathsf{rank}\,(X(p)),$$

where $X(p)$ is the Poisson matrix at p referred to the chart U. The skew-symmetry of X implies that the rank of a Poisson structure at a point is always an even number.

- If a Poisson manifold has maximal rank on M then its dimension n is necessarily even.

► Let us present some examples of Poisson structures.

Example 4.21 (*The canonical Hamiltonian phase space* $\mathbb{R}^{2n}$)

Any constant skew-symmetric matrix on $\mathbb{R}^{2n}$ is the matrix of a Poisson structure in terms of its coordinates $x := (x_1, \ldots, x_{2n})$, as follows from (4.22).

- In particular, the non-degenerate *canonical Poisson structure* on $\mathbb{R}^{2n}$ defined by

$$X = \mathbb{J} := \begin{pmatrix} \mathbb{O}_n & \mathbb{1}_n \\ -\mathbb{1}_n & \mathbb{O}_n \end{pmatrix}, \tag{4.27}$$

is an example of constant Poisson structure on $\mathbb{R}^{2n}$ with maximal rank $2\,n$.

- Defining coordinates $(x_1, \ldots, x_{2n}) := (q_1, \ldots, q_n, p_1, \ldots, p_n)$ we can write the Poisson 2-vector corresponding to (4.27) as

$$\mathbf{X} = \sum_{i=1}^{n} \frac{\partial}{\partial q_i} \wedge \frac{\partial}{\partial p_i},$$

which implies the following canonical Poisson brackets

$$\{\,p_i, p_j\,\} = \{\,q_i, q_j\,\} = 0, \qquad \{\,q_i, p_j\,\} = \delta_{ij}, \qquad i,j = 1, \ldots, n.$$

- If $H \in \mathscr{F}(\mathbb{R}^n)$ then, by using (4.19), we can construct the corresponding canonical Hamiltonian vector field by contracting the Poisson 2-vector $\mathbf{X}$ with the differential of H,

$$\mathsf{d}H = \sum_{i=1}^{n} \left(\frac{\partial H}{\partial q_i} \mathsf{d}q_i + \frac{\partial H}{\partial p_i} \mathsf{d}p_i \right).$$

From formula (4.19) we obtain:

$$\begin{aligned} \mathbf{v}_H &= \widetilde{\mathbf{X}}(\mathsf{d}H) = \mathbf{X}(\cdot, \mathsf{d}H) = -\mathbf{X}(\mathsf{d}H, \cdot) \\ &= \sum_{i=1}^{n} \left(\frac{\partial H}{\partial p_i} \frac{\partial}{\partial q_i} - \frac{\partial H}{\partial q_i} \frac{\partial}{\partial p_i} \right), \end{aligned}$$

which is nothing but that the canonical Hamiltonian vector field (3.38) on $\mathbb{R}^{2n}$.

Example 4.22 (*A Poisson structure on* $\mathbb{R}^3$)

Define the following 3×3 real skew-symmetric matrix:

$$X := \begin{pmatrix} 0 & f_3(x) & -f_2(x) \\ -f_3(x) & 0 & f_1(x) \\ f_2(x) & -f_1(x) & 0 \end{pmatrix},$$

where $f_i \in \mathscr{F}(\mathbb{R}^3)$, $i = 1,2,3$, and $x := (x_1, x_2, x_3)$.

- It can be proved that X defines a Poisson matrix of a Poisson structure on $\mathbb{R}^3$ if and only if there holds
$$\langle \operatorname{rot}_x f(x), f(x) \rangle = 0 \qquad \forall\, x \in \mathbb{R}^3. \tag{4.28}$$
Here $f(x) := (f_1(x), f_2(x), f_3(x))$. Note that (4.28) is a partial differential equation.
- For $A = A(x)$ and $B = B(x)$ arbitrary polynomials in $\mathbb{R}^3$, the function defined by
$$f(x) := A(x)\, \operatorname{grad}_x B(x)$$
is a particular solution of (4.28). Such a function leads to a large number of regular Poisson structures on $\mathbb{R}^3$. Explicitly one finds
$$X = A(x) \begin{pmatrix} 0 & \dfrac{\partial B}{\partial x_3} & -\dfrac{\partial B}{\partial x_2} \\ -\dfrac{\partial B}{\partial x_3} & 0 & \dfrac{\partial B}{\partial x_1} \\ \dfrac{\partial B}{\partial x_2} & -\dfrac{\partial B}{\partial x_1} & 0 \end{pmatrix}.$$
In coordinates we get the following Poisson brackets:
$$\{x_1, x_2\} = A(x)\, \frac{\partial B}{\partial x_3}, \quad \{x_2, x_3\} = A(x)\, \frac{\partial B}{\partial x_1}, \quad \{x_3, x_1\} = A(x)\, \frac{\partial B}{\partial x_2}.$$
- Generically, the matrix X has rank two, while the dimension of the phase space is three. Indeed one can verifies that the function B is a Casimir function of the Poisson structure: the matrix X annihilates the gradient of B:
$$X\, \operatorname{grad}_x B(x) \equiv 0.$$

Example 4.23 (***The three-dimensional Euler top***)

Consider the Poisson structure of Example 4.22 and define
$$A(x) := 1, \qquad B(x) := \frac{1}{2}\left(x_1^2 + x_2^2 + x_3^2\right).$$

- The resulting Poisson matrix is
$$X = \begin{pmatrix} 0 & x_3 & -x_2 \\ -x_3 & 0 & x_1 \\ x_2 & -x_1 & 0 \end{pmatrix}, \tag{4.29}$$
which corresponds to the Poisson 2-vector (see (4.26))
$$\mathbf{X} = x_1 \frac{\partial}{\partial x_2} \wedge \frac{\partial}{\partial x_3} + x_2 \frac{\partial}{\partial x_3} \wedge \frac{\partial}{\partial x_1} + x_3 \frac{\partial}{\partial x_1} \wedge \frac{\partial}{\partial x_2}.$$
Note that X defines the structure constants (4.17) of the three-dimensional orthogonal Lie algebra described in Example 4.17.
- Now consider the smooth function
$$H(x) := \frac{1}{2}\left(a_1\, x_1^2 + a_2\, x_2^2 + a_3\, x_3^2\right),$$
with $a_1, a_2, a_3 \in \mathbb{R}$. We can construct a Hamiltonian vector field in terms of the action of X on the gradient of H:
$$X\, \operatorname{grad}_x H(x) = ((a_2 - a_3)\, x_2\, x_3, (a_3 - a_1)\, x_3\, x_1\, (a_1 - a_2)\, x_1\, x_2). \tag{4.30}$$

In a more geometric language the vector field (4.30) is constructed by contracting the Poisson 2-vector corresponding to X with the differential of H,

$$\mathrm{d}H = a_1\,x_1\,\mathrm{d}x_1 + a_2\,x_2\,\mathrm{d}x_2 + a_3\,x_3\,\mathrm{d}x_3.$$

Indeed, by using (4.19) we obtain

$$\begin{aligned}\mathbf{v}_H &= \widetilde{\mathbf{X}}(\mathrm{d}H) = \mathbf{X}(\cdot,\mathrm{d}H) = -\mathbf{X}(\mathrm{d}H,\cdot)\\ &= (a_2-a_3)\,x_2\,x_3\,\frac{\partial}{\partial x_1} + (a_3-a_1)\,x_3\,x_1\,\frac{\partial}{\partial x_2} + (a_1-a_2)\,x_1\,x_2\,\frac{\partial}{\partial x_3}.\end{aligned}$$

Such a vector field defines exactly the *Euler top* (4.1):

$$\begin{cases}\dot{x}_1 = (a_2-a_3)\,x_2\,x_3,\\ \dot{x}_2 = (a_3-a_1)\,x_3\,x_1,\\ \dot{x}_3 = (a_1-a_2)\,x_1\,x_2.\end{cases} \tag{4.31}$$

This is probably the most famous Hamiltonian system described in terms of an odd number of ODEs. Indeed, the configuration manifold of the three-dimensional Euler top is the Lie group $\mathbf{SO}(3)$, which is usually parametrized in terms of three *Euler angles.*

- Note that we can obtain (4.31) also by using (4.24). For instance,

$$\begin{aligned}\mathbf{v}_H\,[x_1] &\equiv \dot{x}_1 = \{\,x_1, H(x)\,\} = x_{11}\frac{\partial H}{\partial x_1} + x_{12}\frac{\partial H}{\partial x_2} + x_{13}\frac{\partial H}{\partial x_3}\\ &= (a_2-a_3)x_2\,x_3.\end{aligned}$$

- The function H is the Hamiltonian of the system, an integral of motion. Indeed, using (4.25) we get

$$\mathbf{v}_H[H(x)] = (\mathfrak{L}_{\mathbf{v}_H}H)\,(x) = \dot{x}_1\frac{\partial H}{\partial x_1} + \dot{x}_2\frac{\partial H}{\partial x_2} + \dot{x}_3\frac{\partial H}{\partial x_3} = 0$$

The function B is a Casimir, namely another integral of motion, functionally independent on H, which generates a trivial dynamics. We conclude by saying that the Euler top admits two functionally independent integrals of motion which are in Poisson involution w.r.t. the Poisson structure (4.29).

Example 4.24 (*A family of quadratic Poisson structure on* $\mathbb{R}^4$)

A family of quadratic Poisson structures on $\mathbb{R}^4$ is the following. Let x_1, x_2, x_3, x_4 be coordinates on $\mathbb{R}^4$ and define the skew-symmetric matrix

$$X := \begin{pmatrix} 0 & b_1\,x_3\,x_4 & b_2\,x_2\,x_4 & b_3\,x_2\,x_3\\ -b_1\,x_3\,x_4 & 0 & a_3\,x_1\,x_4 & a_2\,x_1\,x_3\\ -b_2\,x_2\,x_4 & -a_3\,x_1\,x_4 & 0 & a_1\,x_1\,x_2\\ -b_3\,x_2\,x_3 & -a_2\,x_1\,x_3 & -a_1\,x_1\,x_2 & 0\end{pmatrix},$$

where $a_1, a_2, a_3, b_1, b_2, b_3 \in \mathbb{R}$ are parameters.

- Four checks of the Jacobi identity suffice to show that this is a Poisson matrix if and only if

$$a_1\,b_1 - a_2\,b_2 + a_3\,b_3 = 0.$$

- Except for the trivial structure the remaining parametric Poisson structures are all of rank 2 and two Casimirs are given by the quadratic functions

$$a_1\,x_2^2 - a_2\,x_3^2 + a_3\,x_4^2, \qquad a_1\,x_1^2 - b_3\,x_3^2 + b_2\,x_4^2.$$

4.4 Hamiltonian mechanics on symplectic manifolds

► We now introduce a special regular Poisson manifold which plays a central role in the theory of Hamiltonian dynamics when the phase space has maximal rank for any point. It turns out that such a manifold is locally diffeomorphic to the canonical Hamiltonian phase space $\mathbb{R}^{2n}$ considered in Chapter 3.

Definition 4.4

Let M be a real r-dimensional smooth manifold.

1. *If M is equipped with a closed non-degenerate 2-form $\omega \in \Omega^2(M)$ then the pair (M, ω) is called* **symplectic manifold**.
2. *Let (M, ω) be a symplectic manifold and let $H \in \mathscr{F}(M)$. The vector field $\mathbf{v}_H$ defined as*
$$\mathbf{v}_H \lrcorner\, \omega = \omega(\mathbf{v}_H, \cdot) := \mathrm{d}H(\cdot), \tag{4.32}$$
is called the **Hamiltonian vector field** *associated with the Hamiltonian H. Its local flow $\Phi_t : I_\varepsilon \times M \to M$, $I_\varepsilon := \{t \in \mathbb{R} : |t| < \varepsilon\}$, is a* **Hamiltonian flow** *on M.*

► From the above definition we can derive the general features of Hamiltonian dynamics on symplectic manifolds. We will see that a symplectic manifold is, in a natural way, an even-dimensional regular Poisson manifold with maximal rank on M.

- Let us consider a coordinate chart U of M where a point $p \in M$ can be expressed by local coordinates $x := (x_1, \dots, x_r)$. Then ω can be locally expressed as
$$\omega = \sum_{1 \leqslant i < j \leqslant r} f_{ij}(x)\, \mathrm{d}x_i \wedge \mathrm{d}x_j,$$
for some smooth functions $f_{ij} \in \mathscr{F}(U)$.
  (a) The closure condition on ω, i.e., $\mathrm{d}\omega = 0$, means that locally there holds
$$\frac{\partial f_{ij}}{\partial x_k} + \frac{\partial f_{jk}}{\partial x_i} + \frac{\partial f_{ki}}{\partial x_j} = 0 \qquad \forall\, i, j, k = 1, \dots, r,$$
and it assures that the induced Poisson bracket satisfies the Jacobi identity and the Hamiltonian flows define Poisson maps.
  (b) The non-degeneracy condition on ω means that locally there holds
$$\mathrm{rank}\big(f_{ij}(x)\big)_{1 \leqslant i < j \leqslant r} = r.$$

Such a condition, together with the skew-symmetry property of ω, implies that any symplectic manifold has even dimension, say $r = 2n$. Furthermore, the differential $2n$-form

$$\omega^{2n} := \underbrace{\omega \wedge \cdots \wedge \omega}_{n \text{ times}} \tag{4.33}$$

defines a volume form. In particular, every symplectic manifold is orientable.

- Formula (4.32) defining a Hamiltonian vector field on (M, ω), implies that the 1-form $\mathbf{v}_H \lrcorner\, \omega$ is closed (even exact) for any vector field $\mathbf{v}_H$, i.e.,

$$\mathrm{d}(\mathbf{v}_H \lrcorner\, \omega) = 0. \tag{4.34}$$

- The closure condition on ω allows a natural definition of a Poisson bracket on M.

  (a) Let $F, H \in \mathscr{F}(M)$ and define the corresponding Hamiltonian vector fields $\mathbf{v}_F, \mathbf{v}_H$ by using (4.32). Then we define a skew-symmetric biderivation on $\mathscr{F}(M)$ by

$$\{F, H\} := \omega(\mathbf{v}_F, \mathbf{v}_H) = \mathrm{d}F(\mathbf{v}_H) = \mathbf{v}_H[F]. \tag{4.35}$$

  The above formula shows that the definition of a Hamiltonian vector field on M, see (4.32), is consistent with the definition of a Hamiltonian vector field on a Poisson manifold, see (4.18).

  (b) We now see that non-degeneracy and closure conditions of ω imply that (4.35) is a well-defined Poisson structure on M. Using $\mathrm{d}\omega = 0$ and formulas (4.15), (4.32) and (4.35) we get

$$\begin{aligned} [\mathbf{v}_F, \mathbf{v}_H] \lrcorner\, \omega &= \mathrm{d}(\mathbf{v}_F \lrcorner\, \mathbf{v}_H \lrcorner\, \omega) = -\mathrm{d}(\omega(\mathbf{v}_F, \mathbf{v}_H)) \\ &= -\mathrm{d}\{F, H\} = -\mathbf{v}_{\{F,H\}} \lrcorner\, \omega. \end{aligned}$$

  Since ω is non-degenerate this shows that

$$[\mathbf{v}_F, \mathbf{v}_H] = -\mathbf{v}_{\{F,H\}}. \tag{4.36}$$

  Applying (4.36) to an arbitrary $G \in \mathscr{F}(M)$ we find

$$\mathbf{v}_F\,[\mathbf{v}_H[G]] - \mathbf{v}_H\,[\mathbf{v}_F[G]] + \mathbf{v}_{\{F,H\}}[G] = 0,$$

  which is precisely the Jacobi identity, as follows from (4.35).

  (c) A more explicit expression for the non-degenerate Poisson bracket induced by ω can be easily derived. Picking local coordinates $(x_1, \ldots, x_r)$,

$r = 2n$, on a chart U we can introduce a skew-symmetric $r \times r$ matrix $\Omega := (\Omega_{ij})_{1 \leqslant i < j \leqslant r}$ by setting

$$\Omega_{ij} := \omega\left(\frac{\partial}{\partial x_i}, \frac{\partial}{\partial x_j}\right). \tag{4.37}$$

Let $F, H \in \mathscr{F}(M)$. Denote by $[\mathbf{v}_F]$ the column matrix whose elements are the coefficients of $\mathbf{v}_F$ w.r.t. the basis $\{\partial/\partial x_1, \ldots, \partial/\partial x_r\}$ and by $[\mathrm{d}F]$ the column matrix whose elements are the coefficients of $\mathrm{d}F$ w.r.t. the basis $\{\mathrm{d}x_1, \ldots, \mathrm{d}x_n\}$. Then (4.32) says that $\Omega[\mathbf{v}_F] = -[\mathrm{d}F]$ and $\omega(\mathbf{v}_F, \mathbf{v}_G) = [\mathbf{v}_F]^\top \Omega[\mathbf{v}_G]$, so that

$$\begin{aligned}\{F, G\} &= [\mathbf{v}_F]^\top \Omega[\mathbf{v}_G] = \left(\Omega^{-1}[\mathrm{d}F]\right)^\top \Omega\left(\Omega^{-1}[\mathrm{d}G]\right) \\ &= -[\mathrm{d}F]^\top \Omega^{-1}[\mathrm{d}G],\end{aligned}$$

which compared with (4.23) gives the Poisson matrix X corresponding to (4.35):

$$X = -\Omega^{-1}. \tag{4.38}$$

- Any flow of a Hamiltonian vector field $\mathbf{v}_H$ on (M, ω) leaves ω invariant (as well as the induced Poisson bracket, in view of (4.35), and the volume form (4.33)). Equivalently, if Φ_t is the Hamiltonian flow of $\mathbf{v}_H$ then $\mathfrak{L}_{\mathbf{v}_H}\omega = 0$. Indeed, by using Cartan formula (4.13) and $\mathrm{d}\omega = 0$, we have:

$$\mathfrak{L}_{\mathbf{v}_H}\omega = \mathrm{d}\left(\mathbf{v}_H \lrcorner \omega\right) = 0,$$

in view of (4.34). An equivalent way to claim that any Hamiltonian vector field $\mathbf{v}_H$ on (M, ω) leaves ω invariant is to say that any Hamiltonian flow Φ_t is a *symplectic morphism* (cf. Theorems 3.16 and 3.18).

► Let us present some examples of symplectic manifolds.

Example 4.25 (*The canonical Hamiltonian phase space* $\mathbb{R}^{2n}$)

In fact, we already know that the canonical Hamiltonian phase space

$$\left(\mathbb{R}^{2n}, \omega\right), \qquad \omega := \sum_{j=1}^{n} \mathrm{d}q_j \wedge \mathrm{d}p_j$$

is a symplectic vector space (of dimension $r = 2n$). Here ω is the canonical symplectic 2-form given in Definition 3.12.

- If $H \in \mathscr{F}(\mathbb{R}^{2n})$ then, by using (4.32), we can construct the corresponding canonical Hamiltonian vector field. The differential of H is

$$\mathrm{d}H = \sum_{i=1}^{n}\left(\frac{\partial H}{\partial q_i}\mathrm{d}q_i + \frac{\partial H}{\partial p_i}\mathrm{d}p_i\right),$$

and a vector field on M can be written as

$$\mathbf{v} = \sum_{i=1}^{n} \left(f_i(q,p)\,\frac{\partial}{\partial q_i} + g_i(q,p)\,\frac{\partial}{\partial p_i} \right).$$

Then, $\mathbf{v}$ is the Hamiltonian vector field corresponding to H is and only if

$$\begin{aligned} \mathrm{d}H &= \sum_{i=1}^{n} \left(\frac{\partial H}{\partial q_i}\mathrm{d}q_i + \frac{\partial H}{\partial p_i}\mathrm{d}p_i \right) \\ &= \left(\sum_{i=1}^{n} \left(f_i(q,p)\,\frac{\partial}{\partial q_i} + g_i(q,p)\,\frac{\partial}{\partial p_i} \right) \right) \lrcorner \left(\sum_{k=1}^{n} \mathrm{d}q_k \wedge \mathrm{d}p_k \right), \end{aligned}$$

which gives

$$f_i(q,p) = \frac{\partial H}{\partial p_i}, \qquad g_i(q,p) = -\frac{\partial H}{\partial q_i}, \qquad i = 1,\ldots,n,$$

as expected. In particular, we find

$$\mathbf{v}_{q_i} = -\frac{\partial}{\partial p_i}, \qquad \mathbf{v}_{p_i} = \frac{\partial}{\partial q_i}, \qquad i = 1,\ldots,n.$$

- The *canonical Poisson structure* on $\mathbb{R}^{2n}$ is defined in terms of (4.37):

$$\{ q_i, p_j \} := \omega\left(\mathbf{v}_{q_i}, \mathbf{v}_{p_j}\right) = \left(\sum_{k=1}^{n} \mathrm{d}q_k \wedge \mathrm{d}p_k \right) \left(-\frac{\partial}{\partial p_i}, \frac{\partial}{\partial q_j} \right) = \delta_{ij},$$

for $i,j = 1,\ldots,n$. Similarly, $\{q_i, q_j\} = \{p_i, p_j\} = 0$ for any $i,j = 1,\ldots,n$. Therefore, see (4.38), the (non-degenerate) Poisson matrix is

$$X = -\mathbb{J}^{-1} = \mathbb{J}.$$

Example 4.26 (*The unit sphere* $\mathbb{S}^2$)

The 2-sphere $\mathbb{S}^2$, regarded as the set of unit vectors in $\mathbb{R}^3$, has tangent vectors at $p \in \mathbb{S}^2$ identified with vectors orthogonal to the vector connecting the origin with p (see Example 4.6).

- The standard symplectic form on $\mathbb{S}^2$ is induced by the standard scalar and vector products:

$$\omega_p(v,w) := \langle\, p, v \times w \,\rangle, \tag{4.39}$$

where $p \in \mathbb{S}^2$ and $v, w \in T_p\mathbb{S}^2$. This is the standard *area form* on $\mathbb{S}^2$ with total area $4\,\pi$.

- In terms of cylindrical polar coordinates $(\theta, h) \in [0, 2\,\pi) \times [-1, 1]$ away from the poles, it is written one finds $\omega = \mathrm{d}\theta \wedge \mathrm{d}h$. The vector fiels $\partial/\partial\theta$ is a Hamiltonian vector field with Hamiltonian $H = h$:

$$\frac{\partial}{\partial\theta} \lrcorner (\mathrm{d}\theta \wedge \mathrm{d}h) = \mathrm{d}h.$$

The motion generated by this vector field is rotation about the vertical axis, which of course preserves both area and height.

It can be proved that any oriented two-dimensional smooth manifold with an area form is a symplectic manifold.

Example 4.27 (*The cotangent bundle of a smooth manifold*)

Let M be a n-dimensional smooth manifold and let T^*M be its ($2\,n$-dimensional) cotangent bundle. It can be proved (but the proof is rather tricky) that T^*M is a symplectic manifold. Note that this implies that T^*M is an orientable manifold even if M is not.

- If $(q_1, \ldots, q_n, p_1, \ldots, p_n)$ are local coordinates on T^*M (note that $(q_1, \ldots, q_n)$ are local coordinates for points on M and $(p_1, \ldots, p_n)$ are local coordinates for tangent vectors on M, i.e., $p_i \equiv \mathrm{d}q_i$) one can prove that the symplectic 2-form can be always written as the canonical symplectic 2-form

$$\omega = \sum_{j=1}^{n} \mathrm{d}q_j \wedge \mathrm{d}p_j.$$

- Indeed, a very standard way to present classical mechanics on symplectic manifolds is to start with the assumption that the *configuration space* of a system is a n-dimensional smooth manifold M. Then the natural *Lagrangian phase space* is the $2n$-dimensional tangent bundle TM, while the natural *Hamiltonian phase space* is the $2n$-dimensional cotangent bundle T^*M. In this setting we can formulate both Lagrangian and Hamiltonian mechanics in a very natural way:

  (a) A *Lagrangian system* is defined in terms of a smooth Lagrangian

  $$\mathscr{L} : TM \to \mathbb{R}.$$

  In any coordinate chart of TM with coordinates $(q_1, \ldots, q_n, \dot{q}_1, \ldots, \dot{q}_n)$ the equations of motion are given by the *Euler-Lagrange equations*

  $$\frac{\mathrm{d}}{\mathrm{d}t}\frac{\partial \mathscr{L}}{\partial \dot{q}_i} = \frac{\partial \mathscr{L}}{\partial q_i},$$

  with $i = 1, \ldots, n$.

  (b) A *Hamiltonian system* is defined in terms of a smooth Hamiltonian

  $$\mathscr{H} : T^*M \to \mathbb{R}.$$

  In any coordinate chart of T^*M with coordinates $(q_1, \ldots, q_n, p_1, \ldots, p_n)$ the equations of motion are given by the *Hamilton equations*

  $$\begin{cases} \dot{q}_i = \dfrac{\partial \mathscr{H}}{\partial p_i}, \\ \dot{p}_i = -\dfrac{\partial \mathscr{H}}{\partial q_i}, \end{cases}$$

  with $i = 1, \ldots, n$.

▶ As anticipated, any symplectic manifold is locally diffeomorphic to the canonical Hamiltonian phase space $\mathbb{R}^{2n}$ presented in Example 4.25. This is the content of the following Theorem, for which we omit the proof and for which we will give a generalization in Theorem 4.3.

Theorem 4.1 (*Darboux*)

Let (M, ω) be a $2n$-dimensional symplectic manifold. Then for every $p \in M$ there exists a coordinate chart U around p and a diffeomorphism $\Psi : U \to V$, where V is an open subset of $\mathbb{R}^{2n}$ parametrized by $(q_1, \ldots, q_n, p_1, \ldots, p_n)$ such that

$$\Psi^* \left(\sum_{j=1}^{n} \mathrm{d}q_j \wedge \mathrm{d}p_j \right) = \omega$$

on U.

No Proof.

4.5 Foliation of a Poisson manifold

► Before considering the problem of the foliation of a Poisson manifold equipped with a degenerate Poisson structure we consider the simpler case when the Poisson manifold has maximal rank everywhere, which implies that its dimension is necessarily even and that $\mathsf{Ham}_p(M) = T_pM$ for all points $p \in M$. In view of what we know about symplectic manifolds the next Theorem is probably not surprising.

Theorem 4.2

Let $(M, \{\cdot, \cdot\})$ be a regular Poisson manifold with maximal rank on M. Then M is a symplectic manifold.

Proof. We proceed by steps.

- Due to the maximal rank assumption on M the map $\widetilde{\mathbf{X}} : \Omega(M) \to \mathfrak{X}(M)$ induced by the Poisson 2-vector field $\mathbf{X}$, $\mathbf{X}(\mathrm{d}F, \mathrm{d}H) := \{F, H\}$, $F, H \in \mathscr{F}(M)$, is everywhere invertible.

- We define ω to be the 2-form which corresponds to its inverse $\widetilde{\mathbf{X}}^{-1}$. Since $\widetilde{\mathbf{X}}(\mathrm{d}H) = \mathbf{v}_H$, for any $H \in \mathscr{F}(M)$, see (4.19), we define the 2-form

$$\omega(\mathbf{v}_F, \mathbf{v}_H) := \left(\widetilde{\mathbf{X}}^{-1}(\mathbf{v}_F)\right)(\mathbf{v}_H) = \mathrm{d}F(\mathbf{v}_H) = \{F, H\}, \tag{4.40}$$

  for any $F, H \in \mathscr{F}(M)$.

- Obviously, ω is non-degenerate due to the non-degeneracy of the Poisson structure on M. The closure condition is a consequence of the Jacobi identity. Indeed, applying formula (4.9) to three Hamiltonian vector fields $\mathbf{v}_F, \mathbf{v}_G, \mathbf{v}_H$ we get

$$\begin{aligned}
\mathrm{d}\omega(\mathbf{v}_F, \mathbf{v}_G, \mathbf{v}_H) &= \mathbf{v}_F[\omega(\mathbf{v}_G, \mathbf{v}_H)] + \omega(\mathbf{v}_F, [\mathbf{v}_G, \mathbf{v}_H]) + \circlearrowleft (\mathbf{v}_F, \mathbf{v}_G, \mathbf{v}_H) \\
&= \mathbf{v}_F[\{G, H\}] - \omega(\mathbf{v}_F, \mathbf{v}_{\{G,H\}}) + \circlearrowleft (F, G, H) \\
&= \{\{G, H\}, F\} - \{F, \{G, H\}\} + \circlearrowleft (F, G, H) \\
&= 2(\{\{G, H\}, F\} + \circlearrowleft (F, G, H)) \\
&= 0,
\end{aligned}$$

  where we used (4.20), (4.40) and the Jacobi identity. Since $T_pM = \mathsf{Ham}_p(M)$ for any $p \in M$ (by the maximal rank assumption) this shows that ω is closed and hence that ω is a symplectic form.

The Theorem is proved ■

► Note that combining Theorem 4.2 with Theorem 4.1 we get as a corollary that if $(M, \{\cdot,\cdot\})$ is a regular Poisson manifold with maximal rank on M then M is locally diffeomorphic to the canonical Hamiltonian phase space with symplectic form given by

$$\omega := \sum_{j=1}^{n} \mathrm{d}q_j \wedge \mathrm{d}p_j.$$

► We now describe the Poisson structure in the neighborhood of any point of a Poisson manifold and deduce from it a (global) decomposition, called *symplectic foliation* of the Poisson manifold into *symplectic leaves* (of varying dimensions).

- Roughly speaking, a foliation of a smooth manifold M looks locally like a decomposition of M into the union of "parallel" submanifolds of smaller dimension. More formally, a m-dimensional *foliation* of an n-dimensional smooth manifold M is a countable collection of subsets M_α of M (called *leaves*) such that:

  1. There holds
  $$\bigcup_\alpha M_\alpha = M, \qquad M_\alpha \cap M_\beta = \emptyset, \quad \alpha \neq \beta.$$
  2. Each leaf M_α is pathwise connected (i.e., if $p, q \in M_\alpha$ then there exists a continuous curve $\gamma : [0,1] \to M_\alpha$ such that $\gamma(0) = p$, $\gamma(1) = q$).
  3. For each point $p \in M$ there exists a local chart U containing p with coordinates $(x_1, \ldots, x_n) : U \to \mathbb{R}^n$ such that the connected components of the intersections of the leaves with U are the level sets of $(x_{m+1}, \ldots, x_n) : U \to \mathbb{R}^{n-m}$.

  Note that coordinates $(x_1, \ldots, x_m)$ provide local coordinates on M_α, which are therefore images of injective immersions. In particular, the leaves have well defined m-dimensional tangent spaces at each point. Therefore any foliation of dimension m defines an m-dimensional distribution.

- We know from Subsection 4.2.4 that there is a close relationship between foliations and integrability of a distribution. For instance, given a vector field on M that is never zero, its integral curves will give a one-dimensional foliation. This observation is generalized by "Frobenius Theorem" saying that the necessary and sufficient conditions for a distribution to be tangent to the leaves of a foliation, is that the set of vector fields tangent to the distribution are closed under the Lie bracket.

► The foliation of a Poisson manifold is described by the next Theorem, which is a generalization of Theorem 4.2 valid for regular Poisson manifolds.

Theorem 4.3 (*Weinstein*)

Let $(M, \{\cdot,\cdot\})$ be a n-dimensional Poisson with

$$\mathsf{rank}_p\{\cdot,\cdot\} = 2r, \qquad p \in M.$$

Then there exists a coordinate chart U of p with coordinates

$$(q_1,\ldots,q_r,p_1,\ldots,p_r,z_1,\ldots,z_s), \qquad s := n - 2r,$$

such that the Poisson 2-vector field on U takes the form

$$\mathbf{X} = \sum_{i=1}^{r} \frac{\partial}{\partial q_i} \wedge \frac{\partial}{\partial p_i} + \sum_{1 \leqslant i,j \leqslant s} f_{ij}(z_1,\ldots,z_s) \frac{\partial}{\partial z_i} \wedge \frac{\partial}{\partial z_j}, \tag{4.41}$$

where any $f_{k\ell}$ is a smooth function which vanishes at p.

No Proof.

► Remarks:

- The rank of $\{\cdot,\cdot\}$ at p is $2r$ but is not necessarily constant on a neighborhood U of p. When the rank is constant and equal to $2r$ on U, then U can be chosen in such a way that such that the functions f_{ij} vanish on U. This leads to the situation described by Theorem 4.2 and the Poisson 2-vector field is the canonical one:
$$\mathbf{X} = \sum_{i=1}^{r} \frac{\partial}{\partial q_i} \wedge \frac{\partial}{\partial p_i}.$$

- If the rank of $\{\cdot,\cdot\}$ at p is constant and equal to $2r$ on a neighborhood U of p, then the Hamiltonian vector fields on U define a distribution of rank $2r$. Such a distribution is integrable because $[\mathbf{v}_F, \mathbf{v}_G] = -\mathbf{v}_{\{F,G\}}$ for all $F, G \in \mathscr{F}(U)$. Thus, we have a (regular) foliation of U, where each leaf has dimension $2r$. The leaves inherit a Poisson structure from $\{\cdot,\cdot\}$. Since the rank of this Poisson structure is $2r$ each leaf carries a symplectic form and U admits a decomposition into *symplectic leaves*.

- Surprisingly, the decomposition into symplectic leaves exists in the neighborhood of any point p of a Poisson manifold. In other words, the rank needs not be constant in a neighborhood of p, but the dimension of the leaves will not be constant in general. Indeed, one can prove that if $(M, \{\cdot,\cdot\})$ is a Poisson manifold then the (singular) distribution on M defined by the Hamiltonian vector fields is integrable in the sense that every $p \in M$ has a coordinate neighborhood U which is, in a unique way, a disjoint union of symplectic manifolds M_α which are Poisson submanifolds of U. The resulting foliation is called the *symplectic foliation* of M and each of its leaves (of varying dimensions) is called *symplectic leaf*.

- Let $M_\alpha \subset M$ be a symplectic leaf of M.

  (a) Theorem 4.2 guarantees that we can construct a symplectic 2-form ω on M_α by setting
  $$\omega(\mathbf{v}_F, \mathbf{v}_H) := \{ F, H \},$$
  for any two Hamiltonian vector fields $\mathbf{v}_F, \mathbf{v}_H$ on M_α. Note that two points belong to the same symplectic leaf if they can be joined by a path whose tangent vector at any point is a Hamiltonian vector field.

  (b) If $F \in \mathscr{F}(M)$ is a Casimir function then F is constant on M_α. Indeed, if F were not locally constant on M_α, then there would exist a point $p \in M_\alpha$ such that $(\mathrm{d}F(\mathbf{v}))_p \neq 0$ for some $\mathbf{v} \in T_pM_\alpha$. But T_pM_α is spanned by $\mathbf{v}_H$ for $H \in \mathscr{F}(M)$ and hence $(\mathbf{v}_H[F])_p = \{F, H\}(p) = 0$ which implies $(\mathrm{d}F(\mathbf{v}))_p = 0$ which is a contradiction. Thus F is locally constant on U and hence constant by connectedness of the leaf M_α. In good cases most or all of the symplectic leaves are level sets of the Casimir functions, but this is not true in general.

Example 4.28 (*The three-dimensional Euler top*)

The three-dimensional Euler top described in Example 4.23 is a simple but non-trivial example of a mechanical system on a Poisson manifold whose phase space admits a symplectic foliation.

- We know that the system is governed by the ODEs
$$\begin{cases} \dot{x}_1 = (a_2 - a_3)\, x_2\, x_3, \\ \dot{x}_2 = (a_3 - a_1)\, x_3\, x_1, \\ \dot{x}_3 = (a_1 - a_2)\, x_1\, x_2, \end{cases} \tag{4.42}$$
where $a_1, a_2, a_3 \in \mathbb{R}$.

- This system is a Hamiltonian system w.r.t. the Poisson structure corresponding to the Poisson 2-vector field
$$\mathbf{X} = x_1 \frac{\partial}{\partial x_2} \wedge \frac{\partial}{\partial x_3} + x_2 \frac{\partial}{\partial x_3} \wedge \frac{\partial}{\partial x_1} + x_3 \frac{\partial}{\partial x_1} \wedge \frac{\partial}{\partial x_2}.$$
In other words, the Poisson manifold on which the system is defined is $\mathbb{R}^3$, coordinatized by x_1, x_2, x_3, with Poisson bracket
$$\{ x_i, x_j \} = \varepsilon_{ijk}\, x_k. \tag{4.43}$$
This also implies that
$$\{ F(x), G(x) \} = \langle x, \mathsf{grad}_x F(x) \times \mathsf{grad}_x G(x) \rangle, \qquad F, G \in \mathscr{F}(\mathbb{R}^3).$$

- The vector field $\mathbf{v}_H$ defining the system (4.42) is obtained as $\mathbf{v}_H = -\mathbf{X}(\mathrm{d}H, \cdot)$ with Hamiltonian given by
$$H(x) := \frac{1}{2}\left(a_1\, x_1^2 + a_2\, x_2^2 + a_3\, x_3^2\right).$$

- We know that the Poisson structure is degenerate. We have $\mathbf{X}(\mathrm{d}B, \cdot) \equiv 0$ where
$$B(x) := \frac{1}{2}\left(x_1^2 + x_2^2 + x_3^2\right)$$
is a Casimir of the Poisson structure.

- The level sets
$$S_b := \left\{ x \in \mathbb{R}^3 \,:\, B(x) = b \right\},$$
with $b > 0$ are the symplectic leaves of the Poisson manifold. These are two-dimensional spheres which can be equipped with a symplectic structure. On each symplectic leaf S_b the Hamiltonian dynamics is locally canonical.
- Excluding the origin (i.e., $b = 0$), which provides a singular leaf, the symplectic 2-form on S_b is given by (see (4.39))
$$\omega_p(v, w) := \frac{1}{b^2} \langle\, p, v \times w \,\rangle ,$$
where $p \in S^2$ and $v, w \in T_p S_b$. Note that the Poisson bracket (4.43) induces on each S_b a well-defined symplectic Hamiltonian vector field.
- Since both H and B are integrals of motion, the motion takes place along the intersections of the level surfaces of the energy H (ellipsoids embedded in $\mathbb{R}^3$) and the angular momentum B (spheres embedded in $\mathbb{R}^3$). The centers of the energy ellipsoids and the angular momentum spheres coincide. The two sets of level surfaces in $\mathbb{R}^3$ develop collinear gradients (for example, tangencies) at pairs of points that are diametrically opposite on an angular momentum sphere. At these points, collinearity of the gradients of H and B implies stationary rotations, that is, fixed points.

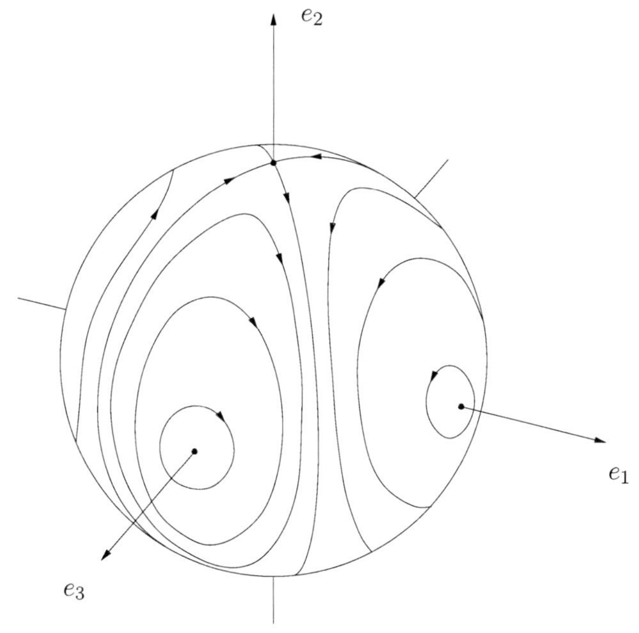

Fig. 4.5. Representation of Euler top flow on the angular momentum sphere.

4.6 Completely integrable systems on symplectic manifolds

► We conclude the Chapter sketching an introduction to the theory of *completely in-*

tegrable systems of classical mechanics on symplectic manifolds. Integrable systems play a crucial role in classical mechanics and their theoretical formulation and investigation is a wide and important branch of the modern theory of dynamical systems. In the simplest (finite-dimensional) instance, the term "integrable system" refers to a mechanical system, described in terms of a Lagrangian or Hamiltonian formulation, whose equations of motion are *solvable by quadratures*, i.e., in terms of a finite number of algebraic operations, the "Inverse function Theorem" and integration.

► We start with an example with one degree of freedom, where we show how the equations of motion can be integrated explicitly, by using only algebraic operations, the process of taking inverse functions and integration.

Example 4.29 (***A particle moving in*** $\mathbb{R}$ ***under the influence of a potential energy***)

We consider a Hamiltonian system on the canonical phase space $\mathbb{R}^2$ defined by the Hamiltonian

$$\mathscr{H}(q,p) := \frac{p^2}{2} + U(q). \tag{4.44}$$

The corresponding Hamilton equations describe the motion of a point-like particle with unit mass moving on the line $\mathbb{R}$ under the influence of a smooth potential energy U.

- We fix a point $(q_0, p_0) \in \mathbb{R}^2$ for which $\mathrm{d}\mathscr{H}(q_0, p_0) \neq 0$ and we denote the value of $\mathscr{H}$ at (q_0, p_0) by h_0. Using $\dot{q} = p$, the Hamiltonian (4.44) implies that the integral curve which starts at (q_0, p_0) satisfies the differential equation

$$\mathrm{d}t = \frac{\mathrm{d}q}{\sqrt{2(h_0 - U(q))}}. \tag{4.45}$$

  The above denominator does not vanish in a neighborhood of q_0, except maybe at q_0, because $\mathrm{d}\mathscr{H}(q_0, p_0) \neq 0$.

- Integrating both sides of (4.45) we find ($t_0 = 0$):

$$t = \int_{q_0}^{q(t)} \frac{\mathrm{d}\xi}{\sqrt{2(h_0 - U(\xi))}},$$

  which defines q (and hence also p) implicitly as a function of t. The obtained functions $(q(t), p(t))$ define, for $|t|$ small, the integral curve of the canonical Hamiltonian vector field $\mathbf{v}_{\mathscr{H}}$, starting at (q_0, p_0), hence they integrate the equations of motion for the initial condition $(q(0), p(0)) = (q_0, p_0)$. Notice that q is obtained by using only algebraic operations, inverting a function and integration.

- A couple of cases where the integration can be carried out in a closed form are the following ones:

  1. $U(q) := \alpha^2 q^{-2}$, $\alpha \in \mathbb{R}, h_0 > 0$. The motion takes place in the region $q_0 < q < \infty$, where $q_0 := \sqrt{\alpha^2/h_0}$. One finds:

$$q(t) = \sqrt{q_0^2 + p_0^2 t^2}, \qquad p_0^2 := 2h_0.$$

  2. $U(q) := \alpha^2 \mathrm{e}^{-2q}$, $\alpha \in \mathbb{R}, h_0 > 0$. The motion is unbounded and takes place in the region $q_0 < q < \infty$, where $q_0 := -(1/2)\log(h_0/\alpha^2)$. One finds:

$$q(t) = \log(\cosh(\beta t)) + q_0, \qquad \beta := \sqrt{2h_0}.$$

► We give the following definition.

Definition 4.5

Let M be a $2n$-dimensional symplectic manifold and let $\mathbf{v}_H$, $H \in \mathscr{F}(M)$, be a Hamiltonian vector field on M. Then $\mathbf{v}_H$ defines a **completely integrable system** *on M if there exists n integrals of motion $F_1 \equiv H, F_2, \ldots, F_n \in \mathscr{F}(M)$ such that:*

1. *$F_1 \equiv H, F_2, \ldots, F_n$ are functionally independent, i.e. the set*

$$U := \left\{ p \in M \, : \, (\mathrm{d}F_1)_p \wedge \cdots \wedge (\mathrm{d}F_n)_p \neq 0 \right\}$$

   is a dense open subset of M.

2. *$F_1 \equiv H, F_2, \ldots, F_n$ are in Poisson involution, i.e.,*

$$\{ F_i, F_j \} = 0 \qquad \forall i, j = 1, \ldots, n.$$

► As a matter of fact, Definition 4.5 is quite restrictive and it imposes both topological restrictions on the Hamiltonian phase space and dynamical restrictions on the Hamiltonian motion generated by the Hamiltonian vector field $\mathbf{v}_H$. The original (purely mechanical and not topological) Liouville's formulation of the Theorem which carries his name is the following: *A Hamiltonian system defined on a $2n$-dimensional symplectic manifold which admits n functionally independent and Poisson involutive integrals of motion is solvable by quadratures.* A more modern (but not the most general) formulation of such a statement also includes some constraints on the topology of the phase space.

Theorem 4.4 (*Arnold-Liouville*)

Let $\mathbf{v}_H$, $H \in \mathscr{F}(M)$, be a Hamiltonian vector field defining a completely integrable system on a $2n$-dimensional symplectic manifold M. If the connected components of the level sets

$$S_{f_1,\ldots,f_n} := \{ p \in M \, : \, F_1(p) = f_1, \ldots, F_n(p) = f_n \},$$

with $f_1, \ldots, f_n \in \mathbb{R}$, are compact, then they are diffeomorphic to n-dimensional tori, invariant under the Hamiltonian flow of $\mathbf{v}_H$. The flow of $\mathbf{v}_H$ on these tori is a linear flow for an appropriate choice of the coordinates.

No Proof.

Example 4.30 (*The n-dimensional harmonic oscillator*)

Consider the canonical symplectic vector space

$$\left(\mathbb{R}^{2n}, \omega \right), \qquad \omega := \sum_{j=1}^{n} \mathrm{d}q_j \wedge \mathrm{d}p_j,$$

with canonical coordinates $(q,p) := (q_1,\dots,q_n,p_1,\dots,p_n)$ and with the Poisson structure coming from the canonical symplectic structure:

$$\{ p_i, p_j \} = \{ q_i, q_j \} = 0, \qquad \{ q_i, p_j \} = \delta_{ij}, \qquad i,j = 1,\dots,n. \tag{4.46}$$

- The *n-dimensional harmonic oscillator* is defined by the Hamiltonian vector field corresponding to the Hamiltonian

$$\mathscr{H}(q,p) := \frac{1}{2}\sum_{i=1}^{n}\left(p_i^2 + \nu_i q_i^2\right), \qquad \nu_i > 0.$$

- It is easy to check that the functions $F_1,\dots,F_n$, defined by

$$F_i(q,p) := \frac{1}{2}\left(p_i^2 + \nu_i q_i^2\right), \qquad i = 1,\dots,n,$$

  are n integrals of motion. They are functionally independent and in Poisson involution w.r.t. the canonical brackets (4.46). Notice that $\mathscr{H} = F_1 + \cdots + F_n$.

- The fibers of the map $(F_1,\dots,F_n)$ over $(f_1,\dots,f_n)$, with $f_i > 0$, are products of circles $p_i^2 + \nu_i q_i^2 = f_i$, hence they are n-dimensional tori.

- When all ν_i are equal one speaks of an *isotropic oscillator*. Notice that in this case each of the functions $q_i p_j - q_j p_i$, $i,j = 1,\dots,n$, is an integral of motion, but these functions are not all in involution.

Example 4.31 (*The rational Calogero-Moser model*)

Consider the canonical symplectic vector space

$$\left(\mathbb{R}^{2n}, \omega\right), \qquad \omega := \sum_{j=1}^{n} \mathrm{d}q_j \wedge \mathrm{d}p_j,$$

with canonical coordinates $(q,p) := (q_1,\dots,q_n,p_1,\dots,p_n)$ and with the Poisson structure coming from the canonical symplectic structure.

- The *rational Calogero-Moser model* (1975) is defined by the Hamiltonian vector field corresponding to the Hamiltonian

$$\mathscr{H}(q,p) := \frac{1}{2}\sum_{i=1}^{N} p_i^2 + \frac{g^2}{2}\sum_{\substack{i,j=1\\ i\neq j}}^{N} \frac{1}{(q_i - q_j)^2}, \qquad g \in \mathbb{R}. \tag{4.47}$$

- The Newton equations of motion are

$$\ddot{q}_i = 2g^2 \sum_{\substack{k=1\\ k\neq i}}^{N} \frac{1}{(q_i - q_k)^3}, \qquad i = 1,\dots,N.$$

  These equations of motion describe a N-particle system with non-local pairwise rational interactions. The mass of each (point-like) particle is equal to one.

- It is possible to show (and it is a non-trivial proof) that the canonical Hamiltonian vector field corresponding to (4.47) provides a completely integrable system. The n functionally independent and Poisson involutive integrals of motion can be chosen among functions of the form

$$F_k(q,p) := \sum_{j=1}^{N} p_j^k + \text{terms of lower degree in momenta}, \qquad k \in \mathbb{N}.$$

Example 4.32 (*The Toda lattice*)

Consider the canonical symplectic vector space

$$\left(\mathbb{R}^{2n},\omega\right), \qquad \omega := \sum_{j=1}^{n} \mathrm{d}q_j \wedge \mathrm{d}p_j,$$

with canonical coordinates $(q,p) := (q_1,\dots,q_n,p_1,\dots,p_n)$ and with the Poisson structure coming from the canonical symplectic structure.

- The *Toda lattice* (1967) is defined by the Hamiltonian vector field corresponding to the Hamiltonian
$$\mathscr{H}(q,p) := \sum_{i=1}^{N}\left(\frac{p_i^2}{2} + \mathrm{e}^{q_{i+1}-q_i}\right).$$
- The corresponding canonical Hamilton equations are
$$\begin{cases} \dot{q}_i = p_i, \\ \dot{p}_i = \mathrm{e}^{q_{i+1}-q_i} - \mathrm{e}^{q_i-q_{i-1}}, \end{cases} \tag{4.48}$$
which are equivalent to Newton equations
$$\ddot{q}_i = \mathrm{e}^{q_{i+1}-q_i} - \mathrm{e}^{q_i-q_{i-1}},$$
subject to one of the two types of boundary conditions: periodic, $q_i \equiv q_{i+N}$ and free, $q_0 = \infty$, $q_{N+1} = -\infty$. These equations of motion describe a lattice of particles interacting with nearest neighbors via forces exponentially depending on distances. The mass of each (point-like) particle is equal to one.
- It is possible to show (and it is a non-trivial proof) that the Hamiltonian vector field defining (4.48) provides a completely integrable system. The n functionally independent and Poisson involutive integrals of motion can be chosen among functions of the form
$$F_k(q,p) := \sum_{j=1}^{N} p_j^k + \text{terms of lower degree in momenta}, \qquad k \in \mathbb{N}.$$

Example 4.33 (*Solvability of the three-dimensional Euler top*)

The three-dimensional Euler top, governed by the system of ODEs

$$\begin{cases} \dot{x}_1 = (a_2 - a_3)\, x_2\, x_3, \\ \dot{x}_2 = (a_3 - a_1)\, x_3\, x_1, \\ \dot{x}_3 = (a_1 - a_2)\, x_1\, x_2, \end{cases}$$

where $a_1, a_2, a_3 \in \mathbb{R}$ is an example of integrable system on a Poisson manifold (see Examples 4.23 and 4.28). For such a system Definition 4.5 and Theorem 4.4 do not apply and we would need a more general notion of complete integrability (on Poisson manifolds). Indeed we have two functionally independent and Poisson involutive integrals of motion but a three-dimensional configuration space which is Poisson manifold equipped with a degenerate Poisson structure. Nevertheless we can show that the Euler top is explicitly solvable in terms of *elliptic functions*.

- We write the system of ODEs of the Euler top in the form
$$\begin{cases} \dot{x}_1 = \alpha_1\, x_2\, x_3, \\ \dot{x}_2 = \alpha_2\, x_3\, x_1, \\ \dot{x}_3 = \alpha_3\, x_1\, x_2, \end{cases} \tag{4.49}$$
where
$$\alpha_1 := \frac{1}{I_2} - \frac{1}{I_3}, \qquad \alpha_2 := \frac{1}{I_3} - \frac{1}{I_1}, \qquad \alpha_3 := \frac{1}{I_1} - \frac{1}{I_2}.$$

We assume for definiteness that $I_1 > I_2 > I_3 > 0$, so that $\alpha_1 < 0$, $\alpha_2 > 0$, $\alpha_3 < 0$.

- System (4.49) admits two functionally independent integrals of motion. In particular, the following three functions are integrals of motion:

$$F_1(x_1, x_2, x_3) := \alpha_2\, x_3^2 - \alpha_3\, x_2^2,$$
$$F_2(x_1, x_2, x_3) := \alpha_3\, x_1^2 - \alpha_1\, x_3^2,$$
$$F_3(x_1, x_2, x_3) := \alpha_1\, x_2^2 - \alpha_2\, x_1^2.$$

Clearly, only two of them are functionally independent because of $\alpha_1 F_1 + \alpha_2 F_2 + \alpha_3 F_3 = 0$.

- One easily sees that the coordinates x_j satisfy the following ODEs with the coefficients depending on the integrals of motion:

$$\begin{cases} \dot{x}_1^2 = \left(F_3 + \alpha_2\, x_1^2\right)\left(\alpha_3\, x_1^2 - F_2\right), \\ \dot{x}_2^2 = \left(F_1 + \alpha_3\, x_2^2\right)\left(\alpha_1\, x_2^2 - F_3\right), \\ \dot{x}_3^2 = \left(F_2 + \alpha_1\, x_3^2\right)\left(\alpha_2\, x_3^2 - F_1\right). \end{cases}$$

The fact that the polynomials on the right-hand sides of these equations are of degree four implies that the solutions are given by elliptic functions. Note that in the above ODEs one regards F_1, F_2, F_3 as fixed real values.

- Let us introduce the four *Jacobi theta functions*. Let $\tau \in \mathbb{C}$ with $\Im\mathfrak{m}\, \tau > 0$, $q := e^{\pi i \tau}$ and u be a complex variable. The we define:

$$\begin{aligned}
\vartheta_1(u) &:= 2\sum_{n=0}^{\infty}(-1)^n q^{\left(n+\frac{1}{2}\right)^2} \sin((2n+1)\pi u) \\
&= -i\sum_{n\in\mathbb{Z}}(-1)^n q^{\left(n+\frac{1}{2}\right)^2} e^{(2n+1)\pi i u}, \\
\vartheta_2(u) &:= 2\sum_{n=0}^{\infty} q^{\frac{1}{4}(2n+1)^2} \cos((2n+1)\pi u) \\
&= \sum_{n\in\mathbb{Z}} q^{\left(n+\frac{1}{2}\right)^2} e^{(2n+1)\pi i u}, \\
\vartheta_3(u) &:= 1 + 2\sum_{n=1}^{\infty} q^{n^2} \cos(2n\pi u) \\
&= \sum_{n\in\mathbb{Z}} q^{n^2} e^{2n\pi i u}, \\
\vartheta_4(u) &:= 1 + 2\sum_{n=1}^{\infty}(-1)^n q^{n^2} \cos(2n\pi u) \\
&= \sum_{n\in\mathbb{Z}}(-1)^n q^{n^2} e^{2n\pi i u}.
\end{aligned}$$

We list some important properties of Jacobi theta functions.

(a) Argument $u = 0$ is usually omitted, the corresponding values are called *theta-constants*:

$$\vartheta_2 := \sum_{n\in\mathbb{Z}} q^{(n+\frac{1}{2})^2}, \qquad \vartheta_3 := \sum_{n\in\mathbb{Z}} q^{n^2}, \qquad \vartheta_4 := \sum_{n\in\mathbb{Z}} (-1)^n q^{n^2}.$$

They satisfy the identity $\vartheta_2^2 + \vartheta_4^2 = \vartheta_3^2$.

(b) The four theta functions are quasi double-periodic functions of u and the play a fundamental role in the classical theory of elliptic functions. In particular they are related to the *Jacobi elliptic functions* via the formulas

$$\mathrm{sn}(u,k) = \frac{\vartheta_3}{\vartheta_2}\frac{\vartheta_1(u/\omega_1)}{\vartheta_4(u/\omega_1)},$$
$$\mathrm{cn}(u,k) = \frac{\vartheta_4}{\vartheta_2}\frac{\vartheta_2(u/\omega_1)}{\vartheta_4(u/\omega_1)},$$
$$\mathrm{dn}(u,k) = \frac{\vartheta_4}{\vartheta_3}\frac{\vartheta_3(u/\omega_1)}{\vartheta_4(u/\omega_1)},$$

where $\omega_1 := \pi\,\vartheta_3^2$. The quantity $k^2 := \vartheta_2^4/\vartheta_3^4$ is called *modulus*, while $(k')^2 = 1 - k^2 = \vartheta_4^4/\vartheta_3^4$ is the *complementary modulus*.

(c) The following *relations between squares* hold (there are two linearly independent ones):

$$\vartheta_1^2(u)\vartheta_4^2 + \vartheta_2^2(u)\vartheta_3^2 - \vartheta_3^2(u)\vartheta_2^2 = 0, \tag{4.50}$$
$$\vartheta_1^2(u)\vartheta_3^2 + \vartheta_2^2(u)\vartheta_4^2 - \vartheta_4^2(u)\vartheta_2^2 = 0, \tag{4.51}$$
$$\vartheta_1^2(u)\vartheta_2^2 + \vartheta_3^2(u)\vartheta_4^2 - \vartheta_4^2(u)\vartheta_3^2 = 0, \tag{4.52}$$
$$\vartheta_2^2(u)\vartheta_2^2 - \vartheta_3^2(u)\vartheta_3^2 + \vartheta_4^2(u)\vartheta_4^2 = 0. \tag{4.53}$$

(d) The following *formulas for derivatives* hold:

$$\begin{aligned}
\vartheta_1'(u)\vartheta_2(u) - \vartheta_1(u)\vartheta_2'(u) &= \vartheta_3(u)\vartheta_4(u)\vartheta_2^2,\\
\vartheta_1'(u)\vartheta_3(u) - \vartheta_1(u)\vartheta_3'(u) &= \vartheta_2(u)\vartheta_4(u)\vartheta_3^2,\\
\vartheta_1'(u)\vartheta_4(u) - \vartheta_1(u)\vartheta_4'(u) &= \vartheta_2(u)\vartheta_3(u)\vartheta_4^2,\\
\vartheta_3'(u)\vartheta_2(u) - \vartheta_3(u)\vartheta_2'(u) &= \vartheta_1(u)\vartheta_4(u)\vartheta_4^2,\\
\vartheta_4'(u)\vartheta_2(u) - \vartheta_4(u)\vartheta_2'(u) &= \vartheta_1(u)\vartheta_3(u)\vartheta_3^2,\\
\vartheta_4'(u)\vartheta_3(u) - \vartheta_4(u)\vartheta_3'(u) &= \vartheta_1(u)\vartheta_2(u)\vartheta_2^2.
\end{aligned}$$

In particular, from the above formulas there follows:

$$\frac{\mathrm{d}}{\mathrm{d}u}\frac{\vartheta_1(u)}{\vartheta_4(u)} = \frac{\vartheta_2(u)}{\vartheta_4(u)}\frac{\vartheta_3(u)}{\vartheta_4(u)}\vartheta_4^2, \tag{4.54}$$
$$\frac{\mathrm{d}}{\mathrm{d}u}\frac{\vartheta_2(u)}{\vartheta_4(u)} = -\frac{\vartheta_1(u)}{\vartheta_4(u)}\frac{\vartheta_3(u)}{\vartheta_4(u)}\vartheta_3^2, \tag{4.55}$$
$$\frac{\mathrm{d}}{\mathrm{d}u}\frac{\vartheta_3(u)}{\vartheta_4(u)} = -\frac{\vartheta_1(u)}{\vartheta_4(u)}\frac{\vartheta_2(u)}{\vartheta_4(u)}\vartheta_2^2. \tag{4.56}$$

- We can now construct the solution of system (4.49), whose form is indeed analog to formulas (4.54–4.56).
- *First Ansatz* ($I_2 > 0$). We consider the Ansatz

$$x_1 = a\,\frac{\vartheta_2(\nu t)}{\vartheta_4(\nu t)}, \qquad x_2 = b\,\frac{\vartheta_1(\nu t)}{\vartheta_4(\nu t)}, \qquad x_3 = c\,\frac{\vartheta_3(\nu t)}{\vartheta_4(\nu t)}. \tag{4.57}$$

Formulas (4.54–4.56) yield:

$$\dot{x}_1 = -\frac{a\nu}{bc}\,\vartheta_3^2\,x_2\,x_3, \qquad \dot{x}_2 = \frac{b\nu}{ca}\,\vartheta_4^2\,x_3\,x_1, \qquad \dot{x}_3 = -\frac{c\nu}{ab}\,\vartheta_2^2\,x_1\,x_2.$$

Therefore the Ansatz gives a solution, if $abc\nu > 0$ and

$$\alpha_1 = -\frac{a\nu}{bc}\,\vartheta_3^2, \qquad \alpha_2 = \frac{b\nu}{ca}\,\vartheta_4^2, \qquad \alpha_3 = -\frac{c\nu}{ab}\,\vartheta_2^2.$$

These are three equations for five unknowns (a,b,c,ν,q). We use them to express the three amplitudes a,b,c through the still unknown values ν,q (that is, through ν and theta-constants):

$$a^2 = -\frac{\nu^2\,\vartheta_2^2\,\vartheta_4^2}{\alpha_2\,\alpha_3}, \qquad b^2 = \frac{\nu^2\,\vartheta_2^2\,\vartheta_3^2}{\alpha_1\,\alpha_3}, \qquad c^2 = -\frac{\nu^2\,\vartheta_3^2\,\vartheta_4^2}{\alpha_1\,\alpha_2}. \tag{4.58}$$

The missing two equations come from the integrals. To find them, we use the relations between squares (three of them involving $\vartheta_4^2(x)$, that is, (4.51–4.53); of course, only two of them are independent). Use (4.52), Ansatz (4.57) and then (4.58) to express the amplitudes:

$$\begin{aligned} 1 &= \frac{\vartheta_1^2(\nu t)}{\vartheta_4^2(\nu t)}\frac{\vartheta_2^2}{\vartheta_3^2}+\frac{\vartheta_3^2(\nu t)}{\vartheta_4^2(\nu t)}\frac{\vartheta_4^2}{\vartheta_3^2} = \frac{x_2^2}{b^2}\frac{\vartheta_2^2}{\vartheta_3^2}+\frac{x_3^2}{c^2}\frac{\vartheta_4^2}{\vartheta_3^2} \\ &= x_2^2\frac{\alpha_1\alpha_3}{\nu^2\vartheta_2^2\vartheta_3^2}\frac{\vartheta_2^2}{\vartheta_3^2}-x_3^2\frac{\alpha_1\alpha_2}{\nu^2\vartheta_3^2\vartheta_4^2}\frac{\vartheta_4^2}{\vartheta_3^2} = x_2^2\frac{\alpha_1\alpha_3}{\nu^2\vartheta_3^4}-x_3^2\frac{\alpha_1\alpha_2}{\nu^2\vartheta_3^4} \\ &= (\alpha_3 x_2^2-\alpha_2 x_3^2)\frac{\alpha_1}{\nu^2\vartheta_3^4}\,. \end{aligned}$$

We see that the following quantity is an integral of motion:

$$F_1 = \alpha_2 x_3^2 - \alpha_3 x_2^2 = -\frac{\nu^2\vartheta_3^4}{\alpha_1}\,. \tag{4.59}$$

There holds $I_1 > 0$, due to $\alpha_2 > 0$, $\alpha_3 < 0$. Similarly, starting with (4.51), we find:

$$\begin{aligned} 1 &= \frac{\vartheta_1^2(\nu t)}{\vartheta_4^2(\nu t)}\frac{\vartheta_3^2}{\vartheta_2^2}+\frac{\vartheta_2^2(\nu t)}{\vartheta_4^2(\nu t)}\frac{\vartheta_4^2}{\vartheta_2^2} = \frac{x_2^2}{b^2}\frac{\vartheta_3^2}{\vartheta_2^2}+\frac{x_1^2}{a^2}\frac{\vartheta_4^2}{\vartheta_2^2} \\ &= x_2^2\frac{\alpha_1\alpha_3}{\nu^2\vartheta_2^2\vartheta_3^2}\frac{\vartheta_3^2}{\vartheta_2^2}-x_1^2\frac{\alpha_2\alpha_3}{\nu^2\vartheta_2^2\vartheta_4^2}\frac{\vartheta_4^2}{\vartheta_2^2} = x_2^2\frac{\alpha_1\alpha_3}{\nu^2\vartheta_2^4}-x_1^2\frac{\alpha_2\alpha_3}{\nu^2\vartheta_2^4} \\ &= (\alpha_1 x_2^2-\alpha_2 x_1^2)\frac{\alpha_3}{\nu^2\vartheta_2^4}\,. \end{aligned}$$

Thus, we have found one more integral of motion,

$$F_3 = \alpha_1 x_2^2 - \alpha_2 x_1^2 = \frac{\nu^2\vartheta_2^4}{\alpha_3}\,, \tag{4.60}$$

with the value $I_3 < 0$, because of $\alpha_1 < 0$, $\alpha_2 > 0$. Finally, we perform a similar computation starting with (4.53):

$$\begin{aligned} 1 &= \frac{\vartheta_3^2(\nu t)}{\vartheta_4^2(\nu t)}\frac{\vartheta_3^2}{\vartheta_4^2}-\frac{\vartheta_2^2(\nu t)}{\vartheta_4^2(\nu t)}\frac{\vartheta_2^2}{\vartheta_4^2} = \frac{x_3^2}{c^2}\frac{\vartheta_3^2}{\vartheta_4^2}-\frac{x_1^2}{a^2}\frac{\vartheta_2^2}{\vartheta_4^2} \\ &= -x_3^2\frac{\alpha_1\alpha_2}{\nu^2\vartheta_3^2\vartheta_4^2}\frac{\vartheta_3^2}{\vartheta_4^2}+x_1^2\frac{\alpha_2\alpha_3}{\nu^2\vartheta_2^2\vartheta_4^2}\frac{\vartheta_2^2}{\vartheta_4^2} = x_1^2\frac{\alpha_2\alpha_3}{\nu^2\vartheta_4^4}-x_3^2\frac{\alpha_1\alpha_2}{\nu^2\vartheta_4^4} \\ &= (\alpha_3 x_1^2-\alpha_1 x_3^2)\frac{\alpha_2}{\nu^2\vartheta_4^4}, \end{aligned}$$

so that the third integral of motion reads

$$F_2 = \alpha_3 x_1^2 - \alpha_1 x_3^2 = \frac{\nu^2\vartheta_4^4}{\alpha_2}\,. \tag{4.61}$$

From this expression we see that our first Ansatz corresponds to the initial data with $I_2 > 0$. Integrals (4.59–4.61) are, of course, linearly dependent, as they should be by their derivation, namely there holds $\alpha_1 F_1 + \alpha_2 F_2 + \alpha_3 F_3 = 0$. These integrals allow us to determine q (or the theta-constants):

$$k^2 := \frac{\vartheta_2^4}{\vartheta_3^4} = -\frac{\alpha_3 F_3}{\alpha_1 F_1}, \qquad (k')^2 := 1-k^2 = \frac{\vartheta_4^4}{\vartheta_3^4} = -\frac{\alpha_2 F_2}{\alpha_1 F_1}.$$

Both modules k^2 and $(k')^2$ are positive and therefore lie in $(0,1)$. For the frequency ν we have the expression

$$\nu^2\vartheta_3^4 = -\alpha_1 F_1.$$

Finally, formulas (4.58–4.61) allow us to express the amplitudes a, b, c in terms of the integrals:

$$a^4 = \frac{F_2 F_3}{\alpha_2\alpha_3}, \qquad b^4 = -\frac{F_1 F_3}{\alpha_1\alpha_3}, \qquad c^4 = -\frac{F_1 F_2}{\alpha_1\alpha_2}.$$

- *Second Ansatz* ($I_2 < 0$). We consider the Ansatz

$$x_1 = a\,\frac{\vartheta_3(\nu t)}{\vartheta_4(\nu t)}, \qquad x_2 = b\,\frac{\vartheta_1(\nu t)}{\vartheta_4(\nu t)}, \qquad x_3 = c\,\frac{\vartheta_2(\nu t)}{\vartheta_4(\nu t)}.$$

A computation similar to the one performed for the first Ansatz gives

$$a^4 = -\frac{F_2\,F_3}{\alpha_2\,\alpha_3}, \qquad b^4 = -\frac{F_1\,F_3}{\alpha_1\,\alpha_3}, \qquad c^4 = \frac{F_1\,F_2}{\alpha_1\,\alpha_2}.$$

4.7 Exercises

Ch4.E1 (a) Consider the parametric map $F : I \to \mathbb{R}^2$, $I := (-1, \infty)$, defined by

$$F(x) := \left(\frac{3\,a\,x}{1+x^3}, \frac{3\,a\,x^2}{1+x^3} \right), \qquad a \in \mathbb{R} \setminus \{0\}.$$

Show that F is not an embedding of I into $\mathbb{R}^2$, i.e., $F(I)$ is a not a differentiable manifold.

(b) Prove that the map $F : \mathbb{R} \to \mathrm{S}^1$ defined by $F(x) := (\cos(2\,\pi\,x), \sin(2\,\pi\,x))$ is a local diffeomorphism, but not a global diffeomorphism.

(c) Prove that the map $F : \mathbb{R} \to \mathbb{R}^2$ defined by $F(x) := \mathrm{e}^x (\cos x, \sin x)$ is an embedding of $\mathbb{R}$ into $\mathbb{R}^2$, so that $F(\mathbb{R})$ is a one-dimensional differentiable manifold. Is $F(\mathbb{R}) \cup \{(0,0)\}$ a differentiable manifold? Why (not)?

(d) Denote by $\mathfrak{gl}(2,\mathbb{R}) \simeq \mathbb{R}^4$ the vector space of all linear maps from $\mathbb{R}^2$ to $\mathbb{R}^2$ and by $S(2,\mathbb{R}) \simeq \mathbb{R}^3$ the space of 2×2 symmetric matrices. Consider the map $F : \mathfrak{gl}(2,\mathbb{R}) \to S(2,\mathbb{R})$, defined by $F(A) := A^\top J\, A$, where

$$J := \begin{pmatrix} 0 & 1 \\ 1 & 0 \end{pmatrix}.$$

Compute the differential $\mathrm{d}F_A$. At what points A does F have maximal rank?

Ch4.E2 Let $\omega \in \Omega^k(M)$ be a differential k-form on M. Let $\mathbf{v} \in \mathfrak{X}(M)$ be a smooth vector field. Prove that $\mathfrak{L}_{\mathbf{v}}\omega = 0$ if and only if ω is constant on the integral curves of $\mathbf{v}$.

Ch4.E3 In $\mathbb{R}^3$ with coordinates $x := (x_1, x_2, x_3)$ consider the following 2-vector field:

$$\mathbf{X} := x_1 \frac{\partial}{\partial x_2} \wedge \frac{\partial}{\partial x_3} + x_2 \frac{\partial}{\partial x_3} \wedge \frac{\partial}{\partial x_1} + x_3 \frac{\partial}{\partial x_1} \wedge \frac{\partial}{\partial x_2}.$$

(a) Prove that $\mathbf{X}$ is a Poisson 2-vector field.

(b) Construct the Hamiltonian vector field corresponding to the smooth function

$$H(x) := \frac{1}{2}\left(a_1\, x_1^2 + a_2\, x_2^2 + a_3\, x_3^2\right) + b_1\, x_1 + b_2\, x_2 + b_3\, x_3,$$

with $a_1, a_2, a_3, b_1, b_2, b_3 \in \mathbb{R}$.

Ch4.E4 Compute the trajectory of a free (point-like and unit mass) particle moving on the surface of the sphere S^2.

Ch4.E5 Compute the trajectory of a free (point-like and unit mass) particle moving on the surface of the torus. Assume that the radius from the origin in $\mathbb{R}^3$ to the center of the torus tube is $R > 0$ and the radius of the tube is $r \in (0, R)$.

Ch4.E6 Consider a free anisotropic oscillator of mass $m = 1$ with n degrees of freedom. Its Newton equations of motion are:

$$\ddot{q}_i + \omega_i^2 q_i = 0, \qquad \omega_i > 0, \qquad i = 1, \dots, n,$$

where $\omega_j \neq \omega_k$ for $j \neq k$.

(a) Find the constraint force field $f = f(q, \dot{q})$, $q := (q_1, \dots, q_n)$, parallel to q, that will keep the oscillator on the sphere S^n.

(b) Write down the (constrained) equations of motion of the anisotropic oscillator on S^n.

(c) Show that the n functions

$$F_i(q) := q_i^2 + \sum_{j \neq i} \frac{(q_j \dot{q}_i - \dot{q}_j q_i)^2}{\omega_i^2 - \omega_j^2}$$

are integrals of motion of the anisotropic oscillator on S^n.

Ch4.E7 Consider two unit mass point-like particles in $\mathbb{R}^3$. The first particle moves along a circle, in a horizontal plane, parametrized by

$$(x_1, x_2) = (r \cos\theta, r \sin\theta), \qquad r \geqslant 0, \theta \in [0, 2\pi).$$

The second particle is constrained on the curve parametrized by

$$(x_1, x_2, x_3) = (r \cos\phi, r \sin\phi, h \sin\phi), \qquad r \geqslant 0, \phi \in [0, 2\pi), h \geqslant 0.$$

Assume that the two particles interact with an elastic force with constant $k > 0$. The gravitational field is neglected.

(a) Write down the Lagrangian of the system.

(b) Write down the Euler-Lagrange equations.

Ch4.E8 The figure below shows a *planar double pendulum*. The rods (lengths ℓ_1, ℓ_2) connecting the masses m_1 and m_2 are rigid and massless, and the whole system is frictionless.

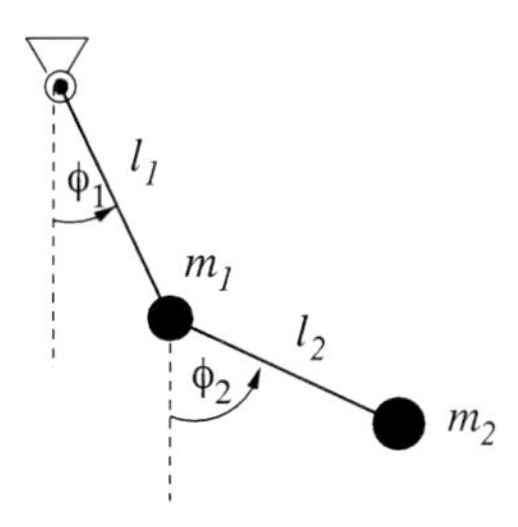

(a) What is the configuration manifold of the system?

(b) Write down the Lagrangian of the system.

(c) Write down the Euler-Lagrange equations.

Index